# Global Dynamics of Social Policy

**Series Editors**
Lorraine Frisina Doetter
University of Bremen
Bremen, Germany

Delia González de Reufels
University of Bremen
Bremen, Germany

Carina Schmitt
Bremen, Germany

Marianne Sandvad Ulriksen
University of Southern Denmark
Odense, Denmark

This open access series welcomes studies on the waves, ruptures and transformative periods of welfare state expansion and retrenchment globally, that is, across nation states and the world as well as across history since the inception of the modern Western welfare state in the nineteenth century. It takes a comprehensive and globalized perspective on social policy, and the approach will help to locate and explain episodes of retrenchment, austerity, and tendencies toward de-welfarization in particular countries, policy areas and/or social risk-groups by reference to prior, simultaneous or anticipated episodes of expansion or contraction in other countries, areas, and risks. One of the aims of this series is to address the different constellations that emerge between political and economic actors including international and intergovernmental organizations, political actors and bodies, and business enterprises. A better understanding of these dynamics improves the reader's grasp of social policy making, social policy outputs, and ultimately the outcomes of social policy.

Gulnaz Isabekova

# Stakeholder Relationships And Sustainability

The Case Of Health Aid To The Kyrgyz Republic

Gulnaz Isabekova
University of Bremen, CRC 1342 "Global Dynamics
of Social Policy" and Research Centre for East European Studies
Bremen, Germany

ISSN 2661-8672          ISSN 2661-8680   (electronic)
Global Dynamics of Social Policy
ISBN 978-3-031-31989-1          ISBN 978-3-031-31990-7   (eBook)
https://doi.org/10.1007/978-3-031-31990-7

This Palgrave Macmillan imprint is published by the registered company Springer Nature Switzerland AG.
The registered company address is: Gewerbestrasse 11, 6330 Cham, Switzerland

Paper in this product is recyclable.

*To my (grand)mother and brother*

# Preface

This book came into being as an outcome of my multisectoral work experience in a nongovernmental organization, private enterprise, and academia. It began with questions, such as what comes after the end of external funding and how relationships among stakeholders matter in the long term. This book aims to expand on these questions in the case of health aid by building a bridge between political science, public health, and development studies. In so doing, it contemplates and reflects on the major issues associated with the long-term sustainability of health aid and the role of the relationships among stakeholders in this matter. Synergizing the perspectives and findings from these disciplines, this book aims to offer a comprehensive analysis of the two phenomena, which may serve as a basis for further studies and as counsel to decision-makers and professionals working in the field.

Bremen, Germany                                                    Gulnaz Isabekova

# Acknowledgments

This book is a product of the research conducted in the Collaborative Research Center 1342 "Global Dynamics of Social Policy" at the University of Bremen. The center is funded by the Deutsche Forschungsgemeinschaft (DFG, German Research Foundation), Project no. 374666841—SFB 1342. The APCs for the Open Access publication were partially funded by the State and University Library Bremen (SuUB).

First of all, I would like to thank Prof. Dr. Heiko Pleines, my *Doktorvater*, who has guided me throughout the whole process with his professionalism, wisdom, trust, and humanity. His dedication to academia and continuous support to colleagues paved the way to opportunities and professional growth for many early stage researchers like me. I would also like to express my gratitude to Prof. Kristina Jönsson and Dr. Monika Ewa Kaminska for their attentiveness, thoughtful suggestions, and support in writing and defending my Ph.D. dissertation, which laid down the basis for this book and commenced with the EU-funded Innovative Training Network "Caspian". I am sincerely thankful for your trust and all the long hours you have spent reading many and, at times, very long pieces from me.

Furthermore, I am grateful to the representatives of civil society and state and international organizations that participated in this research, despite their tight schedules and extensive workload. Your professional insights provided the empirical depth critical to this research by opening the perspectives understudied in the academic literature to this point. I would like to especially thank Venera Toktogonova and the Association of Village Health Committees for their openness and support during the data collection process. I will not name other organizations to maintain the anonymity of research participants; still, I am equally grateful for their time and support.

I would also like to use this opportunity to express my gratitude to my mother, Nazgul Isabekova, for her unconditional love and belief in me. To my big brother, Almaz Isabekov, for his continuous support, investment in my education, and the strong connection we have had since childhood. To my grandmother, Aynek Karymshakova, who, born as a nomad and having experienced many hardships and achievements, always taught us to strive for knowledge. I am also grateful to all other family members, including my sister Aigerim, aunt Gulnara, sister-in-law Aliya, and nieces Perizat, Amira, and Mirana, whom I miss dearly. I am also thankful to my father, Kubanychbek Isabekov, for his support. His determination and professional excellence made him a role model to follow.

Furthermore, I am sincerely grateful to all the wonderful people I have met and collaborated with during this journey. These are my dearest Natalia Zakharchenko and Ohanna Kirakosyan, for their support with data collection process. I am also very grateful to Azhara Kazakbaeva for her support with the transcription of interviews. Separately, I would also like to thank Dr. Ulla Pape, Dr. Esther Somfalvy, and Dr. Ivan Bakalov for their encouraging comments and literature suggestions that found substantial reflection in this book. I am also extremely grateful to multiple friends and colleagues who showed immense support, particularly during the final stages of the writing process: Nathanael Brown, Anastasia Stoll, Irina Wiegand, Stas Gorelik, Benjamin Ahlborn, Dörte Kanis, Jan Matti Dollbaum, Andreas Heinrich, Alesia Kananchuk, Lina Pleines, Aizhan Imanalieva, Tatia Chihkladze, Liva Stupele, Florian Wittmann, Liliana Sanchez, Mareike zum Felde, Karolina Kluczewska, Manuela Putz, Felix Herrmann, Oksana Chorna, and many others. Equally, I am grateful to all participants of the Research Colloquium of the Dept. of Politics and Economics, Research Centre for East European Studies (FSO) at the University of Bremen, for taking their time to comment on the early drafts of this research.

Last but not least, I want to express my gratitude to my fiancé, Dario Landau, for his patience, relentless optimism, and support throughout these years, including during my trips to community-based organizations and health care facilities in Kyrgyzstan. Thank you for your understanding and support during the long working hours over these past years. I am also very thankful to his mother Iris Dorn-Lopez, his brother Luk and his partner Claudi, and other family members, including Domingos Lopez and Peter Dorn, for their warmth, support, and creativity in helping arrange space for me to work in even under highly unusual circumstances.

# Praise for *Stakeholder Relationships and Sustainability*

"Through extensive and careful empirical research Gulnaz Isabekova is able to offer a detailed and systematic examination of her cases, which is of high value not only for students of post-Soviet politics, but also for the broader literature on development aid, health care and infectious diseases."

—Heiko Pleines, *Prof. of Comparative Politics, Research Centre for East European Studies at the University of Bremen*

# Contents

# Abbreviations

| | |
|---|---|
| AFEW | Kyrgyzstan Public Foundation "AIDS Foundation East-West in the Kyrgyz Republic" |
| AIDS | Acquired Immunodeficiency Syndrome |
| ART | Antiretroviral Therapy |
| ARV | Antiretrovirals |
| AVHC/Association of VHCs | The Association of Village Health Committees |
| BMZ | The German Federal Ministry for Economic Cooperation and Development (*das Bundesministerium für wirtschaftliche Zusammenarbeit und Entwicklung*—BMZ) |
| CAH | The "Community Action for Health" Project |
| CBOs | Community-Based Organizations |
| CCM | Country Coordinating Mechanism |
| CDC | The Centers for Disease Control and Prevention |
| COVID-19 | Coronavirus Disease 2019 |
| CSOs | Civil Society Organizations |
| CSW | Commercial Sex Worker |
| DAC | Development Assistance Committee |
| DFID | The United Kingdom's Department for International Development |
| DOTS | Directly Observed Treatment Short-course |
| EECA | Eastern Europe and Central Asia |

| | |
|---|---|
| FMC | Family Medicine Center |
| Gavi | Vaccine Alliance (formerly known as Global Alliance for Vaccines and Immunization) |
| GDF | Global Drug Facility |
| GDP | Gross Domestic Product |
| GHI | Global Health Initiative |
| GIZ | German Corporation for International Cooperation (*die Deutsche Gesellschaft für Internationale Zusammenarbeit*) |
| HIV/AIDS | Human Immunodeficiency Virus Infection and Acquired Immune Deficiency Syndrome |
| HPU | Health Promotion Unit |
| ICRC | International Committee of the Red Cross |
| KfW | German Development Bank (*die Kreditanstalt für Wiederaufbau*) |
| KGS | Kyrgyzstani som |
| KR | Kyrgyz Republic |
| LFA | Local Funding Agent |
| LGBTQ | Lesbian, Gay, Bisexual, Trans, Intersex, and Queer |
| LSG | Local Self-Government |
| M&E | Monitoring and Evaluation |
| MDR-TB | Multidrug-Resistant Tuberculosis |
| MHIF | Mandatory Health Insurance Fund |
| MoH | Ministry of Health |
| MSF | Doctors Without Borders/*Médecins Sans Frontières* |
| MSM | Men Who Have Sex with Men |
| NCPh | National Center of Phthisiology Under the MoH in Kyrgyzstan |
| NGO | Nongovernmental Organization |
| NSP | Needle and Syringe Exchange Program |
| OECD DAC | Organization for Economic Co-operation and Development's Development Assistance Committee |
| ODA | Official Development Assistance |
| OIG | Office of the Inspector General |
| OSF | Open Society Foundation |

| | |
|---|---|
| OST | Opioid Substitution Therapy |
| PEPFAR | President's Emergency Plan for AIDS Relief |
| PHC | Primary Health Care |
| PLHIV | People Living with HIV/Persons Living with HIV |
| PR | Primary Recipient of the Global Fund Grants |
| PRA | Participatory Reflection and Action, formerly known as the Participatory Rural Appraisal |
| PWID | Persons/People Who Inject Drugs |
| RHC | Rayon Health Committees |
| Republican Center | Republican Center for Health Promotion and Mass Communication under the Ministry of Health |
| SDC | Swiss Agency for Development and Cooperation |
| Sida | Swedish International Development Cooperation Agency |
| SR | Sub-recipient of the Global Fund Grants |
| SRC | Swiss Red Cross |
| STIs | Sexually Transmitted Infections |
| SWAp | Sector Wide Approach |
| TB | Tuberculosis |
| The Global Fund | Global Fund to Fight AIDS, Tuberculosis and Malaria |
| UN | United Nations |
| UNAIDS | Joint United Nations Programme on HIV/AIDS |
| UNDP | United Nations Development Programme |
| UNESCO | United Nations Educational, Scientific and Cultural Organization |
| UNICEF | United Nations Children's Fund |
| UNFPA | United Nations Population Fund |
| UNOPS | United Nations Office for Project Services |
| USAID | United States Agency for International Development |
| VHC | Village Health Committee |
| WHO | World Health Organization |

# List of Diagrams

# List of Tables

# 1

# Introduction

Health aid, as it is known today, is a relatively new phenomenon. Before the twentieth century, health assistance was limited mainly to missionaries, who targeted specific geographic areas (Fleßa, 2014). Official development assistance (ODA) for health is a more formalized and structured form of aid. Composed of grants (at least 25% of the total sum) and concessional loans, ODA is provided for development or welfare purposes on a bilateral (country to country) or multilateral (organization to country) basis (OECD, 2009a, p. 180). In contrast to its predecessor, which experiences neither competition among actors nor duplication of efforts (Fleßa, 2014), ODA for health involves multiple bilateral and multilateral actors. According to some estimates, there are up to 61 providers (Knox, 2020, p. 11), each having a specific structure and regulations governing its aid provision and acquisition processes, which may vary considerably across providers.

Notably, ODA for health grew proportionally to the expansion of understanding development as a multidimensional concept not limited to economic growth. First advanced by development theorists and practitioners in the late 1960s, this multidimensional approach to development stressed the various aspects of human welfare, including health,

© The Author(s) 2024
G. Isabekova, *Stakeholder Relationships And Sustainability*, Global Dynamics of Social Policy, https://doi.org/10.1007/978-3-031-31990-7_1

education, and political freedoms (Schafer et al., 2012). Following this approach and growing criticism of the economic focus of development activities (Cornia et al., 1987, 1988), aid to social sectors grew from less than 10% in 1967 to over 40% of total ODA in 2011 (Addison et al., 2015, p. 1356). Health aid also increased from approximately 4% in 1975 (WHO, 2002, p. 12) to 14% of total ODA in 2017 (Knox, 2020, p. 9).[1] It either targeted specific diseases ("vertical" approach) or aimed to strengthen health care systems in general ("horizontal" approach) (Andrews, 2013, p. 130).

Notably, the "vertical" approach subsumes a large share of health aid. Between 2009 and 2018, over half of the health aid was allocated to combatting diseases, with most spending targeting the control of sexually transmitted infections (mainly human immunodeficiency virus infection and acquired immune deficiency syndrome (HIV/AIDS)) and other infectious diseases, such as malaria and tuberculosis (Knox, 2020). This distribution is also due to the establishment of global initiatives focusing on communicable (infectious) diseases. Thus, in the early 2000s, global health initiatives, such as the Vaccine Alliance (Gavi) and the Global Fund to Fight AIDS, Tuberculosis and Malaria (the Global Fund), emerged to facilitate rapid expansion of prevention and treatment services (Biesma et al., 2009). The establishment of these initiatives was also consonant with the global health agenda. The focus on disease control as a global problem corresponded to Goal 6 of the United Nations (UN) Millennium Development Goals (MDGs) (2000–2015) (UN, 2015). The MDGs stressed the role of ODA and a global partnership for development in achieving the stated objectives, but they also noted a slight decrease in the aid flows (ibid.). Following the legacy of the MDGs, the UN Sustainable Development Goals (SDGs) (2015–2030) similarly aim at ending the epidemics of AIDS, tuberculosis (TB), malaria, and other infectious diseases by 2030 (Goal 3, Target 3.3) (UN, n.d.).

However, achievement of the SDGs is jeopardized by the funding gap faced by developing countries (United Nations Development Coordination

---

[1] These data includes members of the Development Assistance Committee and multilateral organizations, which represented 2/3 of the ODA in 2018 (Knox, 2020, p. 9). These estimates do not include the contributions of "emerging" donors, such as Russia, China, and other countries.

Office and Dag Hammarskjöld Foundation, n.d.). Over the decades, ODA has been a reliable source of financing for developing countries (Ahmad et al., 2022), but its share, particularly in certain parts of the world, is decreasing. Although remaining stable in the case of Sub-Saharan Africa, health aid to other regions, such as Latin America, the Caribbean, Eastern Europe, and Central Asia (EECA), has been falling (see Institute for Health Metrics and Evaluation 2023a). Accordingly, Global Health Initiatives in these regions also shrank. For instance, allocations of the Global Fund to Latin America and the Caribbean, Eastern Europe and Central Asia decreased by approximately half, from 7% to 2.9% and from 8% to 2.6%, respectively, of the total investments (see Global Fund, 2011, 2019). Public funding in aid-recipient countries does not necessarily compensate for reductions in ODA to health, as aid is believed to neither facilitate nor hinder public spending on health (WHO, 2019).

This decrease in health aid has been further exacerbated by the implications of the fight against the coronavirus disease of 2019 (COVID-19) pandemic. Indeed, the pandemic has boosted the total ODA to health by 31% compared to the previous year (Brown et al., 2022). This increase, explained by aid providers' allocations to ensure immediate responses to the pandemic, does not necessarily imply more support for areas beyond COVID-19. In fact, the diversion of health care workers and facilities from other areas, often those targeting communicable diseases, decreased access to prevention and treatment services (Economist, 2022). In these circumstances, achieving the UN SDG targets on communicable diseases requires additional funding, a substantial portion of which will aim at catching up to the achievements made before the pandemic.

## 1.1    Sustainability and Relationships in Aid: Problems and Approaches

The reduction in health aid jeopardizes the sustainability of the disease control activities previously covered by it. For instance, there is evidence that countries transitioning[2] from the Global Fund's assistance struggle

---

[2] "Transition" is a process of moving away from donor funding, also referred to as "graduation" or "handover" (Burrows et al., 2016, p. 4).

with reemerging infectious diseases. A 2017 Open Society Foundation (OSF) case study of three countries (Macedonia, Montenegro, and Serbia) suggests service disruptions and an increased HIV burden among key groups (OSF, 2017). The withdrawal of the Global Fund led to similar outcomes in other countries. Civil society organizations in Northern Mexico reported a 60–90% decrease in the distribution of needles and syringes after Global Fund grant program ended (OSF, 2015). In the three years following the end of the Global Fund program in Romania, the rate of HIV positivity among drug users in the country increased from 3% to 30% (OSF, 2014). According to a recent assessment by Gotsadze et al. (2019b), most EECA grant-recipient countries transitioning from the Global Fund's grants face medium- or high-level risks to the continuity of their TB and HIV/AIDS programs after the ending of Global Fund assistance. The authors stress the problems with weak human resources, limited state financing and high dependence on external assistance (ibid.).

In addition to reductions in health aid, the sustainability of the outcomes achieved in disease control activities is further challenged by aid fragmentation. Aid fragmentation refers to "aid that comes in too many small slices from too many donors, creating unnecessary and wasteful administrative costs and making it difficult to target aid where it is needed most" (OECD, 2009b, p. 15). Not specific to health aid but common to development assistance in general, aid fragmentation has multiple repercussions, such as an increased burden on aid recipients and the duplication of efforts.

One example is the large number of meetings between aid providers and recipients. For example, in 2007, Vietnam reported hosting 782 donor missions, each of them demanding "time and attention" from the recipient government (Lawson, 2013, p. 5). In addition to imposing an administrative burden, aid fragmentation is conducive to the duplication of efforts. An extreme example is a case of measles in a little girl in Banda Aceh, Indonesia, after the 2004 Indian Ocean tsunami. The measles symptoms, identified by doctors as unusual, were the outcome of threefold vaccination by three different organizations (Carbajosa, 2005).

The involvement of aid recipients is equally important to the sustainability of disease control efforts. In the measles case described above, coordination among the humanitarian organizations themselves was poor at the time of the disaster, but it was completely ignored in relation to the national government (Susilo, 2010). Indeed, an extensive number of missions do not solve the problem of aid fragmentation, nor does their limited interaction.

This study aims to develop comprehensive analytical frameworks to provide an exhaustive basis for understanding the various forms of relationships between actors and sustainability in the context of health development assistance. It aims to answer the following research question: How do relationships among stakeholders affect the sustainability of health aid?

Multiple parties have sought an answer to this question. Countries providing and receiving development assistance have started multiple initiatives to overcome the problems caused by the duplication of efforts, increase the aid recipient's ownership,[3] and improve the outcomes of development assistance (OECD, 2012, n.d.). The Sector-Wide Approach (SWAp) was one such response. It aimed to unite external assistance under the recipient government's leadership to support its sectoral policy or program (Foster & Leavy, 2001). Thus, the SWAp aimed to overcome the aid fragmentation problem and support the aid recipient's agency. Nevertheless, it demonstrated rather "mixed" performance (Peters et al., 2013, pp. 4–5) and accounted for a small share of aid (Sweeney & Mortimer, 2015), partly because it was bypassed by major aid providers, including those targeting specific diseases.

Similarly, the academic literature on development assistance has equally stressed the inclusion of multiple actors in the aid realization process and the importance of increasing aid recipients' ownership to improve the sustainability of development assistance (e.g., Jerve et al., 2008; Kindornay, 2014; Paine-Andrews et al., 2000; Swedlund, 2017). This discussion of aid recipients' ownership resulted in a distinction between two approaches to development assistance. In a "top-down" approach, assistance was planned by "experts" of donor organizations, whereas a "bottom-up" approach emphasized aid recipients' participation in

---

[3] Ownership is defined as "the control of recipients over the process and outcome of aid negotiations" (Whitfield & Fraser, 2010, pp. 342–343).

defining the objectives and means of development assistance (Kaiser, 2020, pp. 94–95). Andrews (2013) suggests that a concentration on "lone champions" instead of the "broader engagement" of relevant actors leads to the failure of reforms promoted by development programs. He argues that engaging multiple actors is essential for ensuring compliance with the suggested reforms in the local context, as well as local stakeholders' commitment to these reforms (ibid.). Following this logic, aid providers have incorporated some elements of the bottom-up approach in their top-down development programs by emphasizing the aid recipients' involvement in their assistance (Kaiser, 2020, p. 101).

Despite the rhetorical embrace, the actual fulfillment of the bottom-up approach to development assistance has been jeopardized by unequal power dynamics between providers and recipients of development aid (see Hinton & Groves, 2004). Furthermore, the project benchmarks and performance criteria set by donor organizations also affect the terms of aid-recipient participation (Power et al., 2002). Overall, with whom and how to interact in development assistance remain unclear. The inclusion of actors without addressing potential issues related to hierarchy, compatibility, and mutual understanding does not guarantee the desired outcome.

The academic literature on relationships in development assistance is scattered, examining selected forms or relationships between actors in general without nuanced consideration of their types. Some studies have examined coordination (e.g., Aldasoro et al., 2010; Bigsten & Tengstam, 2015; Bourguignon & Platteau, 2015); others, cooperation (Degnbol-Martinussen & Engberg-Pedersen, 2003; Torsvik, 2005; Zimmermann & Smith, 2011) or partnerships (Del Biondo, 2020; Nabyonga Orem et al., 2013). Other research has focused on understanding aid relationships (Eyben, 2006; Hinton & Groves, 2004) or interactions (Lamothe, 2010; Villanger, 2004). However, the link between studies focusing on specific forms or the general notion of relationships between actors has rarely been examined.

Similarly, the meaning of sustainability in the context of development assistance is unclear due to the fragmentation of relevant literature, as the majority of studies focus either on systematic literature review or on empirical analysis of interventions (e.g., case studies), often without theoretical underpinnings (Proctor et al., 2015). This situation contributes to conceptual ambiguity (see Giovannoni & Fabietti, 2013; Shigayeva & Coker, 2015) and inconsistent use of sustainability as a term (Blanchet et al., 2014; Oberth & Whiteside, 2016). The fragmentation of the literature and conceptual ambiguity offer limited implications for a broader understanding of the sustainability of development assistance and the relevant factors.

## 1.2   Research Aims of This Book

First, by combining and systematizing the relevant literature, this book offers an analytical framework for analyzing the relationships between stakeholders. That is, instead of merely stressing the importance of aid relationships, this research analyzes the underlying issues related to the structure of development aid and actors' roles in it. Highlighting the multiplicity of stakeholders involved in aid, it examines the relationships between providers (donor–donor), between providers and the recipient government (donor–recipient state), and between providers, the recipient government, and civil society organizations (CSOs; donor–CSO and recipient state–CSO relationships).

This book aims to synergize the discussion of aid relationships in the development aid literature with a discussion of power and its sources in political theory to provide a more refined analytical framework for analyzing aid relationships (Chap. 2). Differentiating between conventional and alternative perspectives on relationships and power in development assistance, it examines recipients' roles, their potential interdependence, and the (changing) nature of power throughout the assistance.

The analytical framework, composed of four steps, is intended to provide an exhaustive basis for this examination. (1) The discussion of power and its associated terms (resources, consensus/conflict, and interests) provides a necessary conceptual basis for understanding and differentiating between the types of power and its attributes. (2) Further discussion of stakeholders and the context of development aid following the agent-structure approach expand on the relevance of individual and collective agency corresponding to abstract categories (e.g., donor, CSO, recipient state). This approach also places the frequent issues associated with inequality among actors, namely, aid dependency, capacity, aid flexibility, and volatility, into structures that may vary depending on the context/ case but nevertheless remain important to relationships. (3) Analyzing stakeholders' roles throughout the project life cycle by differentiating between the initiation, design, implementation, and evaluation phases is essential to grasp the roles assigned to each actor empirically. (4) Linking the empirical insights from step 3 and the conceptual basis for defining stakeholders, power, and the context in the first two steps leads to a theorization of power dynamics and aid relationships. This step is necessary to place the empirical cases in a broader theoretical framework. This step combines the seven ways of creating power suggested by Haugaard (2003) with the "ideal" types of aid relationships defined by the author of this book in Chap. 2.

Second, this book offers an equally comprehensive analytical framework for understanding the sustainability of health aid. To operationalize sustainability in a consistent and comprehensive manner, it elaborates on the empirical and conceptual definitions of the term. In empirical terms, it defines "what, how or by whom, how much, and by when" to sustain (Iwelunmor et al., 2016, p. 2). In conceptual terms, the book aims to balance donors' and recipients' perspectives on sustainability; for this reason, it adopts a broader definition of sustainability as a continuity of project activities, the maintenance of benefits, and community capacity-building (Shediac-Rizkallah & Bone, 1998).

This book complements Shediac-Rizkallah and Bone's (1998) definition with three further extensions. (1) Acknowledging the relevance of the analysis of both ongoing and complemented projects, it approaches state commitment in terms of necessary legislative amendments and financing as indicative of the sustainability of ongoing initiatives. (2) It complements the operationalization of community capacity-building with an adaptation of Laverack's framework (see Labonte & Laverack, 2001a, 2001b) by focusing on participation, leadership, and resource mobilization. Furthermore, it introduces an aspect that is absent in two previous frameworks, namely, the survival of CSOs beyond the period of development assistance provision. (3) Based on a comprehensive review of research on the sustainability of health care interventions, it also lists factors relevant to the latter. These factors are financing, accounting for the influence of general factors (e.g., political and economic situation in the aid-recipient country), integration into the local context, and organizational factors relevant to the project and the actors implementing it.

Third, systematic operationalization of the relationships, sustainability, and related factors in the context of development aid is followed by the examination of a possible causal link between aid relationships and sustainability. To this end, this book uses the concept of a social mechanism as "a constellation of entities or activities that are linked to one another in such a way that they regularly bring about a particular type of outcome" (Hedström, 2005, p. 11). In addition to providing insights into health aid in the EECA region, this study aims to contribute to the general literature by defining the mechanisms through which the interaction between stakeholders affects the sustainability of health aid. Although specific, these mechanisms are, to a certain extent, generalizable beyond the context of the selected health care programs.

In addition to outlining the underlying issues and main features of social mechanisms, this book emphasizes their role in the formulation of explanatory theories. Thus, by defining social mechanisms, it aims not only to show how but also to explain why the interaction among stakeholders matters to the sustainability of aid. In doing so, it seeks to theorize the relationships between the two phenomena and highlight the conditions under which these relationships are likely to take place and shape sustainability by using Rohlfing's (2012) integrative framework for case studies and causal inferences.

Overall, comprehensive and concise analytical frameworks, based on the extensive literature review and findings from the field, allows for a systematic analysis of sustainability and interaction, including the relevant factors. This mid-range approach extends beyond the alleged universal paradigms and detailed single-case studies by offering a thorough analysis of development projects to identify issues and opportunities applicable to similar initiatives in similar contexts.

## 1.3   Case Selection

Empirically, this book focuses on projects pursuing the bottom-up approach in the developing country context as most favorable for changing the unequal power dynamics between providers and recipients of assistance. Known as a "pioneer" of health care reforms (Ancker et al., 2013), Kyrgyzstan (also referred to as the Kyrgyz Republic) is one of the few countries worldwide to have fully implemented the SWAp. Presuming aid providers' compliance with the national policy and procedures of the aid-recipient governments, the SWAp provides the most favorable environment for altering the conventional power dynamics between donors and recipient governments. The presence of the SWAp in other regions also means that the lessons learned from Kyrgyzstan are equally applicable to other countries implementing this approach to health aid.

Another reason for selecting this country context is that Kyrgyzstan is part of the post-Soviet region—an understudied region in the literature on development aid. Shortly after the collapse of the Soviet Union in 1991, newly independent countries received significant financial and technical assistance from international organizations. However, except for the number of articles discussing the conditions (e.g., Pleines, 2021; Stubbs et al., 2020), assumptions (Wilkinson, 2014), and implications of international support (Ancker & Rechel, 2015; Kim et al., 2018), the post-Soviet region is overlooked in literature on development aid (Leitch, 2016), which largely focuses on Sub-Saharan Africa, Latin America, and Southeast Asia. This book offers insights into health aid in an overlooked context based on case studies in Kyrgyzstan.

The two case studies investigated in this book are the Swiss Agency for Development and Cooperation's (SDC) "Community Action for Health" project and the Global Fund grants to Kyrgyzstan.

SDC represents a traditional bilateral donor whose activities in the health context depend on the geopolitical interests of Switzerland in an aid-recipient country. Unlike the Global Fund, stationed in Geneva, SDC has its representations in aid-recipient countries. Although still accountable to headquarters, these local SDC offices enjoy a relatively high level of autonomy in their policy dialog with recipient governments, budget management, and other areas, which provides them with the flexibility to allocate finances according to the recipients' priorities (OECD, 2005).

Raising approximately US $4 billion annually (Global Fund, 2023a), the Global Fund is among the largest financiers of TB, HIV/AIDS, and malaria programs in the world. It offers grants to countries fulfilling the eligibility criteria (e.g., income status, burden of disease) based on their applications, in which countries indicate how they are going to fight the disease/diseases in question and strengthen their health care systems. As a multilateral donor organization, the Global Fund represents "a new breed of players in global health" that uses a "common blueprint or strategy" across countries to target specific diseases and health challenges (Hanefeld, 2014, p. 54).

Thus, both organizations not only formally acknowledge the importance of ownership but also provide the possibility for aid recipients to define the objectives and activities of the assistance offered by them. In so doing, they embody the "bottom-up" approach to health aid, as their goals and activities are defined by aid recipients.

In addition, the projects differ in their benchmarks and performance criteria. The recipients of the Global Fund projects are expected to comply with its regulations and demonstrate a "good" performance to receive financing continuously. The Community Action for Health project, on the contrary, does not specify the performance criteria and other regulations with which aid recipients need to comply. In this way, the Global Fund projects and the SDC's Community Action for Health project are vivid examples of the bottom-up approach to health aid with and without donor conditionalities. This difference offers another layer of complexity beneficial to understanding the various facets of the bottom-up approach in practice.

Analysis at the project level is essential to understanding how power dynamics and different types of interaction between providers and recipients of health aid form throughout the project life cycle (i.e., its initiation, design, implementation, and evaluation). The focus on the project level also facilitates credible and yet feasible analysis of what sustainable health aid and the relevant factors mean in practice. To ensure the comparability of projects, this book focuses on the TB and HIV/AIDS activities of the Global Fund projects and the Community Action for Health project.[4] The Global Fund grants refer to eight grants that are nevertheless being approached as an ongoing long-term project combatting TB and HIV/AIDS because the objectives of the grants are built on each other. Thus, the administrative division of grants into three- to six-year-long periods corresponds to the length of financial commitments offered by the Global Fund. In contrast, the SDC's Community Action for Health project lasted for nearly seventeen years. It comprised seven phases, from an early pilot to countrywide implementation, which were one continuous project.

## 1.4    Contextual Information on Kyrgyzstan

Kyrgyzstan (also known as the Kyrgyz Republic) is a lower low-middle-income (Global Fund, n.d.-a) Central Asian country with a gross domestic product per capita (current US$) of 1275.9 and a population of 6,700,000 as of 2021 (World Bank Group, 2023). With a total size of 199,900 sq. km, the country is administratively divided into seven regions (*oblasts*) and 40 districts (*rayons*) (National Statistical Committee of the Kyrgyz Republic, 2021).[5] The population is young, with a median age of 27.9 years (ibid.). A large part of the population is ethnic Kyrgyz (73.8%), followed by Uzbek (14.9%), Russian (5.2%), Dungan (1.1%), and other ethnic groups (ibid.). It is estimated that over 80% of the population identifies itself as Muslim, followed by Orthodox Christians (7%) and other religious groups (Usenov, 2022). It should, however, be noted that the extent of religiosity on the individual level, also among those considering themselves as Muslims, varies considerably (ibid.).

The political system of Kyrgyzstan can be characterized as a hybrid regime. The country has been referred to as the "Switzerland of Central Asia" due to its mountainous landscape (see Dorji, 2012) and the "island

---

[4] Unlike the Global Fund projects, the Community Action for Health project included but was not limited to activities targeting TB and HIV/AIDS. For more information on the projects, see Chaps. 6 and 9.

[5] The 7 regions (Batken, Osh, Jalal-Abad, Talas, Chi, Naryn, and Issykkul) are further divided into 40 districts. The capital of the country is Bishkek.

of democracy" because of its initial democratic aspirations (Anderson, 1999). During the three revolutions in 2005, 2010, and 2020, the political system underwent multiple changes, from presidential to parliamentary rule and back to presidential rule. The country used to have a higher level of freedom of speech than other countries in the region. However, there are substantial issues with the rights of sexual minorities (as elsewhere in the former Soviet Union) and concerns with the growing censorship of media and civil society organizations.

Kyrgyzstan gained its independence in 1991 after the collapse of the Soviet Union. Similar to other countries in the region, Kyrgyzstan inherited the *Semashko* health care system, known for its curative rather than preventive approach to diseases. This system is also characterized by the state's paternalistic role as a financer, provider, and regulator of health care services. Overall, the government has remained the main actor in regulating health care systems and defining citizens' entitlement to services (see Isabekova, 2019a). The country has also retained vertical provision of TB and HIV/AIDS services by specialized state agencies, although there have been considerable changes and ongoing reforms in this regard.

Following the collapse of the Soviet Union, Kyrgyzstan struggled with political, social, and economic crises that contributed to the outbreak of tuberculosis. The country's gross domestic product (GDP) declined by half after the collapse of the Soviet Union (Wolfe, 2005, p. 13), impoverishing nearly half (43.5%) of the population (UNDP and ILO, 2008, pp. 25–26). Between 1990 and 2001, the estimated mortality rate related to TB tripled from 9.1 to 29 per 100,000 population (van den Boom et al., 2015, p. 2). The number of TB cases grew from 52 to 88 per 100,000 population; although the actual number of cases was at least two times higher (ibid.). The country also struggled with limited access to testing and poor infection control in medical facilities (WHO/Europe, 2011). Inadequate treatment contributed to the development of multidrug-resistant tuberculosis (MDR-TB).

The incidence of HIV (i.e., new cases reported) increased after the dissolution of the Soviet Union. Situated along one of the three main drug-trafficking routes from Afghanistan to Russia and Europe (Government of KR, 2006), Kyrgyzstan was especially vulnerable to HIV transmission through the use of injection drugs. In the period 1991–1995, the number of persons who injected drugs (PWID) increased by 25% annually and represented 85% of all new HIV cases (Government of KR, 1997). The first HIV case among Kyrgyz citizens was registered in 1996 (ibid.).[6] Improved surveillance and the worsening HIV situation in the country increased the recorded HIV incidence in the 2000s (Ancker et al., 2013), although the official statistics still did not reflect the magnitude of the problem (International Charitable Organization "East Europe and Central Asia Union of People Living with HIV," n.d.). Due to limited testing (Government of KR, 1997), only a third (approximately 30%) of HIV cases were detected (Mansfeld et al., 2015, p. 1).

Overall, health aid has contributed to the prevention, diagnosis, and treatment of TB and HIV/AIDS in the EECA (see Acosta et al., 2016). Nonetheless, despite decreasing ODA to health,[7] the region still has the fastest-growing HIV epidemic and the highest level of MDR-TB in the world (Global Fund, n.d.-b). Kyrgyzstan is on the World Health Organization (WHO) list of 27 countries with a high burden of MDR-TB (WHO, 2015). Fifty-five percent of previously treated patients in the country had MDR-TB (van den Boom et al., 2015, p. 5). Drug resistance is 2.5 times higher among labor migrants than among the general population (Babamuradov et al., 2017, p. 1688). A large proportion of Kyrgyz labor migrants work in Russia and Kazakhstan. While a bilateral agreement with Kazakhstan improved Kyrgyz labor migrants' access to TB services in Kazakhstan (ibid.), their access to health care in Russia remains limited (see Isabekova, 2019b).

Kyrgyzstan has a concentrated form of the HIV epidemic.[8] HIV transmission via intravenous drug injection, initially the prevailing means of infection (Government of KR, 2006, 2012), declined (European Centre for Disease Prevention and Control and WHO/Europe, 2019) and was replaced by heterosexual sex as the main avenue of transmission

---

[6] Fifteen cases of HIV were registered during 1987–1991 among foreign nationals who resided in the country (Godinho et al., 2005, p. 64).

[7] That is, a decrease in the absolute numbers, not per capita. The estimates are based on 2019 US Dollars (see Institute for Health Metrics and Evaluation, 2023).

[8] The HIV epidemic is classified as "concentrated" among the key groups (e.g., people who inject drugs (PWID), men who have sex with men (MSM), and commercial sex workers (CSWs)) if HIV prevalence among the general population is less than 1%; otherwise, the HIV epidemic is classified as "generalized" (see Boily et al., 2015).

(Maytiyeva et al., 2015). Fifty-one percent of HIV incidence in Kyrgyzstan in 2016 resulted from heterosexual intercourse (Government of KR, 2017). This change in the primary mode of transmission shifted the concentrated form of the HIV epidemic from an early to an advanced stage (see World Bank, 2015).

Although concentrated among the key groups, namely, PWID, men who have sex with men (MSM), and commercial sex workers (CSWs) (Government of KR, 2017), HIV infection has been expanding to the general population. Mother-to-child transmission of HIV is still a problem (Maytiyeva et al., 2015), along with HIV infection of children through nosocomial (hospital-acquired) outbreaks. For instance, between 2007 and 2009, 143 children were infected in three hospitals (Ancker et al., 2013, pp. 70–71). Another issue of equal importance to both countries is labor migration toward countries with high HIV prevalence. Labor migrants (mostly seasonal workers) engage in unprotected sex with casual partners (State Partner 4) and, unaware of their HIV status, infect their sexual partners back home upon their return (CSO 6). Working in countries with high HIV/AIDS prevalence (e.g., Russia and Kazakhstan), labor migrants from Kyrgyzstan are at risk of becoming infected with HIV (Government of KR, 2006). There are no accurate estimates of HIV prevalence among this group (Ancker et al., 2013); however, it is estimated that the sexual partners of eight out of twelve HIV-positive pregnant women were associated with labor migration (State Partner 4).

## 1.5 Data Collection

The major sources of data that inform this book are interviews, national legislation on TB and HIV/AIDS, project-related documents, descriptive statistics, and academic and gray literature relevant to the subject.

First, during her fieldwork in 2016 and 2018, the author of this book conducted fifty-two semi-structured interviews with representatives of donor organizations, state authorities, and civil society organizations working on TB and HIV/AIDS. The interviews were conducted in Russian, Kyrgyz, and English. The interviewees were selected based on their availability and responsiveness; to increase outreach, the author collaborated with a research assistant to contact and follow up with the interviewees. A large proportion of the interview transcription was outsourced to an assistant but cross-checked and analyzed by the author using MaxQDA. The interviews were analyzed with thematic content analysis: the interview questions were the basis for the initial categorization of the interview content, followed by a more detailed content-driven analysis and categorization based on the content itself (Kuckartz, 2014, pp. 70–88). This approach ensured the accuracy and comprehensiveness of the interview analysis.

In addition, the national legislation on TB and HIV/AIDS and descriptive statistics obtained from the state structures and WHO were essential to understanding the commitments and contributions of the donors and the national governments in combatting these diseases. These sources complemented the interviews by providing official data about the actors' commitments.[9] Furthermore, this book relied on project-related documents relevant to the selected health care programs. Information on the Global Fund grants is available online (see Global Fund, n.d.-c), and information about the Community Action Health project in Kyrgyzstan was requested from the Swiss Development Cooperation.[10] Finally, analysis of the selected case studies relied on academic and gray literature on TB and HIV/AIDS in Kyrgyzstan.

Despite using various sources, this study acknowledges potential problems of coverage and bias (see Rohlfing, 2012). The use of interviews as primary sources presents a limited picture of sustainability and interactions that is based on interviewees' experiences and perspectives. The secondary sources could be similarly biased by research interest (ibid.) in the case of the academic literature or organizational interest in the case of the gray literature. This study seeks to overcome potential issues with selected coverage and bias by means of triangulation (ibid.).

All interviewees received and signed the "Interview consent form," which outlined the objectives of the research, rules for quotations, terms, and conditions for access to and use of interview material. This form also

---

[9] The author followed the BGN/PCGN transliteration system for referencing Kyrgyz sources and the ISO9 transliteration system for referencing Russian sources.

[10] The author received some information from the SDC about the project, which was supplemented by the interviews as well as by relevant publications by Dr. Tobias Schütz.

provided the researcher's contact details to ask further questions or to request withdrawal from the study at any point. The consent form was provided before the interview, but the participants were asked to sign the forms by the end of the interview to ensure their awareness of the information they provided. All interviewees included in this study expressed their consent to participate in the research by signing the "Interview consent form," agreeing to record their consent, or providing oral consent. Interviewees who chose the third option explained their reluctance with the need to confirm their participation in the research and their answers with higher authorities. By providing unrecorded oral consent, they did not have to participate in the bureaucratic procedures required to obtain such consent. An equally important factor was the general reluctance to sign any document or provide recorded consent, which is common in the post-Soviet region.[11]

## 1.6    Book Structure

The introduction to this book is followed by two analytical chapters expanding on the theoretical underpinnings of aid relationships and sustainability advanced in this book. Chapter 4 demonstrates the application of some of these theoretical considerations by showing how aid dependency, capacity, aid volatility, and flexibility manifest in the selected case studies. The following four chapters in turn offer a thorough analysis of stakeholders' roles throughout the two health projects as well as an assessment of the sustainability of these projects. Chapters 7 and 10 further discuss the aid relationships formed in these projects by linking the empirical findings to the theoretical underpinnings. Chapter 11 discusses

---

[11] The average duration of the interviews was approximately an hour. With the respondents' permission, most interviews were recorded by the author; in other cases, the author took notes. The author provided all respondents with a consent form explaining the objectives of the research, the funding, and the terms and conditions for the use of data. Most respondents signed these agreements, although in some cases, the consent was recorded instead due to interviewees' hesitation to sign a document. All interviewees were anonymized. The interview transcripts are available at the Research Center for East European Studies at the University of Bremen based on the conditions defined by the respondents. For more information about the selection process, list of interviewees, and interview questions, see Isabekova (2023).

the "missing link" or how stakeholder relationships affect the sustainability of health aid. Finally, the conclusion summarizes the major findings by discussing their implications for the broader academic literature and health projects.

# References

Acosta, C., Dara, M., Langins, M., & Kluge, H. (2016). *Good practices in strengthening health systems for the prevention and care of tuberculosis and drug-resistant tuberculosis*. Retrieved February 17, 2023, from http://www.euro. who.int/__data/assets/pdf_file/0010/298198/Good-practices-strengthening-HS-prevention-care-TBC-and-drug-resistant-TBC.pdf?ua=1

Addison, T., Niño-Zarazúa, M., & Tarp, F. (2015). Aid, social policy and development. *Journal of International Development, 27*(8), 1351–1365. https:// doi.org/10.1002/jid.3187

Ahmad, Y., Bosch, E., Carey, E., & Donnell, I. M. (2022). *Six decades of ODA: insights and outlook in the COVID-19 crisis*. Retrieved February 14, 2023, from https://www.oecd-ilibrary.org/sites/5e331623-en/images/pdf/dcd-2019-2159-en.pdf

Aldasoro, I., Nunnenkamp, P., & Thiele, R. (2010). Less aid proliferation and more donor coordination? The wide gap between words and deeds. *Journal of International Development, 22*(7), 920–940. https://doi.org/10.1002/jid.1645

Ancker, S., & Rechel, B. (2015). 'Donors are not interested in reality': The interplay between international donors and local NGOs in Kyrgyzstan's HIV/ AIDS sector. *Central Asian Survey, 34*(4), 516–530. https://doi.org/10.108 0/02634937.2015.1091682

Ancker, S., Rechel, B., McKee, M., & Spicer, N. (2013). Kyrgyzstan: Still a regional 'pioneer' in HIV/AIDS or living on its reputation? *Central Asian Survey, 32*(1), 66–84. https://doi.org/10.1080/02634937.2013.771965

Anderson, J. (1999). *Kyrgyzstan: Central Asia's Island of democracy?* (1st ed.). Routledge.

Andrews, M. (2013). *The limits of institutional reform in development: Changing rules for realistic solutions* (Illustrated Ed.). Cambridge University Press.

Babamuradov, B., Trusov, A., Sianozova, M., & Zhandauletova, Z. (2017). Reducing TB among Central Asia labor migrants. *Health Affairs, 36*(9), 1688. https://doi.org/10.1377/hlthaff.2017.0794

Biesma, R. G., Brugha, R., Harmer, A., Walsh, A., Spicer, N., & Walt, G. (2009). The effects of global health initiatives on country health systems: A review of the evidence from HIV/AIDS control. *Health Policy and Planning, 24*(4), 239–252. https://doi.org/10.1093/heapol/czp025

Bigsten, A., & Tengstam, S. (2015). International coordination and the effectiveness of aid. *World Development, 69*, 75–85. https://doi.org/10.1016/j.worlddev.2013.12.021

Blanchet, K., Palmer, J., Palanchowke, R., Boggs, D., Jama, A., & Girois, S. (2014). Advancing the application of systems thinking in health: Analysing the contextual and social network factors influencing the use of sustainability indicators in a health system—A comparative study in Nepal and Somaliland. *Health Research Policy and Systems, 12*(46), 1–11. https://doi.org/10.1186/1478-4505-12-46

Boily, M.-C., Pickles, M., Alary, M., Baral, S., Blanchard, J., Moses, S., et al. (2015). What really is a concentrated HIV epidemic and what does it mean for west and Central Africa? Insights from mathematical modeling. *Journal of Acquired Immune Deficiency Syndromes, 68*(Suppl 2), 74–82. https://doi.org/10.1097/QAI.0000000000000437

Bourguignon, F., & Platteau, J.-P. (2015). The hard challenge of aid coordination. *World Development, 69*, 86–97. https://doi.org/10.1016/j.worlddev.2013.12.011

Brown, G. W., Tacheva, B., Shahid, M., Rhodes, N., & Schäferhoff, M. (2022). *Global health financing after COVID-19 and the new Pandemic Fund.* Brookings. Retrieved February 14, 2023, from https://www.brookings.edu/blog/future-development/2022/12/07/global-health-financing-after-covid-19-and-the-new-pandemic-fund/

Burrows, D., Oberth, G., Parsons, D., & McCallum, L. (2016). *Transitions from donor funding to domestic reliance for HIV responses: Recommendations for transitioning countries.* Retrieved February 15, 2023, from https://www.globalfundadvocatesnetwork.org/wp-content/uploads/2016/04/Aidspan-APMG-2016-Transition-from-Donor-Funding.pdf

Carbajosa, A. (2005). Demasiado dinero en Banda Aceh. *El País.* Retrieved February 15, 2023, from https://elpais.com/diario/2005/04/13/internacional/1113343204_850215.html

Cornia, G. A., Jolly, R., Stewart, F., Cornia, G. A., Jolly, R., & Stewart, F. (Eds.). (1987). *Adjustment with a human face: Volume 1, protecting the vulnerable and promoting growth.* Oxford University Press.

Cornia, G. A., Jolly, R., Stewart, F., Cornia, G. A., Jolly, R., & Stewart, F. (Eds.). (1988). *Adjustment with a human face: Volume 2, ten country case studies.* Oxford University Press.

Degnbol-Martinussen, J., & Engberg-Pedersen, P. (2003). *Aid: Understanding international development cooperation.* Zed Books.

Del Biondo, K. (2020). Moving beyond a donor-recipient relationship? Assessing the principle of partnership in the joint Africa–EU strategy. *Journal of Contemporary African Studies, 38*(2), 310–329. https://doi.org/10.108 0/02589001.2018.1541503

Dorji, T. (2012). The Switzerland of Central Asia pushes for mountain issues at Rio+20 Summit. *Earth Journalism Network.* Retrieved February 17, 2023, from https://earthjournalism.net/stories/the-switzerland-of-central-asia-pushes-for-mountain-issues-at-rio20-summit

Economist. (2022). How one pandemic made another one worse. Covid-19 set back the battle against tuberculosis. But it also points the way forward. *The Economist.* Retrieved February 14, 2023, from https://www.economist.com/international/2022/10/27/how-one-pandemic-made-another-one-worse

European Centre for Disease Prevention and Control, & WHO/Europe. (2019). *HIV/AIDS surveillance in Europe 2019. 2018 data* (pp. 1–95). Retrieved February 17, 2023, from https://www.ecdc.europa.eu/sites/default/files/documents/HIV-annual-surveillance-report-2019.pdf

Eyben, R. (2006). Introduction. In R. Eyben (Ed.), *Relationships for aid* (pp. 1–17). Earthscan.

Fleßa, S. (2014). Health-related development aid: What comes after it? *The European Journal of Health Economics, 15*(6), 563–566. https://doi.org/10.1007/s10198-013-0551-7

Foster, M., & Leavy, J. (2001). The choice of financial aid instruments. *ODI Working Papers, 158*, 1–36.

Giovannoni, E., & Fabietti, G. (2013). What is sustainability? A review of the concept and its applications. In C. Busco, M. L. Frigo, A. Riccaboni, & P. Quattrone (Eds.), *Integrated reporting* (pp. 21–40). Springer International Publishing. https://doi.org/10.1007/978-3-319-02168-3_2

Global Fund. (2011). *The global fund annual report 2010.* Retrieved February 14, 2023, from https://www.theglobalfund.org/media/1336/corporate_2010annual_report_en.pdf

Global Fund. (2019). *Overview of the 2020–2022 allocations and catalytic investments.* Retrieved February 14, 2023, from https://www.theglobalfund.org/media/9225/fundingmodel_2020-2022allocations_overview_en.pdf

Global Fund. (2023a). *About the global fund*. Retrieved February 15, 2023, from https://www.theglobalfund.org/en/about-theglobal-fund/

Global Fund. (n.d.-a). *Eligibility list 2022*. https://www.theglobalfund.org/media/11712/core_eligiblecountries2022_list_en.pdf

Global Fund. (n.d.-b). *Turning the tide against HIV and tuberculosis. Global fund investment guidance for eastern Europe and Central Asia* (pp. 1–19). Retrieved February 17, 2023, from https://www.globalfundadvocatesnetwork.org/wp-content/uploads/2015/03/Global-Fund-Investment-Guidance-for-EECA_en.pdf

Global Fund. (n.d.-c). *Data explorer*. Retrieved February 17, 2023, from https://data.theglobalfund.org/location/KGZ/grants?components=Tuberculosis,HIV,TB/HIV,Multicomponent,RSSH

Godinho, J., Renton, A., Vinogradov, V., Novotny, T., & Rivers, M.-J. (2005). Reversing the tide: Priorities for HIV/AIDS prevention in Central Asia. *World Bank Working Paper, 54*, 1–186.

Gotsadze, G., Chikovani, I., Sulaberidze, L., Gotsadze, T., Goguadze, K., & Tavanxhib, N. (2019b). Supplement table 2. Comparative table of risk to transition across the countries. *Global Health: Science and Practice, 7*(2), 1–3.

Government of KR. (1997). Nacional''naâ Programma Kyrgyzskoj Respubliki po profilaktike SPIDa i boleznej, peredaûŝihsâ polovym putem, na 1997–2000 gody [National Program of the Kyrgyz Republic on the prevention of AIDS and sexually transmitted diseases for 1997–2000]: Postanovlenie Pravitel'stva Kyrgyzskoj Respubliki ot 1 sentâbrâ 1997 goda № 507 [Decree of the Government of the Kyrgyz Republic dated September 1, 1997 No. 507]. Retrieved February 17, 2023, from http://cbd.minjust.gov.kg/act/view/ru-ru/34692

Government of KR. (2006). Gosudarstvennaâ programma po predupreždeniû èpidemii VIČ/SPIDa i ee social'no-èkonomičeskih posledstvij v Kyrgyzskoj Respublike na 2006–2010 gody [State Programme on prevention of HIV/AIDS epidemic and social and economic consequences in the Kyrgyz Republic for 2006–2010]: Postanovlenie Pravitel'stva Kyrgyzskoj Respubliki ot 6 iûlâ 2006 goda N 498 [Decree of the Government of the Kyrgyz Republic dated July 6, 2006 N 498]. Retrieved February 17, 2023, from http://cbd.minjust.gov.kg/act/view/ru-ru/57612

Government of KR. (2012). Gosudarstvennaâ Programma Po stabilizacii èpidemii VIČ-infekcii v Kyrgyzskoj Respublike na 2012–2016 gody [State Program to Stabilize the HIV Epidemic in the Kyrgyz Republic for 2012–2016]. Retrieved February 17, 2023, from http://cbd.minjust.gov.kg/act/view/ru-ru/93959/20?cl=ru-ru

Government of KR. (2017). Programma Pravitel'stva Kyrgyzskoj Respubliki po preodoleniû VIČ-infekcii v Kyrgyzskoj Respublike na 2017–2021 gody [The Government of the Kyrgyz Republic Program on Overcoming HIV Infection in the Kyrgyz Republic for 2017–2021]: Priloženie 1 Utverždeno postanovleniem Pravitel'stva Kyrgyzskoj Respubliki ot 30 dekabrâ 2017 goda № 852 [Annex 1 approved by the Decree of the Government of the Kyrgyz Republic dated December 30, 2017 No. 852]. Retrieved February 17, 2023, from http://cbd.minjust.gov.kg/act/view/ru-ru/11590

Hanefeld, J. (2014). The Global Fund to Fight AIDS, Tuberculosis and Malaria: 10 years on. *Clinical Medicine, 14*(1), 54–57. https://doi.org/10.7861/clinmedicine.14-1-54

Haugaard, M. (2003). Reflections on seven ways of creating Power. *European Journal of Social Theory, 6*(1), 87–113. https://doi.org/10.1177/1368431003006001562

Hedström, P. (2005). *Dissecting the social: On the principles of analytical sociology* (Illustrated Ed.). Cambridge University Press.

Hinton, R., & Groves, L. (2004). The complexity of inclusive aid. In L. C. Groves & R. B. Hinton (Eds.), *Inclusive aid: Changing power and relationships in international development / edited by Leslie Christine Groves and Rachel Barbara Hinton* (pp. 3–20). Earthscan.

Institute for Health Metrics and Evaluation. (2023). *Financing global health.* Health Focus Areas of Development Assistance for Health (DAH) [The Data Was Filtered by Regions]. Retrieve February 14, 2023, from http://vizhub.healthdata.org/fgh

International Charitable Organization "East Europe and Central Asia Union of People Living with HIV." (n.d.). *Otsenka sovremennoy situatsii: Kyrgyzstan: Analiz vtorichnykh dannykh [Assessment of the current situation: Kyrgyzstan. Analysis of secondary data].* n.p. Retrieved February 17, 2023, from http://ecuo.org/wp-content/uploads/2016/12/Otsenka-situatsii-analiz-vtorichnyh-dannyh-13-MB.pdf

Isabekova, G. (2019a). Diverse health care developments in the PostSoviet space: The role of national and international actors. In S. An, T. Chubarova, & B. Deacon (Eds.), *Social policy, poverty, and inequality in central and Eastern Europe and the former Soviet Union: Agency and institutions in flux* (pp. 238–262). Ibidem Press.

Isabekova, G. (2019b). The contribution of vulnerability of labour migrants to drug resistance in the region: Overview and suggestions. *The European Journal of Development Research, 31*(3), 620–642. https://doi.org/10.1057/s41287-018-0172-1

Isabekova, G. (2023, March 16). *Interviews used to analyze the stakeholders' relationships and the sustainability of selected health projects in the Kyrgyz Republic.* Retrieved March 16, 2023, from https://discuss-data.net/dataset/ adbf1730-5f62-4156-8cb7-b80c7d0489ba/

Iwelunmor, J., Blackstone, S., Veira, D., Nwaozuru, U., Airhihenbuwa, C., Munodawafa, D., et al. (2016). Toward the sustainability of health interventions implemented in sub-Saharan Africa: A systematic review and conceptual framework. *Implementation Science, 11*, 1–27. https://doi.org/10.1186/ s13012-016-0392-8

Jerve, A. M., Lakshman, W. D., & Ratnayake, P. (2008). Sri Lanka: Exploring 'ownership' of aid-funded projects: A comparative study of Japanese, Norwegian and Swedish project aid. In A. M. Jerve, Y. Shimomura, & A. S. Hansen (Eds.), *Aid relationships in Asia: Exploring ownership in Japanese and Nordic aid* (pp. 83–115). Palgrave Macmillan.

Kaiser, M. S. (2020). Are bottom-up approaches in development more effective than top-down approaches? *Journal of Asian Social Science Research, 2*(1), 91–109. https://doi.org/10.15575/jassr.v2i1.20

Kim, E., Myrzabekova, A., Molchanova, E., & Yarova, O. (2018). Making the 'empowered woman': Exploring contradictions in gender and development programming in Kyrgyzstan. *Central Asian Survey, 37*(2), 228–246. https:// doi.org/10.1080/02634937.2018.1450222

Kindornay, S. (2014). Post-2015 partnerships: Shared benefits with the private sector? In B. Tomlinson (Ed.), *Rethinking partnerships in a Post-2015 world: Towards equitable, inclusive and sustainable development: Reality of aid 2014 report* (pp. 69–76). IBON International.

Knox, D. (2020). *Aid spent on health: ODA data on donors, sectors, recipients— Factsheet.* Retrieved February 14, 2023, from https://reliefweb.int/report/ world/aid-spent-health-oda-data-donors-sectors-recipients-factsheet-july-2020

Kuckartz, U. (2014). *Qualitative text analysis: A guide to methods, practice & using software / udo Kuckartz.* SAGE.

Labonte, R., & Laverack, G. (2001a). Capacity building in health promotion, part 1: For whom? And for what purpose? *Critical Public Health, 11*(2), 111–127. https://doi.org/10.1080/09581590110039838

Labonte, R., & Laverack, G. (2001b). Capacity building in health promotion, part 2: Whose use? And with what measurement? *Critical Public Health, 11*(2), 129–138. https://doi.org/10.1080/09581590110039847

Lamothe, H. D. (2010). Re-conceptualizing the international aid structure: Recipient donor interactions and the rudiments of a feedback mechanism. *CEPAL—Serie Financiamiento del desarrollo, 234*, 1–29.

Lawson, M. L. (2013). *Foreign aid: International donor coordination of development assistance.* Congressional Research Service. Retrieved February 28, 2023, from https://sgp.fas.org/crs/row/R41185.pdf

Leitch, D. (2016). *Assisting reform in post-communist Ukraine 2000–2012: The illusions of donors and the disillusion of beneficiaries.* Ibidem Press.

Mansfeld, M., Ristola, M., & Likatavicius, G. (2015). *HIV/AIDS Programme in Kyrgyzstan.* Evaluation report (pp. 1–84). WHO/Europe; Centre for Health and Infectious Disease Research. Retrieved February 17, 2023, from http://www.euro.who.int/__data/assets/pdf_file/0005/273308/HIV-Programme-Review-in-Kyrgyzstan.pdf?ua=1

Maytiyeva, V. S., Chokmorova, U. Zh., Ismailova, A. D., Asybaliyeva, N. A., Yanbukhtina, L. F., Sarybayeva, M. E., et al. (2015). *Stranovoy otchet o dostignutom progresse v osushchestvlenii global'nykh mer v otvet na vich-infektsiyu za 2014 god Kyrgyzskaya Respublika [Country Report on Progress in Implementation of the 2014 Global Response to HIV the Kyrgyz Republic]* (pp. 1–29). Retrieved February 17, 2023, from https://www.unaids.org/sites/default/files/country/documents/KGZ_narrative_report_2015.pdf

Nabyonga Orem, J., Marchal, B., Mafigiri, D., Ssengooba, F., Macq, J., Da Silveira, V. C., & Criel, B. (2013). Perspectives on the role of stakeholders in knowledge translation in health policy development in Uganda. *BMC Health Services Research, 13*(1), 324. https://doi.org/10.1186/1472-6963-13-324

National Statistical Committee of the Kyrgyz Republic. (2021). *Kyrgyzstan v cifrah 2021 [Kyrgyzstan in numbers 2021].* Retrieved February 15, 2023, from http://www.stat.kg/ru/publications/sbornik-kyrgyzstan-v-cifrah/

Oberth, G., & Whiteside, A. (2016). What does sustainability mean in the HIV and AIDS response? *African Journal of AIDS Research, 15*(1), 35–43. https://doi.org/10.2989/16085906.2016.1138976

OECD. (2005). Retrieved February 15, 2023, from http://www.oecd.org/dac/peer-reviews/35297586.pdf

OECD. (2009a). *Managing aid: Practices of DAC member countries* (pp. 1–194). OECD. Retrieved February 14, 2023, from https://www.oecd.org/dac/peer-reviews/35051857.pdf

OECD. (2009b). Development co-operation report 2009. *OECD Journal on Development, 10*(1), 1–254.

OECD. (2012). *The Busan partnership for effective development cooperation.* Retrieved February 15, 2023, from https://www.oecd.org/dac/effectiveness/Busan%20partnership.pdf

OECD. (n.d.). *The Paris declaration on aid effectiveness (2005) and Accra Agenda for Action (2008).* Retrieved February 15, 2023, from http://www.oecd.org/dac/effectiveness/34428351.pdf

OSF. (2014). *Undermining the global fight. The disconnect between the global fund's strategy and the real-life implications of the new funding model.* Retrieved February 15, 2023, from https://www.opensocietyfoundations.org/uploads/416f5097-9eed-4113-b411-3ef88ca7c543/undermining-global-fight-20141201.pdf

OSF. (2015). *Ready, willing, and able? Challenges faced by countries losing global fund support.* Retrieved February 15, 2023, from https://www.globalfundadvocatesnetwork.org/wp-content/uploads/2016/04/ready-willing-and-able-20160403.pdf

OSF. (2017). *Lost in transition: Three Case Studies of Global Fund Withdrawal in South Eastern Europe.* Retrieved February 15, 2023, from https://www.opensocietyfoundations.org/uploads/cee79e2c-cc5c-4e96-95dc-5da50ccdee96/lost-in-translation-20171208.pdf

Paine-Andrews, A., Fisher, J. L., Campuzano, M. K., Fawcett, S. B., & Berkley-Patton, J. (2000). Promoting sustainability of community health initiatives: An empirical case study. *Health Promotion Practice, 1*(3), 248–258.

Peters, D. H., Paina, L., & Schleimann, F. (2013). Sector-wide approaches (SWAps) in health: What have we learned? *Health Policy and Planning, 28*(8), 884–890. https://doi.org/10.1093/heapol/czs128

Pleines, H. (2021). The framing of IMF and World Bank in political reform debates: The role of political orientation and policy fields in the cases of Russia and Ukraine. *Global Social Policy, 21*(1), 34–50. https://doi.org/10.1177/1468018120929773

Power, G., Maury, M., & Maury, S. (2002). Operationalising bottom-up learning in international NGOs: Barriers and alternatives. *Development in Practice, 12*(3/4), 272–284.

Proctor, E., Luke, D., Calhoun, A., McMillen, C., Brownson, R., McCrary, S., & Padek, M. (2015). Sustainability of evidence-based healthcare: Research agenda, methodological advances, and infrastructure support. *Implementation Science, 10*, 1–13. https://doi.org/10.1186/s13012-015-0274-5

Rohlfing, I. (2012). *Case studies and causal inference: An integrative framework.* Palgrave Macmillan.

Schafer, J., Haslam, P. A., & Beaudet, P. (2012). Meaning, measurement and morality in international Development. In P. A. Haslam, J. Schafer, & P. Beaudet (Eds.), *Introduction to international development: Approaches, actors, and issues* (2nd ed., pp. 3–27). Oxford University Press.

Shediac-Rizkallah, M. C., & Bone, L. R. (1998). Planning for the sustainability of community-based health programs: Conceptual frameworks and future directions for research, practice and policy. *Health Education Research, 13*(1), 87–108. https://doi.org/10.1093/her/13.1.87

Shigayeva, A., & Coker, R. J. (2015). Communicable disease control programmes and health systems: An analytical approach to sustainability. *Health Policy and Planning, 30*(3), 368–385. https://doi.org/10.1093/heapol/czu005

Stubbs, T., Reinsberg, B., Kentikelenis, A., & King, L. (2020). How to evaluate the effects of IMF conditionality. *The Review of International Organizations, 15*(1), 29–73. https://doi.org/10.1007/s11558-018-9332-5

Susilo, A. (2010). The ineffectiveness of aid in Aceh re-development projects. *Jurnal Global dan Strategis, 3*(1), 33–42.

Swedlund, H. J. (2017). *The development dance: How donors and recipients negotiate the delivery of foreign aid* (1st ed.). Cornell University Press.

Sweeney, R., & Mortimer, D. (2015). Has the Swap influenced aid flows in the health sector? *Health Economics.* https://doi.org/10.1002/hec.3170

Torsvik, G. (2005). Foreign economic aid; should donors cooperate? *Journal of Development Economics, 77*(2), 503–515. https://doi.org/10.1016/j.jdeveco.2004.05.008

UN. (2015). The millennium development goals report. Retrieved February 14, 2023, from https://www.un.org/millenniumgoals/2015_MDG_Report/pdf/MDG%202015%20rev%20(July%201).pdf

UN. (n.d.). *The 17 goals.* Retrieved February 14, 2023, from https://sdgs.un.org/goals

UNDP, & ILO. (2008). *Kyrgyzstan: ekonomicheskiy rost, zanyatost' i sokrashcheniye bednosti [Kyrgyzstan: Economic growth, employment and poverty reduction].* Retrieved February 17, 2023, from https://www.ilo.org/wcmsp5/groups/public/%2D%2D-europe/%2D%2D-ro-geneva/%2D%2D-sro-moscow/documents/publication/wcms_306630.pdf

United Nations Development Coordination Office, & Dag Hammarskjöld Foundation. (n.d.). *Local insights, global ambition.* Unlocking SDG Financing: Good Practices From Early Adopters. Retrieved February 14, 2023, from https://unsdg.un.org/sites/default/files/Unlocking-SDG-Financing-Good-Practices-Early-Adopters.pdf

Usenov, A. (2022). *Religious politics in Kyrgyzstan: Analysis of achievements and issues*. Central Asian Bureau for Analytical Reporting. Retrieved February 15, 2023, from https://cabar.asia/en/religious-politics-in-kyrgyzstan-analysis-of-achievements-and-issues

van den Boom, M., Mkrtchyan, Z., & Nasidze, N. (2015). *Review of tuberculosis prevention and care services in Kyrgyzstan 30 June–5 July 2014 Mission report* (pp. 1–95). Retrieved February 17, 2023, from http://www.euro.who.int/__data/assets/pdf_file/0010/287803/Review-of-tuberculosis-prevention-and-care-services-in-Kyrgyzstan.pdf?ua=1

Villanger, E. (2004). Company influence on foreign aid disbursement: Is conditionality credible when donors have mixed motives? *Southern Economic Journal, 71*(2), 334–351. https://doi.org/10.2307/4135295

Whitfield, L., & Fraser, A. (2010). Negotiating aid: The structural conditions shaping the negotiating strategies of African governments. *International Negotiation, 15*(3), 341–366. https://doi.org/10.1163/157180610X529582

WHO (Ed.). (2002). *International development assistance and health: The report of working group 6 of the commission on macroeconomics and health*. WHO.

WHO. (2015, January 1). *Use of high burden country lists for TB by WHO in the post-2015 era*. Retrieved December 14, 2020, from https://www.who.int/tb/publications/global_report/high_tb_burdencountrylists2016-2020.pdf

WHO. (2019). *Global spending on health: A world in transition*. Retrieved February 14, 2023, from https://apps.who.int/iris/bitstream/handle/10665/330357/WHO-HIS-HGF-HF-WorkingPaper-19.4-eng.pdf?ua=1

WHO/Europe. (2011). *Tuberculosis country work summary*. Retrieved February 17, 2023, from http://www.euro.who.int/__data/assets/pdf_file/0004/185890/Kyrgyzstan-Tuberculosis-country-work-summary.pdf

Wilkinson, C. (2014). Development in Kyrgyzstan: Failed state or failed state-building? In A. Ware (Ed.), *Development in difficult sociopolitical contexts: Fragile, failed, pariah* (pp. 137–162). Palgrave Macmillan.

Wolfe, D. (2005). *Pointing the way: Harm reduction in Kyrgyz Republic* (pp. 1–60). Harm Reduction Association of Kyrgyzstan. Retrieved February 17, 2023, from https://core.ac.uk/download/pdf/11872287.pdf

World Bank. (2015). *Optimizing investments in Kyrgyz Republic's HIV response*. Retrieved February, 2023, from http://documents.worldbank.org/curated/en/264481477977075763/pdf/109601-WP-GHNDRECAKYRGYZReportMarch-PUBLIC-ABSTRAC-SENT.pdf

World Bank Group. (2023). *The World Bank in the Kyrgyz Republic*. Overview. Text/HTML. Retrieved February 15, 2023, from https://www.worldbank.org/en/country/kyrgyzrepublic/overview

Zimmermann, F., & Smith, K. (2011). More actors, more money, more ideas for international development co-operation. *Journal of International Development, 23*(5), 722–738. https://doi.org/10.1002/jid.1796

# 2

# Theorizing Power, Agents, Structures, and Aid Relationships

The analysis of relationships between providers and recipients of aid inevitably leads to the discussion of power. One can identify two approaches in the development aid literature. The conventional perspective on power primarily emphasizes inequality among actors and aid providers' predominance. In contrast, the alternative perspective highlights aid recipients' agency and suggests that inequalities among stakeholders are not constant.

The conventional perspective builds on the discussion of three approaches to power in development aid by Eyben (2008, pp. 36–37), who differentiated between the differences in powers enjoyed by actors, power distribution as a historical legacy, and power as a "process that enables and constrains action." All three approaches outline specific aspects that sum up to the assumption that inequality among stakeholders is inevitable and is led mainly by donors.

First, the differences in actors' powers suggest that providers enjoy more power than recipients. As the source of power, aid provides the means for donors to hold the recipients accountable (Hinton & Groves, 2004). This accountability, however, works only one way (Renzio, 2006). There are cases of development aid used by donors as "sanctions" against

© The Author(s) 2024
G. Isabekova, *Stakeholder Relationships And Sustainability*, Global Dynamics of Social Policy, https://doi.org/10.1007/978-3-031-31990-7_2

the recipients (Feyissa, 2011, p. 801), but the recipients do not hold donors responsible for breaking their promises due to the fear of not receiving assistance (Eyben, 2008).

The second approach views power distribution as an outcome of historical legacy. Unequal settings between the global "north" and the "south" created the basis for development assistance. The meaning of "development" traces back to the colonization period, when the initial ideas of what "development" is and who defines it were established. This is reflected in, for instance, the underdevelopment of recipients and donor obligations to bring "progress" into these countries (e.g., Schafer et al., 2012). "Development," as defined by donors, was imposed on the recipients.

The third approach examines power by viewing development aid as the process that enables and constrains stakeholders' actions. It suggests that aid *per se* implies inequality (Robb, 2004) because it underlies "gift-giving" and "gift-obligation dynamics" (Hinton & Groves, 2004, p. 12). Following this approach, inequalities between actors are unlikely to be changed because development aid defines or even preassigns the roles, responsibilities, and opportunities of each actor. Overall, the three approaches above are cornerstones of what I call conventional power dynamics in aid, characterized by inequality in resources and hierarchical roles.

The alternative perspective on power is based on another strain of the literature highlighting the recipients' roles, the interdependence of the actors involved in giving and receiving aid, and the changing nature of power throughout the development assistance process. Accordingly, recipients may depend on donor assistance, but after receiving it, they weigh the "pros" and "cons" of the objectives of the aid provider and decide accordingly. Following this perspective, they are not "passive" recipients but discuss the terms and conditions of receiving development aid to maximize "their welfare in the face of budgetary constraints" (Lamothe, 2010, p. 5). Recipients may change their behavior if the incentives and benefits offered by donors are higher than the costs of required changes (ibid., p. 19). If not, recipients retain the status quo. Thus, the reforms anticipated and promoted by development aid take place if the recipient is committed to them.

Furthermore, aid relationships between donors and recipients are characterized by interdependence. Development aid involves actors other than only the direct providers and recipients of the assistance, such as parliaments, governments, constituencies, and local municipalities. Both donors and recipients are accountable for the aid they spend. Although the level of accountability varies depending on the role of the public and the political system of the country, it nevertheless ensures the interdependence of donors and recipients. The actors are mutually dependent because the recipients need the donors' financial resources, and the donors need the recipients' support to show the "success" of their activities (Shutt, 2006a, p. 154; Swedlund, 2017, pp. 75–76). This interdependence outweighs the hierarchies, as both actors are interested in maximizing the output of the assistance and, therefore, are interested in interacting with each other.

Last, there is an evolving or changing nature of power at different stages of the assistance process. Although they exercise more power during the allocation process, donors nevertheless have limited influence over the outcomes of an aid project. As they provide the project finances (in some cases also ideas), donors are important during reform initiation, but their role decreases during the implementation stages (Andrews, 2013, pp. 209–210). In contrast, the role and power of the recipient (state, civil society organizations (CSOs)) increase. Although nonachievement of the outcomes could result in aid suspension, this is not always the case, and it could also be justified by domestic politics, the pressure of constituencies, or reform opponents (Swedlund, 2017, pp. 73–96).

Overall, the agency provided to recipients, stakeholders' interdependence, and the evolving nature of power suggest that the inherent inequalities between donors and recipients underlined in the three interpretations of power are not constant. Following these insights, I suggest a framework composed of the following four steps that are intended to provide a comprehensive basis for grasping the aspects highlighted in the alternative approach:

1. Inspired by scholars in political theory, the first step commences with a reflection on the meaning of power and the common terms associated with it, such as resources, consensus/conflict, and interests. It is intended to provide a necessary conceptual basis for understanding the types of power in the context of inequality in development aid. I intentionally focus on classic political theorists, as they, in my opinion, reflect the aid hierarchy best.

2. This book emphasizes the relevance of both stakeholders and the context in which they interact, which is consonant with the agent-structure approach to aid relationships. In so doing, the second step aims to expand on the relevance of individual and collective agency (e.g., organizational level) of abstract categories, such as "donors," "CSOs," and the "recipient state." Finally, in terms of structure, this book focuses on the frequent issues associated with inequality among actors, namely the recipients' capacities, their dependency on aid, and the flexibility and volatility of aid.

3. The third step calls for a project-level analysis differentiating the following phases of the project cycle: initiation, design, implementation, and evaluation. Empirical analysis at this level offers a detailed yet standardized analysis of development projects, which is beneficial to cross-project comparison.

4. The fourth step culminates the analytical framework by linking the empirical insights from step 3 and the conceptual basis defining stakeholders, power, and the context in the first two steps to a theorization of power dynamics and aid relationships. This step is necessary to understand the empirical cases by placing them in a broader theoretical framework. I built on the seven ways of creating power by Haugaard (2003) because they provide a suitable basis for comprehending the roles and means stakeholders use and the types of power they exercise in relation to each other.

## 2.1    Conceptualization of Power

To a certain extent, the discussion of power in the context of development leaves the following two impressions: providers and recipients stand in opposition to each other, and their powers are inversely related because if the recipients have more power, donors are presumed to have less power. These impressions recall the perceptions of power as a "zero-sum game," in which more power for one actor equals less for another. However, this conceptualization of power was criticized by Parsons as early as the 1960s as inapplicable to all cases (Parsons, 1963). Correspondingly, scholars such as Arendt (1970, p. 44) viewed power as "acting in concert" and, therefore, not antagonistic in the relation of one stakeholder to another. Relying on these insights from scholars in political theory, I define the conceptualization and theorization of power that are essential to defining the types of interaction among stakeholders in step 4 of the analytical framework.

The analysis of any complex phenomenon is associated with multiple issues, and power is not an exception (see Dahl, 1957). There are disagreements about its definition, measurement, and nature. While some scholars defined power as a "circulating medium" (Parsons, 1963, p. 236), others denied its existence as an independent entity, viewing power as "a mode of action upon the actions of others" instead (Foucault, 2002, pp. 341–342). Similarly, the essence of this phenomenon, including its directions (bilateral vs. unilateral) (Goldhamer & Shilds, 1973), interpretations (power "over," "to," and "with") (Pansardi & Bindi, 2021), and forms (dispersed or concentrated), remains contested, along with its measurement and feasibility of empirical observations (Dowding, 2017, p. 4). Overall, there is a tendency toward a multidimensional interpretation of this phenomenon that involves synthesizing different approaches (Ledyaev, 2021).

Indeed, the analysis and operationalization of power are inevitably normative (Lukes, 2005, pp. 37–38). Following the focus of this research on the implications of relationships among stakeholders on the sustainability of health aid, I approach power as a socially constructed phenomenon (Dowding, 2017) and a product of a "set of interacting individuals" (Barnes, 1988, p. 61). I differentiate between the power "over," "to," and "with" due to their relevance to understanding the power dynamics between stakeholders. Power over is among the first forms, and is defined as A having power over B or as relations among controlling and dependent units (see Dahl, 1957). The "power over" form is often associated with hierarchical relations, whereas the "power to" form closely relates to altering these relations. This difference was introduced by Pitkin (1972) and further reemphasized by feminist scholars such as Allen (1998). The "power with" form was introduced by Barnes (1988) and is based on the presumption that power is not attributed to a single entity, which was further strengthened by Arendt's (1970, p. 44) view of power as an "act in concert." There is still an ongoing discussion about interpretations of power and the validity of these differentiations (see Pansardi & Bindi, 2021). However, in the context of development aid, these distinctions are relevant, as they are the key to understanding whether the hierarchy among stakeholders, as in conventional power dynamics, remained or was altered in the course of providing aid or was not present at all.

The "power with" form lies at the core of the analytical approach. This book approaches interaction and sustainability both as an individual and as a collective endeavor. This approach coincides with the perception that power is not something that, as Barnes (1988, pp. 61–62) aptly noted, "radiated from heroic figures; they have glowed with it and illuminated everyone else." Individuals or entities may enjoy power, but it is nevertheless "embedded" in society (ibid.), and the supportive group enables the presence and exercise of power (Arendt, 1970). In this way, power is not attributed to a single entity or an individual but to a broader constellation of stakeholders and structures.

I approach "power with" as an overarching perspective toward the interaction of all stakeholders and their joint impact on the sustainability of development aid. However, for precision and practical reasons, I assign interaction in a dyadic manner by delineating two broader categories of stakeholders (donor–recipient state, recipient state–CSOs, donor–CSOs, donor–donor). Identifying the interaction types of all stakeholders at once would be practically challenging but also possibly analytically meaningless, as this would not allow the precision necessary for grasping the power dynamics. I acknowledge that both aid relationships and aid sustainability are the outcomes of the "power with" form and not individual dyadic interaction types defined in this book. However, the dyadic focus provides a meaningful basis for grasping how the "power with" accumulates and potentially changes, although this discussion falls beyond the focus of this book.

In addition to this conceptualization of power, the analytical approach introduced in this book ingrains the following phenomena associated with power and relevant to defining aid relationships: resources, (in)compliance, and interests.

First, resources are commonly associated with power, with the premise that more resources imply more power (Hinton & Groves, 2004). These are not limited to material resources and include knowledge and access to them. This assumption underlies the unequal relationship among actors in development aid. Nevertheless, although closely related, resources do not equate to power (Giddens, 1984, pp. 15–16). The way actors approach their resources makes a difference, as stakeholders with the same resources may use them dissimilarly (Dahl, 2005, pp. 273–276). Accordingly, the empirical analysis in this book shows the relevance of resources in understanding aid relationships, for instance, in relation to the incentives that one stakeholder may offer to another. However, it also demonstrates that the difference in resources does not necessarily equal hierarchical relationships among actors.

Second, the (in)compliance of stakeholders with the recommendations and regulations of the other stakeholders is another aspect that is essential to grasping the power dynamics in aid. Here, I focus on sanctions as a "reprisal for nonconformity with a prior act of power" (Goldhamer & Shilds, 1973, p. 300), following the act of incompliance (Parsons, 1963, p. 238). It is important to note that sanctions can be positive or negative (e.g., Baldwin, 1971), the main difference being if the change in the situation is for the benefit or disadvantage of the stakeholder to whom sanctions are applied (see Parsons, 1963). The empirical analysis in this book mainly showed the presence of negative sanctions following the act of incompliance. One of the reasons was the visibility of the conflict. However, this may not always be the case, as conflicts rooted in contradicting interests may be latent and never realized from the outside (Lukes, 2005, pp. 28–29). Similarly, consensus among stakeholders could be implied and is not always expressed explicitly (Dowding, 2011a). The empirical analysis was limited to visible conflicts due to the objections stakeholders expressed in relation to actions taken by the other stakeholders. Overall, both sanctions and consensus/conflict provided a useful basis for examining acts of (in)compliance.

Third, closely associated with power, interests are also essential to defining and validating aid relationships. Power is often defined in relation to forcing one to act contrary to one's interests or the capacities of stakeholders to realize their interests (see Lukes, 2005). This reference to power and interests further presumes that power is intentional or in pursuit of specific interests. The definition of interests also closely relates to the costs and benefits that those using and are subject to power face and gain (see Dahl, 2005). This accords with the underlying idea of why stakeholders participate in development aid or choose not to do so. However, scholars disagree about the (un)intentional character of power (e.g., Allen, 1998), its relation to objectives (Giddens, 1984), and the ability of stakeholders to comprehend their interests (Lukes, 2005). The empirical analysis encompassed subsections on stakeholders' interests in pursuing a specific aid relationship form. I argue for stakeholders' abilities to define and voice their interests, noting that the emphasis on the opposite may unintentionally cause unnecessary victimization of stakeholders. Indeed, actors vary in their access to information and capacities, and yet, as the empirical analysis shows, they have pursued their interests by explaining their compliance with specific aid relationship forms.

I follow the simplistic definition of interests related to the realization of personal and organizational objectives due to the different levels of abstraction pursued in the theoretical approach to the operationalization of actors. Individuals and organizations representing donors and recipients operate in conditions of uncertainty since they are insecure about each other's actions and the amount as well as the duration of development assistance (Swedlund, 2017). Furthermore, the complexity of development assistance, which is related to a multiplicity of actors, interests, and the areas involved, results in the actors receiving incomplete information. Therefore, I suggest that stakeholders have limited or "bounded" rationality in maximizing their personal as well as organizational interests. "Bounded rationality" means that actors are constrained in their "information-processing" abilities by risks, uncertainty, limited awareness of other options, and the "complexity" of the setting, resulting in an inability to choose "the best course of action" (Simon, 1972, pp. 162–164).

In addition to rationality, interests relate to actors' perceptions of what is "important" and acceptable from their personal and organizational perspectives, as well as in relation to other stakeholders. The actors' "mental image of the world" frames their perceptions of and reactions to the ongoing processes and preferences for certain decisions, or what Scharpf (1997, p. 62) defined as "subjective preferences." Personal perception is also shaped by what is "acceptable and legitimate" from both individual and organizational perspectives (Campbell, 2004, p. 96). This interrelation between individual preferences and acceptability is vivid, particularly in the cases of politically and culturally salient issues. Equally, the actors' choices are guided not only by personal perceptions but also by a "relational" aspect of the actors to each other (Scharpf, 1997, pp. 69–84), which emphasizes actors' responsiveness to ongoing processes and others' reactions to these processes, which also shape their perceptions. This once again reemphasizes the assumption that decisions do not take place in isolation but in the context of not only structural factors but also in relation to other stakeholders involved.

## 2.2 Conceptualizing Agents and Structures

Following the long-standing discussion on the roles of actors and the relevance of the context in development aid, I emphasize the significance of both actors and structures in understanding power dynamics. This approach corresponds with a meso-level theorization of power as a conceptual tool for specific purposes advocated by Haugaard (2002).

In this book, actors and stakeholders refer to organizations and occasionally individuals whom I approach as agents that act depending on incentives provided in the relevant structures and the roles assigned to these agents (Dowding, 2017, p. 22). An action is defined as changing "the pre-existing state of affairs or course of events" (Giddens, 1984, p. 14). I acknowledge the significance of both individual and organizational levels of analysis but largely keep to the organizational level, except for cases in which individual actions explicitly emanated from individuals and their specific backgrounds, contributing to actions beyond the organizational perspective. This attribution to roles relates to practical concerns, namely, "collective" agency is "easier to comprehend" than that of an individual (Dowding, 2011b, p. 9). Furthermore, I believe that individuals are shaped by the organizations they represent as well as the roles they are assigned to, particularly in the context of aid. This assumption accords with Scharpf (1997, p. 12), who suggested that individuals were "much less free in their actions" but represent certain entities and act on behalf of them. Indeed, individuals also pursue personal interests shaped by their comprehension of reality. The impact of self-interest is specifically relevant to leadership positions, where individuals have fewer organizational constraints (ibid., p. 62). However, even with these positions, individuals are censored by their positions and organizations. In addition to individual and organizational perspectives, agency in the context of development aid closely relates to the roles assigned to "providers" and "recipients." Therefore, I link individuals and organizations to broader analytical entities, donors, recipient states, and civil society organizations.

Structures are equally significant to power dynamics. "Recursively organized sets of rules and resources" enable and constrain stakeholders (Dowding, 2011c, p. 10), shaping their action and inaction (Lukes, 2005, p. 26). The structures encompass a number of phenomena, but in the context of development aid, some are regularly of specific relevance. For example, although common to development assistance in general, the inequality between the providers and recipients of aid varies across cases. I suggest that aid dependency and the capacity of the recipient are vital to understanding these variations. Furthermore, actors dealing with development assistance face the problems of aid volatility (uncertainty) and aid (non) flexibility. Similar to inequality, these phenomena are common to development aid, although donor policies on these issues vary; therefore, I attribute these factors to the providers and not the recipients of assistance (Table 2.1).[1]

**Table 2.1** Structures and their relevance to agents

| Recipient | Aid dependency | Capacity |
|-----------|----------------|----------|
| Donor | Aid volatility | Aid flexibility |

[1] For the justification of the relevance of these factors to aid relationships, see Isabekova (2019).

First, aid dependency is critical to understanding the actors receiving aid. A country (in this framework, a recipient state and a CSO) is aid-dependent when it cannot "achieve objective X in the absence of aid for the foreseeable future" (Lensink & White, 1999, p. 13). For example, if stakeholders are interested in conducting specific reforms, aid dependency means that the recipient cannot implement the reforms without the donor. Obviously, financial and institutional constraints may prevent the recipient state or CSOs from implementing the desired reforms or policies independently. However, we need to distinguish between the necessity for "additional" support and the "sole" reliance on it. A recipient country or a CSO seeking donor support in addition to its own resources is not aid-dependent; however, the country fully relying on the assistance is aid-dependent. Although reflected in individual actors, aid dependency remains a structural issue because it is rooted in a broader country/region/global context beyond these actors.

There are different measurements of aid dependence, but this study suggests a sector-specific definition. Glennie and Prizzon (2012) propose a quantitative indicator of dependence, calculated by the ratio of aid to the gross national income of the recipient country. For civil society organizations, this could refer to the ratio of "external" funding to the resources of the organization. These types of indicators are useful for the general ranking of recipient countries/organizations, but they are not helpful for understanding the power dynamics within specific sectors, for example, health care. Generally, a country's dependence on aid is not equal to its sectoral dependence. The state may receive a large amount of aid but no health care aid. The sectoral division of the assistance provides a more accurate picture, but even in this case, the numbers might be misleading. For instance, the share of "external" health expenditure as part of current health expenditure in 2016 in Kyrgyzstan was approximately 4% (World Bank Group, 2023). One may assume that the country is relatively "independent" from aid because public (state) and private (patients) contributions to health care are much higher than those from donors. However, the empirical chapters in this book demonstrate the opposite. Although independent at the general level, the country relies on technical assistance in health care reforms or financing in terms of access to treatment. Therefore, the analytical framework presented here suggests that a more specific sector or subsectoral focus provides a better understanding of the aid dependence of recipients.

The second factor relevant to understanding the differences in power dynamics is capacity. Broadly defined as "the ability of people, organizations, and society as a whole to manage their affairs successfully" (OECD, 2011, p. 2), capacity in a narrow sense refers to the individual, organizational, and systems' abilities/"competencies" to implement their functions (European Centre for Development Policy Management, 2008, p. 2). Based on these definitions, this chapter operationalizes capacity as a recipient's ability to perform its functions and administer its activities with a focus on the availability of human resources. Human resources are essential to negotiations, implementation, and the evaluation of development assistance. A limited capacity, reflected in an insufficient number of staff members and their qualification issues, causes communication problems with donors.

However, again, critical to understanding stakeholders, "capacity" is still an outcome of broader issues in a given sector or country and is not limited to an individual actor. Thus, in her interviews with donor representatives in Sub-Saharan Africa, Swedlund (2017, pp. 92–93) highlights the staff shortages and computer literacy problems of the recipient countries. Limited capacity is related to and caused by a "brain drain" from public institutions. Qualified staff members are often recruited by donors offering better remuneration and advancement policies (Swedlund, 2017; Toornstra & Martin, 2013). Similar issues with staff retention are noticed in the case of CSOs (Frontera, 2007), although there are differences within this group. The level of staff rotation in community-based organizations (CBOs), where members work on a voluntary basis, might be higher than in a nongovernmental organization (NGO), which pays its employees and provides additional nonfinancial incentives, such as training and travel.

Third, stakeholders operate in conditions of uncertainty related to aid appropriation procedures and relatively short development program durations. Aid volatility varies depending on aid modalities, with budget support being more predictable than project-based assistance. There is a general acknowledgment of the need for increased aid predictability (e.g., Menocal & Mulley, 2006). Correspondingly, the emphasis on long-term partnerships in institutional capacity-building in the 2000s, for instance, contrasts with the *ad hoc* assistance provided in the 1990s (Leitch, 2016, p. 195). Nevertheless, donors have different aid appropriation procedures and opportunities to make commitments before their partners. Bilateral aid from OECD countries often depends on the annual appropriations approved by their parliaments on the basis of their governments' proposals (Isabekova, 2019). Multilateral aid, by contrast, depends on the contributions of funding countries and organizations. Overall, making long-term commitments beyond the scope of the specific project is problematic in cases of both bilateral and multilateral donors.

Fourth, aid flexibility is equally important to understanding stakeholders. The flexibility of donors has been emphasized in relation to the ability to adjust to local priorities and contexts (Hirschhorn et al., 2013). Strict regulations from the parliament or the government negatively impact the flexibility of the assistance by assigning it to certain purposes. Thus, driven by the goals defined by the "central" authorities of donor agencies and not necessarily by the recipients on the ground, tied aid is commonly viewed as the opposite of flexibility. However, in addition to preset objectives, flexibility is also closely related to responsiveness to changes taking place on the ground. This relates to what Leitch (2016, p. 215) calls the "institutional factors on the donors side." For this reason, in addition to adherence to recipients' priorities, I explore the decision-making authority held by the field offices of donor organizations. Authorities delegated to the "field" offices of donor organizations contribute to the ability to respond to changes on the ground. In contrast, a highly centralized organizational structure means the concentration of decision-making authorities at relevant institutions or headquarters. As a rule, this affects aid responsiveness due to bureaucratic delays.

Overall, the reflections on the roles of stakeholders as individual and collective agents and structures in the form of key issues attributed to aid relationships complement the conceptual definition of power in the first step. A combination of these two steps offers a valuable guide to the empirical analysis in the following step.

## 2.3   Project Life Cycle

To grasp the power dynamics in development assistance projects, I suggest a project-level analysis by differentiating the four phases of the assistance, namely, initiation, design, implementation, and evaluation. These stages are not consecutive since evaluation, for instance, can take place before, during, or at the end of the project. However, differentiation into phases allows for the roles of actors, as well as the division of labor between them, to be analyzed throughout the assistance process. Following Bachrach and Baratz (1962), power closely relates to agenda-setting and prioritizing the issues one wants resolved, irrespective of their relevance to the subject. Thus, a detailed analysis of project phases allows the analysis of power dynamics, including but not limited to, agenda-setting and observing stakeholder participation throughout the project realization process.

## 2.4   Uniting Theory and Empirical Findings

This step links the empirical insights from step 3 and the conceptual basis defining stakeholders, power, and the context in the first two steps to a theorization of power dynamics and aid relationships. It builds on the seven ways of creating power, defined by Haugaard (2003), as it offers a suitable basis for grasping the "power over" and "power to" forms in aid projects.

His theorization intends to organize and explain the analysis of and insight into power by scholars such as Parsons, Luhmann, Barnes, Clegg, Giddens, Bachrach and Baratz, Foucault, Lukes, Weber, Dahl, Mann, and Poggi (see Haugaard, 2003). In his theorization, Haugaard provides some structured way of understanding the power dynamics between stakeholders. His overview was utilized by Shutt (2006b) to analyze aid relationships, inspiring the inquiry and application of this approach in this book as well.

The first way of creating power is through social order, which derives its essence in the predictability assured through the intended reproduction of meaning accepted and emulated by others (Haugaard, 2003, pp. 90–91). Two practices are highlighted in this regard, namely, "structuring," which occurs through attributing a similar meaning to actions irrespective of time and place, and an agent that intentionally exercises these actions (ibid.). In development assistance, for instance, individuals and organizations aim to demonstrate the ownership of aid recipients by negotiating the objectives and adjusting them accordingly. These actions, conducted in numerous *couloirs,* represent the intended reproduction of this principle. The similarity and impersonalization in the reproduction of meaning contribute to predictability, the foundation of the social order, which also requires its broader acceptability, which is ensured by the second element. The "confirm-structuring" embodies the "public" and "willing" reproduction of the meaning, pointing to the acceptance and consensus regarding the meaning (ibid.). Accordingly, the practice mentioned above is broadly acknowledged and adhered to (more or less) by various stakeholders in multiple countries and sectors. However, consensus is not always the case, or as Haugaard notes, acceptance does not preclude conflict, as the social orders taken for granted today may not have been prevalent but were "fought for" in the past (ibid., 96). Similarly, ownership and actions regarding aid were not common throughout the history of aid.

The second way of creating power, that is, structural bias, is closely connected to the former, as it ensures the predictability and stability of the social order. Nevertheless, system bias represents a different source for creating power through structural constraints imposed by one actor upon another (Haugaard, 2003, p. 94). It is characterized by the process in which stakeholders or initiatives inconsistent with and aiming to change the prevalent social order face the "noncollaboration" of those who reproduce it (ibid.). Thus, the constraints would ensure stakeholder compliance with the principle of ownership and the unacceptance of those unwilling to implement it upon receiving assistance. This supports the premise that structural biases are not necessarily negative but rather essential to the stability of the social system (Haugaard, 2003, pp. 93–96). However, constraints do not empower stakeholders to the same extent. Predictability is ensured at the expense of certain forms of interaction and power that might have benefited stakeholders disadvantaged by the social order (ibid.). However, actors comply with constraints for reasons that are not necessarily "consciously chosen" and could be an outcome of structural and cultural patterns and other reasons (Lukes, 2005, pp. 25–26). In addition to consensus regarding the order, social biases may also involve conflict in the form of destruction or evasion from supporting the system (Haugaard, 2003, pp. 95–96).

The third way of creating power relates to social consciousness, which is in the foreground of the reproduction of specific order and biases. Predictability here derives from stakeholders' perceptions that specific meanings are not "arbitrary" and exist "out there in the world" but are consistent with their interpretations of the matter at stake and the world in general (Haugaard, 2003, pp. 97–99). Thus, the critical aspect of reproducing meaning and biases lies in their relation to the actors' systems of thought. For instance, gender equality and women's empowerment in development assistance relate to the consciousness of gender equality in a broader context. Prevalent among members of the Organization for Economic Co-operation and Development's Development Assistance Committee, this explicit emphasis today may not have been common some time ago or in contexts with different perceptions of gender. Similarly, the reasoning behind some health care intervention programs relating the cognitive abilities and behaviors of individuals to their ethnic background would be unthinkable today (see Morgan, 1993). In this way, changes in the system of thought result in practices common to the previous system becoming obsolete (Haugaard, 2003, pp. 98–99).

The fourth way of creating power is the relationship between tacit and discursive knowledge. It is based on Giddens' (1984, pp. 4–5) differentiation between the "practical" or unconscious knowledge actors use in their social lives and the "discursive" or conscious knowledge underpinned by their reflection, rationale, and reasoning. Haugaard (2003, pp. 100–102) notes that most knowledge in social life is tacit to ensure stakeholders' comprehension of reality. However, a transformation of this tacit knowledge into discursive knowledge allows distance, evaluation, and recognition (ibid.). An instance of this transformation could be, for instance, reflections on the meaning of development, which changed from an initial economic focus to a broader operationalization of development to include social, environmental, and other factors. Haugaard notes that tacit knowledge may benefit those in power, but its transformation into discursive knowledge may change the order if actors realize that the reproduction of social structures disadvantaged them (ibid.). Following the example mentioned above, critical reflections on development, in combination with the evidence for the noticeable shortcomings of a solely economic reform focus (see Cornia et al., 1987, 1988), offered favorable circumstances for targeting social needs neglected before.

The fifth way of creating power, that is, reification, stabilizes existing power relations and structural reproduction on the premise that they embody something "more than social constructs" (Haugaard, 2003, pp. 102–103). This source of power resembles the system of thought but differs from it in one dimension. Stakeholders conform to structures and related practices, not because of their relations to actors' perceptions of reality but for other reasons. This resemblance and distinction are not explicitly stated in the theoretical framework but are nevertheless critical to its valid application. Following Haugaard (2003, pp. 103–105), "non-arbitrariness" in reification could be based on multiple grounds, including religion, nature, truth, and science, with practices incompliant with existing structures as going against these concepts. For example, health care programs frequently appeal to science to support their objectives and activities, which could be a vivid example of reification in practice.

The sixth way of creating power, that is, discipline, is manifested through socialization aimed at establishing a routine based on practical consciousness (Haugaard, 2003, pp. 105–106). This, in a way, follows the logic opposite to the one described in the relationship between tacit and discursive knowledge. It does not result in predictability but originates from it (ibid.). The assumption consonant with what Barnes (1988, p. 58) referred to as social power is at the disposal of those able to judge and decide upon routines. Haugaard (2003, pp. 106–107) notes that as a relatively modern phenomenon, this power is based on the premise that discipline is not arbitrary, and practices inconsistent with it are "irregularities" and foes of the social order. In the context of development assistance, disciplinary power could relate to functions agreed upon and assigned to stakeholders in projects and the relevant training they received. However, closely related to stakeholders' socialization and use of practical knowledge, disciplinary power is also limited to the extent to which stakeholders "internalize" the suggested routines (ibid.). These limitations are explicitly discussed in Chap. 8 of this book.

The seventh way of creating power is coercion, which represents the last resort measure. Haugaard (2003, pp. 108–109) builds an analogy with physical power and suggests that it comes into play when other sources of power fail, approaching coercion as neither an "ultimate" nor "effective" source of power. This perception is consonant with that of Arendt (1970), who defined coercion as the weakest form of power, as well as that of Parsons (1963), who stated that coercion was not equal to power. Coercion is often discussed in relation to freedom. However, presuming that stakeholders have freedom and make decisions depending on their interests and structural factors, I approach coercion as an act of compelling that is contrary to the will and interests of stakeholders. Although associated with physical power, coercion takes other forms in stakeholder relationships, as discussed in Chap. 8 of this book.

Overall, the seven ways of creating power help theorize the roles of stakeholders and their actions in relation to each other, as empirically discussed in the third step. This theorization is necessary to grasp how the conceptualization of power, agents, and structures unfolded in selected cases. More specifically, by defining ways of creating power, I expand on the types of power, how actors used resources, or whether a conflict/consensus accompanied this process. This theorization is critical to validating the empirical findings in relation to the type of power ("over/"to") stakeholders created.

The type of power is essential to understanding whether an alternative approach to stakeholder relationships, characterized by interdependence, recipient agency, and changing power dynamics, occurred. In contrast, it could also be the case that the conventional approach to stakeholder relationships, mainly characterized by inequality, took place. However, I refrain from attributing specific ways of creating power to any of these two approaches or types of aid relationships discussed in the following subsection. The discussion of the seven ways clearly showed that the "power over" or the "power to" forms could equally be produced by some stakeholders, depending on how they use their resources or whether there is a conflict/consensus. Nevertheless, the seven ways of creating power provide a good indication of whether "power over" and/or "power to" emerged between stakeholders. Assuming that the former is a characteristic of the conventional approach to stakeholders and their relationships with each other, I suggest that the types of relationships associated with inequality have "power over" as the prevailing form. In contrast, aid relationships associated with equality have "power to" as the prevailing form of power emerging between actors.

However, given that power may be changing throughout the project cycle, I acknowledge that in some cases, it could be a combination of the "power over" and "power to" forms. Here, the roles of stakeholders throughout the project cycle and the analysis of structures (aid dependency, capacity, aid volatility, and flexibility) may help. If stakeholders were equally engaged throughout the project life cycle and structures changed in favor of the aid recipient, the "power to" prevailed; otherwise, the "power over" form prevailed. Why does this matter? Equal engagement of stakeholders addresses the problem of limited involvement of aid recipients and aid fragmentation (Chap. 1). It also demonstrates the recipients' agency and abilities to raise issues or participate in decision-making. Structures, in turn, demonstrate whether the context in which aid relationships took place was favorable to changing the hierarchy underlined in the conventional approach. I argue that aid volatility and inflexibility of providers, as well as aid dependency and limited capacity on recipients' sides, are favorable to retaining the hierarchy among stakeholders. In contrast, aid predictability and flexibility, accompanied by relatively limited aid dependency but necessary capacity, are beneficial for altering the hierarchy.

I argue that considerations regarding engagement in the project cycle and structures are case-dependent and not attached to a specific type of aid relationship. However, inequality and hierarchy are attributed to certain types, as discussed below.

Aid relationships in this book encompass a variety of relationships between stakeholders, including noncoordination, coordination, unequal cooperation encompassing recipient/donor-driven cooperation, and a "utilitarian approach toward CSOs, and equal forms of cooperation, such as partnerships, and an "empowerment" approach toward CSOs. I acknowledge that my findings in this book demonstrate the power dynamics limited to the time and space covered in the empirical chapters. However, further theorizing regarding the power dynamics in aid relationships helps us abstract from project phases to a broader conceptualization of stakeholder interaction. Therefore, this section operationalizes different types of relationships, building on the findings and filling the gaps in the relevant academic literature and international agreements. The suggested types of interaction are Weber's "ideal" types, which combine the features found and relevant to them in reality but in a manner that is coherent with each type of interaction (Oxford Reference, 2023).

First, noncoordination may range from the noninteraction of actors with each other to the noncompliance of one actor with the priorities of another. Among donors, this means no exchange of information, resulting in unawareness of each other and a subsequent possible duplication of activities. For interactions between donors and recipient states, noncoordination may refer to a donor(s) pursuing activities without exchanging information with the state or without complying with its priorities. Regarding the interaction of donors and the recipient state with CSOs, noncoordination is expressed by the noninvolvement of the latter in development assistance. Since the coordination of activities is time- and resource-consuming, noncoordination could be beneficial to actors from the short-term perspective, as this does not require time and an additional workload in contrast to coordination or cooperation.

I suggest that noncoordination takes place in cases of inequality between donors and recipients. The recipient is too aid-dependent to raise the issue of noncoordination or has no capacity to require/implement the donors' compliance with its requirements. Donors, in turn, are disincentivized by the time and resources needed for coordination to initiate this voluntarily. These disincentives may further be exacerbated by limited aid flexibility and certainty. As coordination and cooperation presume a certain level of adjustment and awareness of aid flows, aid volatility and inflexibility may be accompanying factors of noncoordination. Potential incentives for donors to coordinate with each other could relate to increasing their influence over the recipient. However, as the recipient is aid-dependent, each donor may already have leverage over it and may not understand the benefits of coordination.

Second, coordination among stakeholders is expressed by the parallel implementation of activities with information exchange. It is characterized by an agreement on priorities, but actors pursue their activities toward these priorities without involving each other. This agreement on priorities entails the recipient's capacity and/or the donors' willingness to engage in coordination. The recipient's capacity is essential since the recipient is expected to request and, most importantly, ensure donor coordination and compliance with its priorities. However, coordination could also be the outcome of donor initiatives. The reasoning behind the coordination taking place as a result of the donor's initiative could be (in addition to altruistic motives) the donor's interest in increasing their influence over the recipient. Thus, the recipient may still be aid-dependent, but the influence of the individual donor may not be sufficient or as high as in the case of noncoordination.

Coordination is beneficial to both recipients and providers of aid in the long-term, as it decreases transaction costs. Therefore, similar to cooperation, coordination has been emphasized in a number of international documents (e.g., OECD, n.d.). However, the costs stakeholders face overshadow its long-term benefits in the short term. Its establishment requires staff involvement and time for negotiations that may take up to several years (Lawson, 2013). Facing a trade-off between long- and short-term benefits, the actors may favor the latter.

Third, in the context of development assistance, I define cooperation as a joint realization of aid, which may take unequal and equal forms. The former takes place when one of the actors, be it a provider or a recipient, dominates in the aid realization process, while the latter means that actors are equally engaged in the process. This equality is not only about the stakeholders' responsibilities but is also about their roles and "say" in the process. Although "equal" and "unequal" labels of cooperation inevitably recall normative connotations, I approach them as mere variations of cooperation. Following Weber (1986, pp. 28–29), I acknowledge that domination, in its general meaning, is an essential component of social action and can take different forms in regard to how one imposes their "own will upon the behavior" of others. Dominance in unequal cooperation takes different forms, depending on the stakeholders involved.

Between donors, cooperation means that one donor relies on the operational procedures of another and complies with its regulations and approaches. The donor may have a "leading" role due to a larger share of finances or taking over the responsibility for the outcomes. This emergence of the "dominant" actor could be the outcome of other donors being "less motivated" to compete or "much poorer" (Bueno de Mesquita & Smith, 2016, p. 2). Cooperation presumes a certain level of flexibility and certainty necessary for negotiations among stakeholders and relevant adjustments that are possibly more demanding than that required in coordination, in which stakeholders agree on priorities and not on operational procedures and approaches.

In relationships between donors and recipient states, unequal cooperation takes two forms, namely, cooperation driven by donors and their conditions and aid driven by the recipient state and its priorities. There is an extensive discussion of conditionality in the development aid literature concerning the requirements the recipient was expected to fulfill in economic, political, and other terms (e.g., Crawford & Kacarska, 2019; Molenaers et al., 2015). This type of interaction is probably characterized by aid dependency and capacity issues on the recipient's side that also define the flexibility and predictability possible and provided by the donor. Thus, if the recipient is relatively aid-independent and/or has the capacity, the donor may be forced to be more flexible and predictable to enforce its conditions.

Another form of unequal cooperation between the donor and the recipient is when aid is driven by the priorities of the latter, with donors adjusting their activities accordingly. The idea of the recipient state being the "driver" of interaction complies with the notion of "ownership," as emphasized in the Paris Agenda, by increasing its role in the process and in achieving assistance results. This type of cooperation presumes high capacity and relative independence of recipients from external aid, accompanied by flexibility and predictability of assistance, which may not be immediately offered but are achieved during the negotiation process.

In civil society organizations' interactions with other stakeholders, unequal cooperation refers to a "utilitarian" approach. This approach was initially used to discuss community participation in development aid (Morgan, 2001). The analytical framework presented here extends to explaining the interaction of CSOs with other actors. Following the "utilitarian" perspective, communities (in this book CSOs) are involved in development assistance through "passive means" used to reach the project objectives (Rasschaert et al., 2014, p. 7). According to the "utilitarian" perspective, CSOs are dependent on "external" assistance due to their low capacity and structural barriers (ibid.). These include, for instance, illiteracy (Jana et al., 2004), gender-related biases (WHO, 2008), the political situation in the country, and poverty (Fawcett et al., 1995). In these conditions, donors may not have sufficient incentives to offer the predictability and flexibility regarding their assistance. Moreover, even if provided, CSOs may not be able to negotiate for these characteristics because of the fear of losing access to funding.

Fourth, equal cooperation takes place when all the actors are involved in the aid realization process and have an equal say in it. None of the parties dominate aid realization. This notion of equality recalls partnerships that are founded on equality, trust (Hyden, 2008), nonconditionality (Abrahamsen, 2004), and shared responsibilities and authority. Trust is ensured in partnerships in which stakeholders fulfill their commitments (Del Biondo, 2020); however, this is often problematic, as actors may break their promises in the face of pressure from their constituencies, parliaments, and so forth. Ideally, a partnership has no conditions, meaning that stakeholders fulfill their responsibilities voluntarily. However, conditionality may be inevitable and still be present in a partnership as long as it applies to all parties. Furthermore, I also define shared responsibilities and authorities that are essential to partnerships, as these provide a necessary underpinning for equality in the aid realization process.

In general, equal cooperation or partnership rarely takes place in practice, although there are some exceptions. The most pressing issue for donors in equal cooperation contexts is the adoption of joint procedures, which is essential to the joint implementation and evaluation of the assistance. These aspects require lengthy discussions and a high degree of flexibility and predictability. Given the complex structures of donors and their adherence to their own rules, this might be problematic to implement in practice. Between donors, equal cooperation seldom occurs because of "harmonization" issues, although there are exceptions among medium-sized organizations sharing similar perspectives (see Isabekova, 2019). In the relationships of donors with recipient states, equal cooperation similarly assumes the presence of equality, trust, nonconditionality, and shared responsibilities between the parties. This is problematic due to power dynamics and the inherent inequality between the actors, as discussed in the beginning of this chapter. The recipients might be reluctant to participate or criticize the donor because of the fear of donors cutting funding (Hinton, 2004). As the agenda is still set by donors (Nissanke, 2008), partnerships (or equal cooperation) might be viewed as "little more than rhetoric" (Abrahamsen, 2004, pp. 1455–1456). Because of aid dependence and limited capacity, equal cooperation rarely takes place between recipient countries and their donors.

Finally, the relationship of donors and recipient states with the CSOs' definition of equal cooperation in the analytical framework presented here is based on the "empowerment" approach. Similar to the "utilitarian" perspective described in unequal cooperation, this approach was initially suggested for community involvement (Morgan, 2001, p. 223). This chapter extends it to cover CSOs. Empowerment is a "process of gaining influence over conditions that matter to people" (Fawcett et al., 1995, p. 679). In development aid, CSOs are able to express their concerns, set priorities, and participate in negotiations and the decision-making process. They equally cooperate with other partners by participating throughout the assistance process. Following this approach, CSOs are viewed as the source of initiative rather than "passive" aid recipients (Morgan, 2001, p. 223; Rasschaert et al., 2014, p. 7). However, there is inherent inequality between donors, recipient states, and CSOs because of the differences in resources and the structure of development assistance (see the section on power dynamics). The power dynamics further vary among CSOs. CBOs are relatively aid-dependent and require more capacity-building activities. There is evidence that at the end of a development project, CBOs will continue its activities if it continues to receive funding from another donor; otherwise, they will cease or decrease their activities (Ahluwalia et al., 2010). The dependence of CBOs on donors is clearly illustrated by the statement of one CBO member in Central Asia: "getting a grant is similar to receiving money from God" (Earle et al., 2004, p. 34). In contrast to CBOs, NGOs might also be aid-dependent but have a relatively higher capacity, although there is variation between local, national, and international NGOs. An organization with several branches across the country or in several countries has more human and financial resources than one operating in a village or a town.

Referring to aid relationships, I intended to synergize the narrow and general approaches in the international documents and development aid literature. The classification of the "ideal" types of interactions among stakeholders presented above serves two purposes. First, it provides the level of abstraction necessary to observe the link between interaction among stakeholders and the sustainability of development aid. Second, this level of abstraction is critical to our comprehension of complex relationships and learning beyond specific cases. In other words, how cases are selected for empirical research may help us understand both aid relationships and their connotations of sustainability in other contexts.

# References

Abrahamsen, R. (2004). The power of partnerships in global governance. *Third World Quarterly, 25*(8), 1453–1467. https://doi.org/10.1080/014365904 2000308465

Ahluwalia, I. B., Robinson, D., Vallely, L., Gieseker, K. E., & Kabakama, A. (2010). Sustainability of community-capacity to promote safer motherhood in northwestern Tanzania: What remains? *Global Health Promotion, 17*(1), 39–49. https://doi.org/10.1177/1757975909356627

Allen, A. (1998). Rethinking power. *Hypatia, 13*(1), 21–40.

Andrews, M. (2013). *The limits of institutional reform in development: Changing rules for realistic solutions* (Illustrated Ed.). Cambridge University Press.

Arendt, H. (1970). *On violence.* Houghton Mifflin Harcourt.

Bachrach, P., & Baratz, M. S. (1962). Two faces of power. *The American Political Science Review, 56*(4), 947–952. https://doi.org/10.2307/1952796

Baldwin, D. A. (1971). The power of positive sanctions. *World Politics, 24*(1), 19–38. https://doi.org/10.2307/2009705

Barnes, B. (1988). *The nature of power.* University of Illinois Press.

Bueno de Mesquita, B., & Smith, A. (2016). Competition and collaboration in aid-for-policy deals. *International Studies Quarterly, 60*(3), 413–426. https://doi.org/10.1093/isq/sqw011

Campbell, J. L. (2004). *Institutional change and globalization.* Princeton University Press.

Cornia, G. A., Jolly, R., Stewart, F., Cornia, G. A., Jolly, R., & Stewart, F. (Eds.). (1987). *Adjustment with a human face: Volume 1, protecting the vulnerable and promoting growth.* Oxford University Press.

Cornia, G. A., Jolly, R., Stewart, F., Cornia, G. A., Jolly, R., & Stewart, F. (Eds.). (1988). *Adjustment with a human face: Volume 2, ten country case studies.* Oxford University Press.

Crawford, G., & Kacarska, S. (2019). Aid sanctions and political conditionality: Continuity and change. *Journal of International Relations and Development, 22*(1), 184–214. https://doi.org/10.1057/s41268-017-0099-8

Dahl, R. A. (1957). The concept of power. *Behavioral Science, 2*(3), 201–215. https://doi.org/10.1002/bs.3830020303

Dahl, R. A. (2005). *Who governs?: Democracy and power in an American City* (2nd ed.). Yale University Press.

Del Biondo, K. (2020). Moving beyond a donor-recipient relationship? Assessing the principle of partnership in the joint Africa–EU strategy. *Journal of Contemporary African Studies, 38*(2), 310–329. https://doi.org/10.108 0/02589001.2018.1541503

Dowding, K. (2011a). Consent. In *Encyclopedia of power* (pp. 137–138). SAGE Publications.

Dowding, K. (2011b). Agency. In *Encyclopedia of power* (pp. 6–10). SAGE Publications.

Dowding, K. (2011c). Agency-structure problem. In *Encyclopedia of power* (pp. 10–11). SAGE Publications.

Dowding, K. (2017). *Social and political power.* Oxford Research Encyclopedia of Politics. https://doi.org/10.1093/acrefore/9780190228637.013.198.

Earle, L., Fozilhujaev, B., Tashbaeva, C., & Djamankulova, K. (2004). Community development in Kazakhstan, Kyrgyzstan and Uzbekistan: Lessons learnt from recent experience. *Occasional Papers Series, 40*, 1–63.

European Centre for Development Policy Management. (2008). Capacity change and performance: Insights and implications for development cooperation. *Policy Management Brief, 21*, 1–12.

Eyben, R. (2008). *Power, mutual accountability and responsibility in the practice of international aid: A relational approach.* IDS Working Paper, 305. Retrieved February 20, 2023, from https://opendocs.ids.ac.uk/opendocs/bitstream/handle/20.500.12413/4164/Wp305.pdf?sequence=1&isAllowed=y

Fawcett, S. B., Paine-Andrews, A., Francisco, V. T., Schultz, J. A., Richter, K. P., Lewis, R. K., et al. (1995). Using empowerment theory in collaborative partnerships for community health and development. *American Journal of*

*Community Psychology,* 23(5), 677–697. https://doi.org/10.1007/BF02506987

Feyissa, D. (2011). Aid negotiation: The uneasy "partnership" between EPRDF and the donors. *Journal of Eastern African Studies, 5*(4), 788–817. https://doi.org/10.1080/17531055.2011.642541

Foucault, M. (2002). The subject and power. In J. D. Faubion (Ed.), *Power essential works of Foucault 1954-1984* (Vol. 3, pp. 326–348). Penguin Books.

Frontera. (2007). *Motivating Staff and Volunteers. Working in NGOs in the South.* Prepared for People In Aid by FRONTERA, an international management and development consulting organisation. Retrieved March, 2019, from https://www.refugeerightstoolkit.org/wp-content/uploads/2013/11/Motivating-Staff-and-Volunteers.pdf

Giddens, A. (1984). *The constitution of society. Outline of the theory of structuration.* University of California Press.

Glennie, J., & Prizzon, A. (2012). *From high to low aid: a proposal to classify countries by aid receipt.* Retrieved February 28, 2023, from https://cdn.odi.org/media/documents/7621.pdf

Goldhamer, & Shilds, E. A. (1973). Types of power and status. In L. H. Ross (Ed.), *Perspectives on the social order. Readings in sociology* (pp. 299–306). McGraw-Hill.

Haugaard, M. (2002). *Power: A reader.* Manchester University Press.

Haugaard, M. (2003). Reflections on seven ways of creating power. *European Journal of Social Theory, 6*(1), 87–113. https://doi.org/10.1177/1368431003006001562

Hinton, R. (2004). Enabling inclusive aid: Changing power and relationships in international development. In L. C. Groves & R. B. Hinton (Eds.), *Inclusive aid: Changing power and relationships in international development* (pp. 210–220). Earthscan.

Hinton, R., & Groves, L. (2004). The complexity of inclusive aid. In L. C. Groves & R. B. Hinton (Eds.), *Inclusive aid: Changing power and relationships in international development / edited by Leslie Christine Groves and Rachel Barbara Hinton* (pp. 3–20). Earthscan.

Hirschhorn, L. R., Talbot, J. R., Irwin, A. C., May, M. A., Dhavan, N., Shady, R., et al. (2013). From scaling up to sustainability in HIV: Potential lessons for moving forward. *Globalization and Health, 9*(57), 1–9. https://doi.org/10.1186/1744-8603-9-57

Hyden, G. (2008). After the Paris declaration: Taking on the issue of power. *Development Policy Review, 26*(3), 259–274. https://doi.org/10.1111/j.1467-7679.2008.00410.x

Isabekova, G. (2019). The relationships between stakeholders engaged in development assistance: Towards an analytical framework. *SOCIUM SFB 1342 Working Papers, 3*, 1–28.

Jana, S., Basu, I., Rotheram-Borus, M. J., & Newman, P. A. (2004). The Sonagachi project: A sustainable community intervention program. *AIDS Education and Prevention, 16*(5), 405–414. https://doi.org/10.1521/aeap.16.5.405.48734

Lamothe, H. D. (2010). Re-conceptualizing the international aid structure: Recipient donor interactions and the rudiments of a feedback mechanism. *CEPAL—Serie Financiamiento del desarrollo, 234*, 1–29.

Lawson, M. L. (2013). *Foreign aid: International donor coordination of development assistance.* Congressional Research Service. Retrieved February 28, 2023, from https://sgp.fas.org/crs/row/R41185.pdf

Ledyaev, V. (2021). Conceptual analysis of power: Basic trends. *Journal of Political Power, 14*(1), 72–84. https://doi.org/10.1080/2158379X.2021.1877002

Leitch, D. (2016). *Assisting reform in post-communist Ukraine 2000-2012: The illusions of donors and the disillusion of beneficiaries.* Ibidem Press.

Lensink, R., & White, H. (1999). Aid dependence. Issues and indicators. *Expert Group on Development Issues, 2*, 1–86.

Lukes, S. (2005). *Power: A radical view* (2nd ed.). Palgrave Macmillan.

Menocal, A. R., & Mulley, S. (2006). Learning from experience? A review of recipient- government efforts to manage donor relations and improve the quality of aid. *ODI Working Paper, 268*, 1–25.

Molenaers, N., Dellepiane, S., & Faust, J. (2015). Political conditionality and foreign aid. *World Development, 75*, 2–12. https://doi.org/10.1016/j.worlddev.2015.04.001

Morgan, L. M. (1993). *Community participation in health: Politics of primary Care in Costa Rica.* Cambridge University Press.

Morgan, L. M. (2001). Community participation in health: Perpetual allure, persistent challenge. *Health Policy and Planning, 16*(3), 221–230.

Nissanke, M. (2008). Donor-recipient relationship in the aid: Effectiveness debate. In A. M. Jerve, Y. Shimomura, & A. S. Hansen (Eds.), *Aid relationships in Asia: Exploring ownership in Japanese and Nordic aid* (pp. 22–40). Palgrave Macmillan.

OECD. (n.d.). The Paris Declaration on Aid Effectiveness (2005) and Accra Agenda for Action (2008). Retrieved February 15, 2023, from http://www.oecd.org/dac/effectiveness/34428351.pdf

OECD. (2011). *Perspectives note: The enabling environment for capacity development*. Retrieved February 28, 2023, from https://www.oecd.org/development/accountable-effective-institutions/48315248.pdf

Oxford Reference. (2023). *Ideal type*. Retrieved February 16, 2023, from https://www.oxfordreference.com/display/10.1093/oi/authority.20110803095956574;jsessionid=8CB5C64761000B0277F2E49F5D0C7051

Pansardi, P., & Bindi, M. (2021). The new concepts of power? Power-over, power-to and power-with. *Journal of Political Power, 14*(1), 51–71. https://doi.org/10.1080/2158379X.2021.1877001

Parsons, T. (1963). On the concept of political power. *Proceedings of the American Philosophical Society, 107*(3), 232–262.

Pitkin, H. F. (1972). *Wittgenstein and justice: on the significance of Ludwig Wittgenstein for social and political thought*. University of California Press. Retrieved February 28, 2023, from http://archive.org/details/wittgensteinjust00pitk

Rasschaert, F., Decroo, T., Remartinez, D., Telfer, B., Lessitala, F., Biot, M., et al. (2014). Sustainability of a community-based anti-retroviral care delivery model—A qualitative research study in Tete, Mozambique. *Journal of the International AIDS Society, 17*(18910), 1–10. https://doi.org/10.7448/IAS.17.1.18910

Renzio, P. (2006, January 1). *Promoting mutual accountability in aid relationships*. Synthesis Note. Retrieved March 4, 2020, from https://www.odi.org/sites/odi.org.uk/files/odi-assets/events-documents/3586.pdf

Robb, C. (2004). Changing power relationships in the history of aid. In L. C. Groves & R. B. Hinton (Eds.), *Inclusive aid: Changing power and relationships in international development* (pp. 21–41). Earthscan.

Schafer, J., Haslam, P. A., & Beaudet, P. (2012). Meaning, measurement and morality in international Development. In P. A. Haslam, J. Schafer, & P. Beaudet (Eds.), *Introduction to international development: Approaches, actors, and issues* (2nd ed., pp. 3–27). Oxford University Press.

Scharpf, F. W. (1997). *Games real actors play: Actor-centered institutionalism in policy research*. Westview Press.

Shutt, C. (2006a). Money matters in aid relationships. In R. Eyben (Ed.), *Relationships for aid* (pp. 154–170). Earthscan.

Shutt, C. (2006b). Power in aid relationships: A personal view. *IDS Bulletin, 37*(6), 79–87. https://doi.org/10.1111/j.1759-5436.2006.tb00325.x

Simon, H. A. (1972). Theories of bounded rationality. In C. B. McGuire & R. Radner (Eds.), *Decision and organization: A volume in honor of Jacob Marschak* (pp. 161–176). North-Holland publishing company.

Swedlund, H. J. (2017). *The development dance: How donors and recipients negotiate the delivery of foreign aid* (1st ed.). Cornell University Press.

Toornstra, F., & Martin, F. (2013). Building country capacity for development results: How does the international aid effectiveness agenda address the capacity gaps? In H. Besada & S. Kindornay (Eds.), *Multilateral development cooperation in a changing global order. // multilateral development cooperation in a changing global order* (pp. 89–114). Palgrave Macmillan.

Weber, M. (1986). Domination by economic power and by authority. In S. Lukes (Ed.), *Power* (pp. 28–36). New York University Press.

WHO. (2008). *Community involvement in tuberculosis care and prevention Towards partnerships for health: Guiding principles and recommendations based on a WHO review*. Retrieved February 28, 2023, from http://apps.who.int/iris/bitstream/10665/43842/1/9789241596404_eng.pdf

World Bank Group. (2023). *External health expenditure (% of current health expenditure)—Kyrgyz Republic*. Retrieved February 28, 2023, from https://data.worldbank.org/indicator/SH.XPD.EHEX.CH.ZS?locations=KG

# 3

# Sustainability of Health Assistance

Sustainability is a multidisciplinary and multidimensional phenomenon. From the 1970s to the 1990s, studies on sustainability and sustainable development focused on the impact of human activity on the environment (Giovannoni & Fabietti, 2013). Only in the late 1980s did the perception of sustainability expand beyond ecology, nature conservation, and environmental degradation to include the social and economic aspects of this phenomenon (see Kidd, 1992). The 1987 "Our Common Future" report of the United Nations (UN) World Commission on Environment and Development (the Brundtland Report) and the 1992 Earth Summit in Rio manifested this multidimensionality, contributing to a "three-pillar" (environment, economic, and social) perspective of sustainability (see Purvis et al., 2019). Although the practical feasibility of this approach (Károly, 2011) and simultaneous attainment of all three dimensions to the same extent (Boussemart et al., 2020) remain unsettled, the three-pillar perspective manifested itself in the UN Sustainable Development Goals (SDGs 2015–2030). The SDGs, *per se*, embody the multidisciplinarity and multidimensionality of sustainability as a phenomenon.

At the same time, multidisciplinarity is among the main reasons for the ambiguity of the literature on sustainability. Following the Brundtland Report (1987), multiple authors discussed the interrelation, (in)

© The Author(s) 2024
G. Isabekova, *Stakeholder Relationships And Sustainability*, Global Dynamics of Social Policy, https://doi.org/10.1007/978-3-031-31990-7_3

compatibility, and balanced representation of the three pillars, along with the indicators and factors relevant to sustainability (e.g., Purvis et al., 2019). A multiplicity of studies contributed to the establishment of sustainability science—an interdisciplinary field aimed at identifying indicators and methods for sustainability research (Kajikawa, 2008). Nevertheless, the conceptual and theoretical underpinning behind the integrated approach toward sustainability remained weak, leading to difficulties in defining and characterizing sustainability (Purvis et al., 2019). This opacity is echoed in terminological inconsistency and conceptual ambiguity observed in the literature on the sustainability of development assistance to health care.

Sustainability is often used interchangeably with, among other terms, routinization, institutionalization, adaptability, resilience, and continuity (Gruen et al., 2008; Kiwanuka et al., 2015). However, if routinization and institutionalization focus on the standardization and integration of practices at organizational levels (Pluye et al., 2004; Scheirer & Dearing, 2011), sustainability refers to the integration of a practice or a change in the system as a whole. Similarly, the adaptability or flexibility of a program when faced with situational changes (Shigayeva & Coker, 2015) and the resilience or ability of the system (or a program) "to maintain" itself (Mayer, 2008, pp. 278–279) and continuity describe specific characteristics of sustainability but not the term itself. Depending on internal and external changes and challenges, interventions are likely to be adjusted or to retain certain features. However, it is not clear which aspects and to what extent they need to be continued for an intervention to be "sustainable" (Stirman et al., 2012, p. 10). Thus, all the terms mentioned above denote specific parts of sustainability but not the phenomenon as a whole.

Terminological inconsistency is an outcome of fragmentation of the literature on the sustainability of development assistance to health care. There is a prevalence of case studies contributing to context-specific knowledge but leaving the question of general implications unanswered. On reviewing the literature on the sustainability of health care programs in sub-Saharan Africa, Latin America, and Southeast Asia, it becomes clear that the research focuses on either systematic literature reviews or empirical analysis of interventions (i.e., case studies), often without any theoretical underpinning. This fragmentation and underdevelopment of the literature (Iwelunmor et al., 2016; Stirman et al., 2012) is reflected in the recommendations for further research on sustainability in health care, which emphasize, among other issues, the conceptual clarity and necessity for understanding the link between specific characteristics of interventions and contextual factors to the sustainability of interventions (Proctor et al., 2015).

A genuine understanding of sustainability and related factors also presupposes the awareness of biases behind this phenomenon. Sustainability is an "inherently political" phenomenon (Purvis et al., 2019, p. 692), which has a different meaning for different actors (Morgan, 2001). For donors, it implies the long-term financial costs of the program being taken over by the recipient, while for the recipient, sustainability refers to the freedom to make changes to enable the continuity of a program over time (Walsh et al., 2012). This dissimilarity in interpretation suggests that different actors are unlikely to have a similar understanding of what is sustainable and what is not. However, because of unequal power dynamics, the definition of sustainability and the factors associated with it may reflect the interests of stronger groups and not necessarily the recipients of health aid (see Murphy, 2012). The subsections below expand on the operational framework, which aims to overcome the terminological and conceptual ambiguity of sustainability in the context of health aid. Adapting and extending the existing analytical models, it intends to provide an equitable framework for analyzing various perspectives of sustainability and the factors related to it.

## 3.1 Operationalizing Sustainability

Sustainability in empirical terms requires identifying a matter of interest, relevant actors, a timeframe, and the extent of sustainability—in other words, "what, how or by whom, how much and by when" to sustain (Iwelunmor et al., 2016, p. 2). This book focuses on development assistance for health care and the roles of multiple stakeholders in ensuring sustainability, which is not achieved by a single or limited number of actors (see Chap. 2). As indicated in the introduction to this book, the

**Table 3.1** Sustainability in empirical terms

| | |
|---|---|
| "What" to be sustained | Ongoing and completed health care projects |
| "How/by whom" | Donors and recipients of the assistance |
| "By when" | Up to the present day |
| "How much" | Emphasis on the continuous and not on the categorical (yes/no) nature of this phenomenon |

majority of health care programs are not sustained beyond the end of donor financing. Yet, sustainability does not automatically come along with the end of the financing; it is rather built throughout the realization of a health care aid. The case studies of health care programs selected for this book represent completed and ongoing projects (Table 3.1). Both types of projects are of equal value to understanding the sustainability of health care aid to the present day. Most data I have collected on selected projects range from 1991, when the Kyrgyz Republic gained its independence, to 2018, when I conducted the second fieldwork. The factual data were updated in 2022 to reflect the current state of organizations and selected programs in the face of the global coronavirus disease-19 (COVID-19) pandemic. Therefore, "up to the present day" in this book denotes the state of play at that specific point in time when data were collected (see Sztompka, 1993, p. 12). Last, sustainability is a "matter of degree rather than an all-or-none phenomenon" (Shediac-Rizkallah & Bone, 1998, p. 96). However, I refrain from assigning weak and strong or partial and full ranges to sustainability, as these measurements are inherently subjective (see Savaya et al., 2008). Instead, I define sustainability *vis-à-vis* the three perspectives described in the following subsection.

## 3.2   Conceptual Definition

This book adopts Shediac-Rizkallah and Bone's (1998) conceptualization of sustainability as maintaining benefits, continuing program activities, and building the capacity of a recipient community. First, maintaining benefits refers to services or infrastructure provided within development assistance (Altman, 1995; Torpey et al., 2010). This book focuses on the services received by the population targeted by the assistance (i.e., the project beneficiaries) and, where relevant, on any use of hardware or infrastructure provided to the beneficiaries, continuing beyond the duration of the program. Second, in terms of continuity of project activities, it overviews the activities continued and discontinued by the end of the project. Self-evident in the cases of completed health care projects, these two aspects can, however, also be assessed in ongoing development programs. This book identifies arrangements made by aid recipients to maintain the services and continue project activities at their own expense or through financing from other donors as implications for the sustainability of these services and activities beyond the duration of the initial project funding. Third, community capacity-building presumes, among other issues, empowerment of a recipient community through a development program (see Shediac-Rizkallah & Bone, 1998).

Broad but nevertheless precise, Shediac-Rizkallah and Bone's (1998) operationalization of sustainability is cited by multiple authors. However, studies largely concentrate on one or several of three aspects: maintenance of benefits (e.g., Chambers et al., 2013; Johnson et al., 2004), continuity of project activities (Cassidy et al., 2006; Schell et al., 2013), and community capacity-building (Alexander et al., 2003). This book incorporates all three aspects to characterize the sustainability of development assistance for health care. This ensures a comprehensiveness of the analysis and a balance of donors' and recipients' perspectives on this phenomenon. Some authors elaborate further on these three categories. Scheirer and Dearing (2011), for instance, assess the maintenance of procedures and policies promoted during project implementation, continued attention to the problem, and dissemination of program ideas and activities. Although useful for an in-depth analysis of specific aspects of the assistance, these indicators may be onerous for assessing the sustainability of development projects as a whole.

This book adheres to the original conceptualization by Shediac-Rizkallah and Bone (1998), although with further adaptations to resolve the ambiguity of the third aspect. Not as straightforward as the two others, it requires further consideration of nuances related to the operationalizations of "community" and "capacity-building."

The romanticization of "community" and approaching it as a homogeneous unit is opposite to the success of community-oriented programs (Morgan, 1993, p. 44). Among other qualities, communities have nested hierarchies and power relations. Therefore, it is important to ask who the community is and whom it represents (Yeo, 1993). Men and women, the elderly, and marginalized groups, such as men who have sex with men and others, all have different social statuses. For instance, the status of men and the elderly in Central Asia, their access to resources and decision-making processes, is not comparable to the status of women (Earle et al., 2004). Without addressing these issues, development programs may simply reinforce existing hierarchies (Wells et al., 2012) instead of empowering the community as a whole. Acknowledging the hierarchies inherent in communities, this book focuses on community organizations, including community-based organizations (CBOs) and nongovernmental organizations (NGOs), which target or work with marginalized groups. This specific focus on community organizations, rather than on communities as a whole, contributes to and also assures the empirical feasibility of the assessment of capacity-building.

Capacity-building in this book refers to activities that contribute to the ability of a community organization to formulate and express its concerns and use internal and external resources to achieve its goals. This definition combines the operationalization provided by multiple scholars.[1] Internal resources in this definition include individual and organizational assets, such as skills, experience, and associations with other organizations; external resources refer to the physical assets (e.g., hospitals, social service institutions) that the organization can use in its activities (Mcknight & Kretzmann, 2012). The operationalization of capacity-building used in this book goes beyond Shediac-Rizkallah and Bone's (1998) emphasis on training and its role in supporting the roles of community members as sources of information and expertise. In so

---

[1] Labonte and Laverack (2001a, p. 115) and Sarriot et al. (2004a, p. 28) described community capacity as the ability of communities to define, evaluate, and "act on health (or any other) concerns of importance to their members." In addition, Jackson (2003), Raeburn et al. (2006), and Goldberg and Bryant (2012) characterized community capacity-building as improving general performance and the ability to achieve the stated goals through establishment of a necessary environment for it, including planning, needs assessment, and assessment of resources.

doing, it amplifies (expands) the meaning of capacity-building and shifts the emphasis from health projects to the abilities and agency of community organizations.

To measure community organizations' capacity-building, I adopt the model suggested by Laverack and focus on participation, leadership, and resource mobilization.[2] The emphasis on these three aspects corresponds to the agency of community organizations highlighted in the definition above. Engagement in problem-setting and the ability to influence decisions are inherent to participation (Labonte & Laverack, 2001a), which is critical to the responsiveness of development programs to local concerns. Although concomitant to participation, leadership refers to the ability of the community organization to define the problems, suggest solutions, critically assess the factors contributing to inequalities, and develop relevant strategies to address them (ibid.). Finally, resource mobilization is instrumental to the first two aspects. However, the ability of an organization to mobilize resources in addition to its own assets (ibid.) may take different forms.

This book focuses on development aid, state financing, community fundraising, and liaisons with other organizations as potential sources for resource mobilization. Donor funding is a typical source of financing for community organizations in developing countries. However, reductions in development assistance and unpredictability of funding flows make this source of financing unreliable. Inclusion of costs into the national state budget once the project has ended is another option, for instance, through state provision of services under social contracting, though budget deficits and government prioritization of other areas not targeted by community organizations may exclude this possibility. The third option is a mobilization of resources at the community level, for instance, through collecting donations and in-kind support, introducing membership fees (Paine-Andrews et al., 2000), or conducting income-generating activities (Walsh et al., 2012). However, there are some underwater stones here as well. For example, poverty may increase competition for limited

---

[2] The original source is an unpublished Ph.D. thesis by Laverack (1999), which was expanded further by Labonte and Laverack (2001a, 2001b).

resources (Roussos & Fawcett, 2000) or question the viability of fund-raising activities at all.

Lastly, mobilization of resources also takes place through association with organizations that have similar objectives (Paine-Andrews et al., 2000). In addition to in-kind and political support, these alliances may contribute to strengthening human resources, also through the improvement of skills (Hirschhorn et al., 2013). Through collaboration with medical workers, community organizations working in health care may, for instance, gain supervisory support (Ajayi et al., 2012) and link their activities to existing health care services (WHO, 2008). However, the same organizations may be gatekeepers that are protective of their areas of interest. Thus, medical professionals at times oppose the involvement of community organizations in health care due to the lack of medical training and expertise of the latter (Morgan, 2001).

Overall, resource mobilization is one of the most important yet problematic components of community capacity-building. In addition to the knowledge and skills of community organizations (Sarriot et al., 2004a), it also largely depends on the overall political and economic situation in the relevant region or country. This interplay between organizational aspects and external factors brings to mind an indicator not included in the original model for analyzing capacity-building by Laverack (see Labonte & Laverack, 2001a, 2001b), namely, the survival of community organizations beyond the duration of the project funding.

Both the continuity of project activities and the maintenance of benefits largely depend on the survival of community organizations. For this reason, multiple stakeholders raised this issue during my fieldwork in the Kyrgyz Republic. Participation, leadership, and resource mobilization contribute to but do not necessarily guarantee the survival of a community organization. The organization may take part in the decision-making process, demonstrate leadership, have multiple sources of funding, and still discontinue its activities. For this reason, I include the survival of the organization beyond the duration of project funding as an additional indicator of community capacity-building (Table 3.2).

**Table 3.2** Conceptual definition of sustainability

| | |
|---|---|
| Maintenance of benefits | Continuity of benefits offered to targeted groups after the end of the project (e.g., services and infrastructure) |
| Continuity of project activities | Continuity of project activities after the end of the project funding |
| Community capacity-building | Activities contributing to community organizations' capacity-building<br>1. Leadership, or the ability of these organizations to define the problems, suggest solutions, and critically reflect on the general issues relevant to their work<br>2. Mobilization of resources via donor and state financing, fundraising activities, or liaison with other organizations<br>3. Survival of civil society organizations (CSOs) beyond the end of a donor-financed project |

The source: author's adoption of Shediac-Rizkallah and Bone (1998) and Laverack (see Labonte & Laverack, 2001a, 2001b). Participation in problem-setting and the ability to influence decisions are not included in the analysis of community organizations' capacity-building process, as it is separately evaluated in the analysis of stakeholders' roles throughout the project's life cycle (Chaps. 5 and 8)

## 3.3  Factors Influencing the Sustainability of Health Care Interventions

In addition to the empirical and conceptual definition, the accurate analysis encompasses the related factors, as sustainability does not materialize in a vacuum. Rather, it depends on several internal and external factors that are difficult to predict (Sarriot et al., 2004b). In the case of development assistance to health care, internal factors are technical or program-related elements, such as management, planning, implementation, and achievement of stated goals (Bossert, 1990; Shigayeva & Coker, 2015). External factors are the economic and political situations in the recipient country that shape the system in general, including the number of medical workers in the country, health care coverage (Iwelunmor et al., 2016), and political support for specific programs. All these factors contribute to the uncertainty associated with the sustainability of health care programs. This section summarizes the main factors associated with this

**Table 3.3** The factors relevant to sustainability

| | |
|---|---|
| Financing | At the expense of the aid recipient or through a combination of "innovative" funding methods |
| Accounting for the influence of general factors | • Political and economic situation in the aid-recipient country<br>• Health care system, the epidemiological burden of the disease<br>• Availability of medical workers<br>• National priorities |
| Integration into local context | Consideration for historical, systemic, and cultural specifics, but with the account for the social stigma and discrimination against specific groups |
| Organizational factors | • Project duration<br>• Capacity of the organization implementing health care interventions (managerial, financial, and structural characteristics of the organization and its human resources) |

phenomenon based on the literature on the sustainability of health care interventions (Table 3.3). Awareness of these factors is critical to the sustainability of health aid, although the prominence of each of these conditions may depend on a particular project and country setting.

First of all, donors provide the initial financing for projects, but at the end of the assistance period, or in the best-case scenario, at the beginning, the question of funding continuity arises. This continuity, as a rule, is ensured at the expense of aid recipients or through a mixture of funding mechanisms. A plan for program continuation, including evidence of the recipient's contribution or evidence of a combination of the donor's and recipients' funding sources, is often a precondition for assistance (Schell et al., 2013). This requirement for the recipient's contribution may increase the share of domestic funding to the areas targeted by development aid. UNAIDS (2015, p. 54) notes that in 2005, development assistance accounted for 69% of all HIV-related spending, but by 2014, contributions from domestic sources in low- and middle-income countries represented 57% of all HIV-related expenditure.

However, the recipient's contribution may not be sufficient to cover the costs of all health care programs. For instance, in the case of projects combatting tuberculosis (TB), the course of treatment for drug-resistant forms of this disease ranges between US $1218 and $6313 per patient

(Laurence et al., 2015; Ormerod, 2005). Depending on the burden of disease, these costs may be unaffordable to patients and governments in developing countries. Similarly, a recent study of development assistance for community health workers by Lu et al. (2020) suggests that in the context of low-income countries, domestic public spending alone cannot fully cover national community health worker programs. Nevertheless, the role of domestic resources remains imperative, notably as aid recipients transition from external assistance. As development aid to Eastern Europe, Central Asia, and Latin America is decreasing, national contribution becomes the key subject in sustainability discussions (Burrows et al., 2016).

Continuous financing of health care programs may also be provided through a combination of various other funding mechanisms. These are affordable procurement of medicines via large donors, optimization of spending, and adoption of "innovative" funding methods (Oberth & Whiteside, 2016, p. 3). As seen above, the cost of medicines is a key issue. Procurement via large organizations, such as the Global Drug Facility, provides access to affordable quality-assured TB medications and diagnostics because they purchase large quantities of health products (Stop TB Partnership, 2019). However, the price per item may be higher for individual countries since they procure considerably smaller quantities than these organizations. For this reason, the establishment of procurement mechanisms is essential to the sustainability of health projects involving diagnostic and treatment services (see Chap. 9). Another option for assuring additional funding is the optimization of spending on health care. Yet, given the limited amount of resources and budget deficits of the countries receiving development assistance, this measure may not guarantee significant savings to cover the costs of health care programs.

The third option, introducing "innovative" funding methods, may take different forms, including specific tax mechanisms, fundraising activities, and recruitment of volunteers. For example, additional taxation on mining or cigarette companies for their contribution to the "risky environment and conditions" for the development of TB could be a supplementary source of financing for TB services (Amo-Adjei, 2013, pp. 4–5). Similarly, taxes on airlines, formally employed individuals and companies, and mobile phone usage could increase national spending on

HIV/AIDS-related services (Oberth & Whiteside, 2016). However, there is no guarantee that gathered resources are actually going to be used for public health programs (ibid.). The sustainability of community organizations' activities is associated with community support, volunteering, and local fundraising activities (Abdul Azeez & Anbu Selvi, 2019). Nevertheless, donor financing represents the largest share of funding for community health workers (see Saint-Firmin et al., 2021), and the actual contribution of "innovative" funding methods in economically weak countries remains ambivalent.

Furthermore, the economic, social, and political situation in the recipient country has implications for health care interventions, particularly in the long-term perspective (Bossert, 1990). During the project implementation period, these external factors are at best targeted by project implementers. By the end of the program, however, when the implementers withdraw, the influence of these factors on project outcomes increases. Economic crises or changes in the ruling party may alter domestic politics and government priorities, resulting in cuts to health care spending. Outcomes may also be impacted by issues in the local health care system, such as a shortage of health care professionals, their burn-out, an increased number of patients, and poor record-keeping systems (Harpham & Few, 2002; Iwelunmor et al., 2016). The prioritization of a program by the recipient country is also relevant to the continuity of outcomes. Thus, prevention programs are less likely to be sustained due to the lack of immediate visible effect (Shediac-Rizkallah & Bone, 1998). Short-term visibility of an intervention facilitates continuity of curative (e.g., treatment-oriented) programs over prevention programs. Similarly, the type of activity is also relevant to its continuity. Training programs, for example, offer continuity of outcomes at a low cost. Trained personnel disseminate knowledge further, also via "training of trainers" (ibid., p. 101). In this regard, Kiwanuka et al. (2015) suggest that training women may be particularly beneficial since they are more likely to stay in the community and train others than men. These are only a few examples of the external factors which are relevant to health care programs. Although program managers cannot foresee the influence of every possible aspect, they may nevertheless reflect on known factors during the project implementation/design phase in order to mitigate their impact by the end of the project.

Similarly, integration into the local context is essential for the sustainability of health care interventions. Sarriot et al. (2004a, p. 34) note that sustainability plans are "meaningless" out of context, and they are not alone in this assessment. A systematic evaluation of 125 studies identifies compliance with the local context as one of the most important factors for the sustainability of health care programs (Stirman et al., 2012). The significance of the context is equally acknowledged by practitioners. A vivid example thereof is the Paris Declaration on Aid Effectiveness (2005) and its emphasis on using "country systems and procedures" and "existing capacities" of the recipient country in development assistance (OECD, n.d., pp. 4–5). Although difficult to define, the context, in a general sense, refers to the setting broader than the intervention itself and the discussion of current politics and actors, including systemic, historical, and cultural factors (Andrews, 2013).

Systemic, historical, and cultural factors are reflected in institutional structures and actors' preferences. Closely interrelated, these factors may hinder, promote, or contribute to the mixed results of a health care intervention, as demonstrated in the following examples of reforms targeting primary health care and decreased tea consumption.

Despite the broad acknowledgment and efforts made by domestic and external actors, funding for primary health care in post-Soviet countries remains small. In addition to the medical lobby behind hospital care and political, economic, and social hurdles (Kühlbrandt, 2014), the small funding also corresponds to the structure of the health care system inherited from the Soviet Union. The *Semashko* health care system emphasized a curative approach and little prevention or health care promotion. After the collapse of the Soviet Union, newly independent countries initiated multiple reforms to change this system (Isabekova, 2019). Nevertheless, the emphasis on treatment and curative approaches remained, often at the expense of prevention (Kazatchkine, 2017).

Furthermore, an interplay of cultural and historical factors may contribute to the mixed outcomes of health care interventions. For example, excessive tea consumption is a problem in Central Asian countries. The Swiss health promotion program in the Kyrgyz Republic targeted, among other issues, child nutrition by promoting the use of a micronutrient powder and abstention from black tea consumption during meals. Although tea consumption among pregnant women and children did decrease (Schüth et al., 2014), the rate of anemia reduction among children was still lower than that in other countries (Lundeen et al., 2010). One of the main reasons was the continued tea consumption in the population (see Tobias Schüth, 2011).

Overall, the "context" has controversial implications for programs advocating for changes that are incompliant with mainstream norms. For example, a local population may be open to some projects but not to others. Thus, teachers may be reluctant to introduce sex education based on the belief that condoms could be perceived as an encouragement to have sex (Maticka-Tyndale et al., 2010). As Rashed et al. (1997) note, public attitudes toward bed nets and condoms differ because the first is culturally acceptable, while the second is not. Cultural values defining individual and collective behavior are highly relevant to health care programs (Airhihenbuwa, 1995). However, certain health care interventions, such as harm reduction programs—including needle exchange services and methadone substitution therapy for persons who inject drugs—access to condoms, sexually transmitted diseases for commercial sex workers and men who have sex with men, are controversial in the local context of many countries. Although these programs are also expected to address the context of the recipient country, the extent of their integration into a context that discriminates against the groups targeted by these projects may be limited.

(In)acceptability of specific practices and stigma and discrimination against certain groups is erroneously associated with the "morality" clause related to individual behavior, although it may, in fact, be the result of a reaction to the unknown or mere discrimination of marginalized groups. HIV/AIDS may be perceived as an outcome of or even a punishment for "immoral" behavior, not concomitant with "traditional" values, such as abstinence and fidelity (Hannon, 1990). Similarly, stigma is closely associated with degrading moral status (Goffman, 1963), leading to the discrimination of specific groups of the population affected by certain diseases. There, however, is a difference between persons *affected* and *associated* with diseases. Among persons living with HIV, commercial sex workers and men who have sex with men are more marginalized than children or pregnant women. This discrimination against certain groups also may materialize in the selectivity of the groups entitled to HIV/AIDS-related services (Oberth & Whiteside, 2016). For instance, following a significant reduction in financing from the Global Fund to Fight AIDS, Tuberculosis and Malaria (Global Fund), the government may continue the prevention of mother-to-child transmission of HIV but cut these services for persons who inject drugs (OSF, 2015).

In both cases, (in)acceptability and stigma, the question of whether these two are parts of the moral system of a society or merely indicate the cultural variants remains open. Notably, a cultural variant is selected by an individual based on its popularity in a given environment, whereas the moral system encompasses a set of codes of conduct persisting over time due to its contribution to mutually beneficial social cooperation (Luco, 2014). In addition to the cultural meaning, stigma also may be the result of a "tactical response to perceived threats, real dangers, and fear of the unknown" (Yang et al., 2007, p. 1528). Thus, (in)acceptability of specific practices, stigma, and discrimination against some groups may be the outcome of cultural biases, response to the unknown, or even oppression of less powerful, often marginalized groups.

Elaborated analysis of moral systems and cultural biases goes beyond the scope of this book, but it nevertheless reasserts the necessity for going beyond the romanticized perspective of a "context." Thus, integration into the local context presumes the awareness of power dimensions and hierarchies and considers perceptions of certain practices in the given context and beyond. This resonates with the distinction between the code of conduct followed by an individual or group ("descriptive") against the one that, in certain conditions, would be supported by "all rational people" ("normative") (Gert & Gert, 2002). In other words, a practice pursued in a given context does not necessarily represent a widespread norm or contribute to the benefit of people living in this context. The natural rights of persons, irrespective of their gender, nationality, or sexual orientation, are elaborated in the United Nations Universal Declaration of Human Rights (1948), the International Covenant on Civil and Political Rights (1966), and the International Covenant on Economic, Social, and Cultural Rights (1966) (see Office of the United Nations High

Commissioner for Human Rights, 2023a, 2023b). Although the universal applicability of these documents is part of a continuous discussion, negative implications of stigma and discrimination on public health are evident. The social stigma and discrimination against marginalized groups, such as men who have sex with men and sex workers, jeopardize their access to health care (Oberth & Whiteside, 2016) and contribute to high HIV/AIDS prevalence among these groups (UNAIDS, 2014).

Clearly, interventions based on "established values and practices" are better accepted (Aubel & Samba-Ndure, 1996, pp. 53–54), but what can be done with others who may not comply with the local but a general perception of morality? Awareness of stakeholders' interests and societal hierarchies and going beyond the abstract notion of culture or context is the first step. It may be followed by strategies pursued by some community organizations that learned to build their discourse of providing access to HIV treatment to marginalized groups for the benefit of the general population (see Chap. 8).

In addition to the external factors mentioned above, the sustainability of health care programs is also influenced by internal factors, and this section focuses on two of them: the duration of the project and the capacity of the organization implementing it.

The duration of a project is inherent to its sustainability beyond the period of donor funding. There is a close correlation between time and change because change is closely related to time, and time is associated with change (Sztompka, 1993). Changes promoted by health care programs and their sustainability take time, particularly if these changes contradict the values or habits accepted in the recipient society. Projects with a duration of three to five years are often referred to as "seed funding" or "demonstration" projects that are expected to get financing from elsewhere by the end of those three to five years (Scheirer, 2005, p. 320). Nevertheless, both practitioners and researchers agree that, in practice, most interventions terminate before achieving maturity (Shediac-Rizkallah & Bone, 1998), and the majority of these "premature" projects are discontinued after the end of donor funding (Altman, 1995, p. 527).

Nevertheless, donor organizations vary in their approaches to project duration. Swiss development assistance usually lasts longer than other average development programs. In contrast, organizations such as the Global Fund have a three-year project duration with the possibility of prolonging it; although uncertainty in regard to financing has its implications on relations between the actors (see Chap. 4).

Another factor inherent to the sustainability of health care interventions is the capacity of the organization implementing the relevant project. The OECD (2011, p. 2) defines capacity as the "ability of people, organizations, and society as a whole to manage their affairs successfully." This book focuses on an organizational level. Capacity encompasses the managerial, financial, and structural characteristics of the organization and its human resources (Shigayeva & Coker, 2015). These attributes may suggest that larger organizations with developed structure, networking, and access to resources are preferred over their smaller counterparts. However, this is not necessarily the case. Although beneficial in some instances, increased professionalization may be counterproductive in other instances. Similarly, a large structure also may come with bureaucratization and rigidity of the organization, which decreases its flexibility and responsiveness to the local context. Furthermore, in relation to human resources, it should be noted that in the context of development programs, there is a continuous rotation of staff members on both sides, that is, the donor and the recipients. Human resources are particularly pressing for recipient state agencies and nongovernmental organizations, in contrast to donor institutions offering attractive employment conditions and recruiting the most qualified staff in aid-recipient countries (Swedlund, 2017).

Along with these general features of the organization, another significant factor to sustainability is the leadership and commitment of staff members to the health care program (Scheirer & Dearing, 2011; Shigayeva & Coker, 2015). Although difficult to measure, the dedication of staff members of implementing organizations was apparent in the two case studies covered in this book.

## 3.4    Summary

This section discussed the conceptual and empirical ambiguity of sustainability, and listed the factors relevant to it. As a concept, sustainability is defined in relation to the continuity of project activities once the project has ended, maintaining benefits offered to the targeted population, and building the capacity of the recipient community (Shediac-Rizkallah & Bone, 1998). In addition to conceptualization, this chapter also discussed the empirical operationalization of sustainability: defining what to sustain, by whom, to what extent, and for how long (Iwelunmor et al., 2016, p. 2). As sustainability analysis takes place in the context of uncertainty, this section also presented the factors related to sustainability of health care interventions, namely financing, accounting for general conditions, integration into local contexts, and organizational aspects. The impact of each of these factors is case-specific, though awareness of these conditions contributes to a better understanding of sustainability in health care interventions.

## References

Abdul Azeez, E. P., & Anbu Selvi, G. (2019). What determines the sustainability of community-based palliative care operations? Perspectives of the social work professionals. *Asian Social Work and Policy Review, 13*(3), 334–342. https://doi.org/10.1111/aswp.12185

Airhihenbuwa, C. O. (1995). *Health and culture. Beyond the western paradigm* (New Ed.). SAGE PUBN.

Ajayi, I. O., Jegede, A. S., & Falade, C. O. (2012). Sustainability of intervention for home Management of Malaria: The Nigerian experience. *Journal of Community Medicine and Health Education*, 1–8. https://doi.org/10.4172/2161-0711.1000175

Alexander, J. A., Weiner, B. J., Metzger, M. E., Shortell, S. M., Bazzoli, G. J., Hasnain-Wynia, R., et al. (2003). Sustainability of collaborative capacity in community health partnerships. *Medical Care Research and Review, 60*(4), 130–160. https://doi.org/10.1177/1077558703259069

Altman, D. G. (1995). Sustaining interventions in community systems: On the relationship between researchers and communities. *Health Psychology, 14*(6), 526–536. https://doi.org/10.1037//0278-6133.14.6.526

Amo-Adjei, J. (2013). Perspectives of stakeholders on the sustainability of tuberculosis control programme in Ghana. *Tuberculosis Research and Treatment, 2013*, 1–6. https://doi.org/10.1155/2013/419385

Andrews, M. (2013). *The limits of institutional reform in development: Changing rules for realistic solutions* (Illustrated Ed.). Cambridge University Press.

Aubel, J., & Samba-Ndure, K. (1996). Lessons on sustainability for community health projects. *World Health Forum, 17*(1), 52–57.

Bossert, T. J. (1990). Can they get along without us? Sustainability of donor-supported health projects in Central America and Africa. *Social Science & Medicine, 30*(9), 1015–1023. https://doi.org/10.1016/0277-9536(90)90148-l

Boussemart, J.-P., Leleu, H., Shen, Z., & Valdmanis, V. (2020). Performance analysis for three pillars of sustainability. *Journal of Productivity Analysis, 53*(3), 305–320. https://doi.org/10.1007/s11123-020-00575-9

Burrows, D., Oberth, G., Parsons, D., & McCallum, L. (2016). *Transitions from donor funding to domestic reliance for HIV responses: Recommendations for transitioning countries.* Retrieved February 15, 2023, from https://www.globalfundadvocatesnetwork.org/wp-content/uploads/2016/04/Aidspan-APMG-2016-Transition-from-Donor-Funding.pdf

Cassidy, E. F., Leviton, L. C., & Hunter, D. E. K. (2006). The relationships of program and organizational capacity to program sustainability: What helps programs survive? *Evaluation and Program Planning, 29*(2), 149–152. https://doi.org/10.1016/j.evalprogplan.2005.12.002

Chambers, D. A., Glasgow, R. E., & Stange, K. C. (2013). The dynamic sustainability framework: Addressing the paradox of sustainment amid ongoing change. *Implementation Science, 8*(117), 1–11. https://doi.org/10.1186/1748-5908-8-117

Earle, L., Fozilhujaev, B., Tashbaeva, C., & Djamankulova, K. (2004). Community development in Kazakhstan, Kyrgyzstan and Uzbekistan: Lessons learnt from recent experience. *Occasional Papers Series, 40*, 1–63.

Gert, B., & Gert, J. (2002). *The definition of morality*. Retrieved March, 2023, from https://plato.stanford.edu/entries/morality-definition/

Giovannoni, E., & Fabietti, G. (2013). What is sustainability? A review of the concept and its applications. In C. Busco, M. L. Frigo, A. Riccaboni, & P. Quattrone (Eds.), *Integrated reporting* (pp. 21–40). Springer International Publishing. https://doi.org/10.1007/978-3-319-02168-3_2

Goffman, E. (1963). *Stigma: Notes on the management of spoiled identity* (First Touchstone Ed.). Simon & Schuster, Inc.

Goldberg, J., & Bryant, M. (2012). Country ownership and capacity building: The next buzzwords in health systems strengthening or a truly new approach to development? *BMC Public Health, 12*(531), 1–9. https://doi.org/10.1186/1471-2458-12-531

Gruen, R. L., Elliott, J. H., Nolan, M. L., Lawton, P. D., Parkhill, A., McLaren, C. J., & Lavis, J. N. (2008). Sustainability science: An integrated approach for health-programme planning. *Lancet, 372*(9649), 1579–1589. https://doi.org/10.1016/S0140-6736(08)61659-1

Hannon, P. (1990). Aids: Moral issues. Studies: An Irish. *Quarterly Review, 79*(314), 103–115.

Harpham, T., & Few, R. (2002). The Dar Es Salaam urban health project, Tanzania: A multi-dimensional evaluation. *Journal of Public Health Medicine, 24*(2), 112–119. https://doi.org/10.1093/pubmed/24.2.112

Hirschhorn, L. R., Talbot, J. R., Irwin, A. C., May, M. A., Dhavan, N., Shady, R., et al. (2013). From scaling up to sustainability in HIV: Potential lessons for moving forward. *Globalization and Health, 9*(57), 1–9. https://doi.org/10.1186/1744-8603-9-57

Isabekova, G. (2019). Diverse health care developments in the PostSoviet space: The role of national and international actors. In S. An, T. Chubarova, & B. Deacon (Eds.), *Social policy, poverty, and inequality in central and Eastern Europe and the former Soviet Union: Agency and institutions in flux* (pp. 238–262). Ibidem Press.

Iwelunmor, J., Blackstone, S., Veira, D., Nwaozuru, U., Airhihenbuwa, C., Munodawafa, D., et al. (2016). Toward the sustainability of health interventions implemented in sub-Saharan Africa: A systematic review and conceptual framework. *Implementation Science, 11*, 1–27. https://doi.org/10.1186/s13012-016-0392-8

Jackson, S. F. (2003). Working with Toronto neighbourhoods toward developing indicators of community capacity. *Health Promotion International, 18*(4), 339–350. https://doi.org/10.1093/heapro/dag415

Johnson, K., Hays, C., Center, H., & Daley, C. (2004). Building capacity and sustainable prevention innovations: A sustainability planning model. *Evaluation and Program Planning, 27*(2), 135–149. https://doi.org/10.1016/j. evalprogplan.2004.01.002

Kajikawa, Y. (2008). Research core and framework of sustainability science. *Sustainability Science, 3*, 215–239. https://doi.org/10.1007/s11625-008-0053-1

Károly, K. (2011). Rise and fall of the concept sustainability. *Journal of Environmental Sustainability, 1*(1), 1–13. https://doi.org/10.14448/jes.01. 0001

Kazatchkine, M. D. (2017). Health in the Soviet Union and in the post-soviet space: From utopia to collapse and arduous recovery. *Lancet, 390*(10102), 1611–1612. https://doi.org/10.1016/S0140-6736(17)32383-8

Kidd, C. V. (1992). The evolution of sustainability. *Journal of Agricultural and Environmental Ethics, 5*(1), 1–26. https://doi.org/10.1007/BF01965413

Kiwanuka, S. N., Tetui, M., George, A., Kisakye, A. N., Walugembe, D. R., & Kiracho, E. E. (2015). What lessons for sustainability of maternal health interventions can be drawn from rural water and sanitation projects?: Perspectives from eastern Uganda. *Journal of Management and Sustainability, 5*(2), 97–107. https://doi.org/10.5539/jms.v5n2p97

Kühlbrandt, C. (2014). Primary health care. In B. Rechel, E. Richardson, & M. McKee (Eds.), *Trends in health systems in the former Soviet countries* (pp. 111–128). WHO. https://www.euro.who.int/__data/assets/pdf_file/0019/261271/Trends-in-health-systems-in-the-former-Soviet-countries. pdf%3Fua%3D1

Labonte, R., & Laverack, G. (2001a). Capacity building in health promotion, part 1: For whom? And for what purpose? *Critical Public Health, 11*(2), 111–127. https://doi.org/10.1080/09581590110039838

Labonte, R., & Laverack, G. (2001b). Capacity building in health promotion, part 2: Whose use? And with what measurement? *Critical Public Health, 11*(2), 129–138. https://doi.org/10.1080/09581590110039847

Laurence, Y. V., Griffiths, U. K., & Vassall, A. (2015). Costs to health services and the patient of treating tuberculosis: A systematic literature review. *PharmacoEconomics, 33*(9), 939–955. https://doi.org/10.1007/s40273-015-0279-6

Lu, C., Palazuelos, D., Luan, Y., Sachs, S. E., Mitnick, C. D., Rhatigan, J., & Perry, H. B. (2020). Development assistance for community health workers in 114 low- and middle-income countries, 2007–2017. *Bulletin of the World Health Organization, 98*(1), 30–39. https://doi.org/10.2471/BLT.19.235499

Luco, A. (2014). The definition of morality. *Social Theory and Practice, 40*(3), 361–387. https://doi.org/10.5840/soctheorpract201440324

Lundeen, E., Schueth, T., Toktobaev, N., Zlotkin, S., Hyder, S. M. Z., & Houser, R. (2010). Daily use of sprinkles micronutrient powder for 2 months reduces anemia among children 6 to 36 months of age in the Kyrgyz Republic: A cluster-randomized trial. *Food and Nutrition Bulletin, 31*(3), 446–460. https://doi.org/10.1177/156482651003100307

Maticka-Tyndale, E., Wildish, J., & Gichuru, M. (2010). Thirty-month quasi-experimental evaluation follow-up of a national primary school HIV intervention in Kenya. *Sex Education, 10*(2), 113–130. https://doi.org/10.1080/14681811003666481

Mayer, A. L. (2008). Strengths and weaknesses of common sustainability indices for multidimensional systems. *Environment International, 34*(2), 277–291. https://doi.org/10.1016/j.envint.2007.09.004

Mcknight, J. L., & Kretzmann, J. P. (2012). Mapping community capacity. In M. Minkler (Ed.), *Community organizing and community building for health and welfare* (3rd ed., pp. 171–186). Rutgers University Press.

Morgan, L. M. (1993). *Community participation in health: Politics of primary Care in Costa Rica*. Cambridge University Press.

Morgan, L. M. (2001). Community participation in health: Perpetual allure, persistent challenge. *Health Policy and Planning, 16*(3), 221–230.

Murphy, K. (2012). The social pillar of sustainable development: A literature review and framework for policy analysis. *Sustainability: Science, Practice and Policy, 8*(1), 15–29. https://doi.org/10.1080/15487733.2012.11908081

Oberth, G., & Whiteside, A. (2016). What does sustainability mean in the HIV and AIDS response? *African Journal of AIDS Research, 15*(1), 35–43. https://doi.org/10.2989/16085906.2016.1138976

OECD. (2011). *Perspectives note: The enabling environment for capacity development*. Retrieved February 28, 2023, from https://www.oecd.org/development/accountable-effective-institutions/48315248.pdf

OECD. (n.d.). *The Paris declaration on aid effectiveness (2005) and Accra agenda for action (2008)*. Retrieved February 15, 2023, from http://www.oecd.org/dac/effectiveness/34428351.pdf

Office of the United Nations High Commissioner for Human Rights. (2023a). *Universal declaration of human rights*. Retrieved March 2, 2023, from https://www.ohchr.org/en/human-rights/universal-declaration/translations/english

Office of the United Nations High Commissioner for Human Rights. (2023b). *Human rights instruments*. Retrieved March 2, 2023, from https://www. ohchr.org/en/instruments-listings

Ormerod, L. P. (2005). Multidrug-resistant tuberculosis (MDR-TB): Epidemiology, prevention and treatment. *British Medical Bulletin, 73–74*, 17–24. https://doi.org/10.1093/bmb/ldh047

OSF. (2015). *Ready, willing, and able? Challenges faced by countries losing global fund support*. Retrieved February 15, 2023, from https://www.globalfundadvocatesnetwork.org/wp-content/uploads/2016/04/ready-willing-and-able-20160403.pdf

Paine-Andrews, A., Fisher, J. L., Campuzano, M. K., Fawcett, S. B., & Berkley-Patton, J. (2000). Promoting sustainability of community health initiatives: An empirical case study. *Health Promotion Practice, 1*(3), 248–258.

Pluye, P., Potvin, L., Denis, J. L., & Pelletier, J. (2004). Program sustainability: Focus on organizational routines. *Health Promotion International, 19*(4), 489–500. https://doi.org/10.1093/heapro/dah411

Proctor, E., Luke, D., Calhoun, A., McMillen, C., Brownson, R., McCrary, S., & Padek, M. (2015). Sustainability of evidence-based healthcare: Research agenda, methodological advances, and infrastructure support. *Implementation Science, 10*, 1–13. https://doi.org/10.1186/s13012-015-0274-5

Purvis, B., Mao, Y., & Robinson, D. (2019). Three pillars of sustainability: In search of conceptual origins. *Sustainability Science, 14*(3), 681–695. https:// doi.org/10.1007/s11625-018-0627-5

Raeburn, J., Akerman, M., Chuengsatiansup, K., Mejia, F., & Oladepo, O. (2006). Community capacity building and health promotion in a globalized world. *Health Promotion International, 21*(S1), 84–90. https://doi. org/10.1093/heapro/dal055

Rashed, S., Johnson, H., Dongier, P., Gbaguidi, C. C., Laleye, S., Tchobo, S., et al. (1997). Sustaining malaria prevention in Benin: Local production of bednets. *Health Policy and Planning, 12*(1), 67–76. https://doi.org/10.1093/ heapol/12.1.67

Roussos, S. T., & Fawcett, S. B. (2000). A review of collaborative partnerships as a strategy for improving community health. *Annual Review of Public Health, 21*, 369–402. https://doi.org/10.1146/annurev.publhealth.21.1.369

Saint-Firmin, P. P., Diakite, B., Ward, K., Benard, M., Stratton, S., Ortiz, C., et al. (2021). Community health worker program sustainability in Africa: Evidence from costing, financing, and geospatial analyses in Mali. *Global Health: Science and Practice, 9*(Supplement 1), S79–S97. https://doi. org/10.9745/GHSP-D-20-00404

Sarriot, E. G., Winch, P. J., Ryan, L. J., Bowie, J., Kouletio, M., Swedberg, E., et al. (2004a). A methodological approach and framework for sustainability assessment in NGO-implemented primary health care programs. *The International Journal of Health Planning and Management, 19*(1), 23–41. https://doi.org/10.1002/hpm.744

Sarriot, E. G., Winch, P. J., Ryan, L. J., Edison, J., Bowie, J., Swedberg, E., & Welch, R. (2004b). Qualitative research to make practical sense of sustainability in primary health care projects implemented by non-governmental organizations. *The International Journal of Health Planning and Management, 19*(1), 3–22. https://doi.org/10.1002/hpm.743

Savaya, R., Spiro, S., & Elran-Barak, R. (2008). Sustainability of social programs: A comparative case study analysis. *American Journal of Evaluation, 29*(4), 478–493. https://doi.org/10.1177/1098214008325126

Scheirer, M. A. (2005). Is sustainability possible?: A review and commentary on empirical studies of program sustainability. *American Journal of Evaluation, 26*(3), 320–347. https://doi.org/10.1177/1098214005278752

Scheirer, M. A., & Dearing, J. W. (2011). An agenda for research on the sustainability of public health programs. *American Journal of Public Health, 101*(11), 2059–2067. https://doi.org/10.2105/AJPH.2011.300193

Schell, S. F., Luke, D. A., Schooley, M. W., Elliott, M. B., Herbers, S. H., Mueller, N. B., & Bunger, A. C. (2013). Public health program capacity for sustainability: A new framework. *Implementation Science, 8*(1), 15. https://doi.org/10.1186/1748-5908-8-15

Schüth, T. (2011). From people's mandate to national policy. *Medicus Mundi Schweiz Bulletin, 119*, n.p.

Schüth, T., Jamangulova, T., Aidaraliev, R., Aitmurzaeva, G., Iliyazova, A., & Toktogonova, V. (2014). Community Action for Health in the Kyrgyz Republic: Overview and Results. Sharing Experiences in International Cooperation. Issue Paper on Health Series, (3a), 1–31.

Shediac-Rizkallah, M. C., & Bone, L. R. (1998). Planning for the sustainability of community-based health programs: Conceptual frameworks and future directions for research, practice and policy. *Health Education Research, 13*(1), 87–108. https://doi.org/10.1093/her/13.1.87

Shigayeva, A., & Coker, R. J. (2015). Communicable disease control programmes and health systems: An analytical approach to sustainability. *Health Policy and Planning, 30*(3), 368–385. https://doi.org/10.1093/heapol/czu005

Stirman, W. S., Kimberly, J., Cook, N., Calloway, A., Castro, F., & Charns, M. (2012). The sustainability of new programs and innovations: A review of the empirical literature and recommendations for future research. *Implementation Science: IS, 7*(17), 1–19. https://doi.org/10.1186/1748-5908-7-17

Stop TB Partnership. (2019). *Global Drug Facility (GDF): Increasing global access to quality-assured TB treatments and diagnostics.* Retrieved March 2, 2023, from https://www.stoptb.org/facilitate-access-to-tb-drugs-diagnostics/global-drug-facility-gdf

Swedlund, H. J. (2017). *The development dance: How donors and recipients negotiate the delivery of foreign aid* (1st ed.). Cornell University Press.

Sztompka, P. (1993). *The sociology of social change* (1st ed.). Wiley-Blackwell.

Torpey, K., Mwenda, L., Thompson, C., Wamuwi, E., & van Damme, W. (2010). From project aid to sustainable HIV services: A case study from Zambia. *Journal of the International AIDS Society, 13*(19), 1–7. https://doi.org/10.1186/1758-2652-13-19

UNAIDS. (2014). *The gap report.* Retrieved February 3, 2023, from https://www.unaids.org/sites/default/files/media_asset/UNAIDS_Gap_report_en.pdf

UNAIDS. (2015). *How AIDS changed everything. MDG 6: 15 years, 15 lessons of hope from the AIDS response* (pp. 1–543). Retrieved March 2, 2023, from https://www.unaids.org/sites/default/files/media_asset/MDG6 Report_en.pdf

Walsh, A., Mulambia, C., Brugha, R., & Hanefeld, J. (2012). "The problem is ours, it is not CRAIDS'". Evaluating sustainability of community based organisations for HIV/AIDS in a rural district in Zambia. *Globalization and Health, 8*(1), 40. https://doi.org/10.1186/1744-8603-8-40

Wells, K. J., Preuss, C., Pathak, Y., Kosambiya, J. K., & Kumar, A. (2012). Engaging the community in health research India. *Technology and Innovation, 13*, 305–319. https://doi.org/10.3727/194982412X13292321140886

WHO. (2008). *Community involvement in tuberculosis care and prevention Towards partnerships for health: Guiding principles and recommendations based on a WHO review.* Retrieved February 28, 2023, from http://apps.who.int/iris/bitstream/10665/43842/1/9789241596404_eng.pdf

World Commission on Environment and Development. (1987). *Our common future: Report of the world commission on environment and development.* Retrieved March 2, 2023, from https://sustainabledevelopment.un.org/content/documents/5987our-common-future.pdf

Yang, L. H., Kleinman, A., Link, B. G., Phelan, J. C., Lee, S., & Good, B. (2007). Culture and stigma: Adding moral experience to stigma theory. *Social Science & Medicine, 64*(7), 1524–1535. https://doi.org/10.1016/j.socscimed.2006.11.013

Yeo, M. (1993). Toward an ethic of empowerment for health promotion. *Health Promotion International, 8*(3), 225–235. https://doi.org/10.1093/heapro/8.3.225

# 4

# The Role of Structural Factors in Selected Health Programs

As noted in Chap. 1, this book specifically focuses on two case studies: (1) the "Community Action for Health" project, financed by the Swiss Agency for Development and Cooperation (SDC) (hereinafter the CAH/ Swiss project), and (2) grants from the Global Fund to Fight AIDS, Tuberculosis and Malaria (the Global Fund) targeting tuberculosis and HIV/AIDs in the Kyrgyz Republic (hereinafter the Global Fund project/ grants). This chapter elaborates on the case-specific factors relevant to understanding the interaction and sustainability of these health care initiatives by focusing on the four factors delineated in the analytical framework as essential to health care initiatives. These are predictability and flexibility of assistance on the donors' sides, and dependency and capacity on the recipients' sides (Chap. 2).

## 4.1 Aid Predictability

In the context of health assistance, aid predictability refers to the duration that donor organizations can commit themselves, financially or by other means, to the assistance they offer (Chap. 2). Both SDC and the

© The Author(s) 2024
G. Isabekova, *Stakeholder Relationships And Sustainability*, Global Dynamics of Social Policy, https://doi.org/10.1007/978-3-031-31990-7_4

Global Fund acknowledge the significance of aid predictability and commit themselves to improving it. However, the outcomes are diverse due to the development assistance organizational structure offered by these two actors.

First, the Swiss Development Cooperation structure allows for multiyear predictability. The Swiss Parliament adopts the "Dispatch on International Co-operation" every four years (Federal Department of Foreign Affairs and State Secretariat for Economic Affairs, 2020). This document underpins the country's view of development, such as poverty reduction and sustainable development (2021–2024), and is not limited to official development assistance (ibid.). Three organizations are responsible for implementing the Dispatch: (1) the Swiss Agency for Development and Cooperation, (2) the Human Security Division within the Federal Department of Foreign Affairs, and (3) the Economic Cooperation and Development Division of the State Secretariat for Economic Affairs within the Federal Department of Economic Affairs, Education and Research (OECD, 2019, pp. 13–14). Nevertheless, the SDC manages the most significant part of the Dispatch on International Cooperation program (68%), including technical and financial cooperation and humanitarian assistance (ibid.). The four-year budget planning of the program, combined with sound forecast information and multiyear funding agreements, provides the basis for the reliability of Swiss aid (ibid., p. 18). Moreover, Swiss aid agencies have buffer funds to rely on; this means that Swiss aid organizations can make four- to five-year commitments, and in the case of the SDC, this extends up to ten years (OECD, 2009, p. 213). Accordingly, Switzerland offers more predictable aid from a multiyear perspective.

The "Community Action for Health" Project (CAH) is a vivid example of the predictability of Swiss development assistance in Kyrgyzstan and beyond. With the average duration of Swiss projects in Kyrgyzstan being approximately ten years (see Embassy of Switzerland in KR, 2013), the CAH lasted for seventeen years (2001–2017). Several interviewees stressed that this long-term duration was essential to the project performance (IO Partner 5) because working with existing structures takes time, with the first three years spent on building networks (IO Partner 9). This multiyear predictability is a strong feature of the Swiss Development

Cooperation (OECD, 2014). According to estimates of the Global Partnership for Effective Development Co-operation (n.d.), which is a multistakeholder platform aiming to promote the effectiveness of development efforts, Swiss aid demonstrates high aid predictability. In 2018, for instance, Switzerland performed better (70%) than the average bilateral member of the Organization for Economic Co-operation and Development's Development Assistance Committee (OECD DAC) (53.2%) in terms of the medium-term predictability (two to five years) of its aid (ibid.).

In addition to funding, another significant aspect of predictability in the CAH was the long-term engagement of the project coordinator. Invited by the Swiss Red Cross (SRC) to conduct a pilot study on community involvement in health care in 2001, Dr. Tobias Schütz stayed in Kyrgyzstan for almost thirteen years. He administered most of the project process, from its early initiation to countrywide expansion and further extension of community-based organizations at the national level. According to one interviewee, the absence of short-term consultants benefited the project (IO Partner 5). Indeed, the project coordinator's continuous engagement contributed to building partnerships and uniformity of principles and approaches throughout the CAH. Notably, the long-term presence of key staff members is one of the aspects of Swiss aid appreciated by partner countries as beneficial to collaboration (OECD, 2019). However, thirteen years is probably an exception rather than a rule to a long-term presence.

Second, the Global Fund encourages the initiative of countries, multilateral and nongovernmental organizations, and private foundations willing to unite their efforts against tuberculosis, human immunodeficiency virus infection and acquired immune deficiency syndrome (HIV/AIDS), and malaria worldwide. In this way, the initiative to establish the Global Fund came from Japan, the European Commission, United Nations agencies, participants of the African Summit on HIV/AIDS (2001), the United States, and a number of other stakeholder countries (Global Fund, n.d.-b). However, one could specifically highlight the role of the Bill and Melinda Gates Foundation, which provided the "single largest nongovernment pledge" in the amount of US $100 million in 2001 (ibid., pp. 15–24). Over eighty countries made or pledged contributions to the Global Fund, with the Organization for Economic Cooperation and Development's Development Assistance Committee (OECD DAC) members and the European Commission representing, as of the end of 2021, the leading government donors (Global Fund, 2023a).

Similar to Swiss aid, the Global Fund (2023b) offers relatively predictable assistance by allocating funding to countries on a three-year basis. This period corresponds with the replenishment cycle, during which the governments and organizations supporting the Global Fund (2023c) pledge their financial contributions. Adopted in 2005, this approach was intended to provide "more stable and predictable" financing (ibid.). Benefiting ongoing programs, the three-year period also allowed sufficient time for countries to prepare their applications. In so doing, this approach aimed to eliminate gaps between calls and inflated costs in applications prepared on short-term notice observed during the time when the organization made announcements on an *ad hoc* basis in the past (Global Fund, n.d.-b). Additionally, as part of its 2012–2016 strategy, the organization introduced the New Funding Model, which, among others, stipulated early feedback on proposals, intending to decrease the waiting time and increase their chances of success (Global Fund, 2013). The Global Fund has equally aimed to increase the predictability of ongoing projects. Its Rolling Continuation Channel Initiative, for instance, stipulated up to six years of funding for "high-performing grants," with applications reviewed quarterly instead of on an annual basis (Global Fund, n.d.-b, pp. 37–38). Furthermore, the organization aimed to further increase transparency by announcing the eligibility of countries for grants based on their disease burden in each of the three components supported by the Global Fund (i.e., tuberculosis, HIV/AIDS, and malaria) and income classification (Global Fund, n.d.-c). Overall, the Global Partnership for Effective Development Co-operation (n.d.) estimates that the Global Fund performs better (66.8% in 2016) than the average vertical program (see Chap. 1 for a definition) (42.8%) in terms of medium-term predictability of its assistance.

Kyrgyzstan has been a long-term recipient of Global Fund grants. The country has received tuberculosis (TB) and HIV/AIDS grants since 2004, with the average duration of grants being approximately 4.5 years

(Global Fund, n.d.-a). It should be noted that each grant was built around the objectives and activities of the former. This approach contributed to the continuity of efforts in both areas. This continuity and uniformity laid the foundation for approaching the grants as continuous projects against tuberculosis and HIV/AIDS, respectively.

Nevertheless, the Global Fund's financial commitment remains limited to three years, with financing beyond being dependent on the availability of funds. The Global Fund's dependence on financiers was also visible during the accusations of fraud in the grants. Confirmed in "a very small number" of countries and activities, reports in the mass media about fraud have nevertheless contributed to the perception that the organization lost control over its grant disbursements (Brown & Griekspoor, 2013, p. 139). In response to these allegations, several countries announced the halt of their financing, resulting in a seven to eight billion dollar funding shortfall (The Lancet, 2011). Consequently, at the 25th board meeting in November 2011, the Global Fund suspended the planned call for new grants but assured financing for ongoing programs (Moszynski, 2011). It took several measures to address the problems related to fraud allegations. In addition to replacing several senior managers, it changed its operational model and emphasized "more risk-based supervision" in grant implementation (Brown & Griekspoor, 2013, pp. 139–142).

Overall, both Switzerland and the Global Fund acknowledge and plan to ensure the predictability of their assistance, which is also reflected in their performance compared to an average donor. However, lasting for almost seventeen years, the CAH is a striking example of the predictability of Swiss aid, supported by the long-term presence of the project coordinator. Similarly, lasting 4.5 years on average, the Global Fund grants to Kyrgyzstan demonstrate the commitment and efforts of this organization to aid predictability. Furthermore, building around the preceding objectives, each grant contributed to the continuity of activities, laying the foundation for analytically treating them as ongoing projects against tuberculosis and HIV/AIDS. Nevertheless, the organizational dependency on the replenishment cycles limits its ability to make longer commitments, which will be discussed in the following section.

## 4.2   Aid Flexibility

Aid flexibility in this book denotes the stakeholders' abilities to change the development assistance and the extent to which this change demands specific procedures that may indirectly hinder the stakeholders from initiating this process (Chap. 2). Switzerland and the Global Fund acknowledge and commit to providing flexible assistance consonant with recipients' needs and objectives, albeit with differing success.

Flexibility is a strong feature of Swiss aid within and beyond Kyrgyzstan due to the relevant emphasis and organizational structure. Switzerland allows flexible programming and budgeting adaptable to changing circumstances at country and project levels (OECD, 2019). This emphasis is further supported by the structure of the Swiss development cooperation stipulating decentralization and allowing a certain level of autonomy for field offices. Although part of Swiss embassies, the Swiss Cooperation Offices are still "fairly autonomous" (OECD, 2005, p. 73). They report directly to headquarters in Geneva, have policy dialogs with the recipient governments and other donors, and manage local staff and local budgets (ibid., p. 74). Country directors have some flexibility in allocating funds according to the priorities annually defined in collaboration with partner countries (ibid., p. 218). This flexibility of Swiss aid is also reflected in the SDC's operations in Kyrgyzstan. The organization is among the three organizations providing budget support to assist in the realization of the national health care program. It also provides project-related health assistance, which benefits from organizational flexibility, as the CAH shows.

The autonomy of Swiss aid agencies was conducive to the flexibility of the "Community Action for Health" and its responsiveness to local needs. Driven by specific objectives of empowering the communities to improve their health and to support the partnership between the state health care system and local communities (Schüth, n.d.), the project was nevertheless open to local initiatives. As demonstrated in Chap. 5, the CAH targeted the issues and solutions identified by local communities and those prioritized in national health care. This openness to activities suggests that although covering all significant areas, the project description and funding still provided space for introducing alterations. It also permitted rather unbureaucratic approval of budget changes, including further adjustments of costs and activities (IO Partner 11). This guaranteed the project's responsiveness to the changing circumstances, also in terms of the needs of community-based organizations and the areas of concern highlighted by the local community.

At the same time, Swiss aid faces challenges in balancing the different levels of accountability, which may also affect its flexibility. As demonstrated above, Switzerland ensures the accountability of its aid to recipients by providing flexibility and responsiveness to local needs. Nevertheless, Swiss development organizations, as any others, are primarily accountable to citizens paying taxes for aid or, in practice, the organizations representing these citizens. Switzerland has no political ties to its aid (OECD, 2019), but there are growing voices about conforming official development assistance to national interests. One example thereof is the political pressure to target irregular migration to Switzerland by linking the assistance to migration policies of recipient countries (ibid.). If successful, these initiatives will provide political ties to Swiss aid, which will also affect its flexibility.

However, the link to national interests is not the only pressure on development assistance, as accountability to taxpayers also presumes the achievement of stated objectives and the use of funds accordingly. This may increase the control over finances and, in so doing, restrict the "spontaneity" of allocations. One interviewee noted that the control over the financing and budget specifications in the CAH increased, reducing the initial flexibility of the initiative (IO Partner 11). Although in need of further investigation, in the broader context of increasing pressure on the accountability of aid, this suggestion points to the controversial relationship between accountability and flexibility. In other words, increased control over the assistance is opposite to its flexibility.

Like Switzerland, the Global Fund commits itself to providing flexible assistance. The organization recognizes the problem with requesting project proposals instead of accepting the existing national strategies or applying project cycles instead of adjusting themselves to the cycle of the national program of an aid-recipient country (UNAIDS, 2005a, pp. 14–15). Accordingly, the Global Fund asks applicants to conform their proposals with national strategies (Chap. 8) and allows sending funding requests at any time during the initial three-year allocation period to ensure alignment with national budgeting cycles (Global Fund, 2013).

Nevertheless, as the case of the Global Fund project in Kyrgyzstan shows, the three-year period does not necessarily comply with the duration of national health care programs. According to a state representative, the Global Fund is among the few donors explicitly committing finances. For this reason, the organization is also explicitly stated as a source of financing for specific activities (Government of KR, 2017b). Other organizations may similarly support the national program, but their commitments are not stated anywhere (State Partner 2). Nevertheless, even a three-year commitment does not cover the entire duration of the national program. Another interviewee explained that with national programs (i.e., against TB and HIV/AIDS) being developed for five years, the funding for the remaining two years remains unknown (State Partner 4).

Furthermore, the allowable changes to approved grants seem insignificant. The Global Fund attempts to consider recipients' suggestions and implement relevant changes (CSO 8) by adjusting to unexpected expenditures, savings, cancelations, and transfer of some activities (IO Partner 20). However, these changes are "typically not substantial" (Vujicic et al., 2011, p. 2) and remain within 10–15% of the grant's total amount (IO Partner 20). More substantial changes, such as providing treatment instead of prevention, may be problematic (ibid.) and involve additional bureaucratic hurdles, as suggested by another interviewee (State Partner 4). The respondent noted that the approval might come or not, with different conditions and limitations applied and negotiations lasting months, particularly in the cases involving medications (ibid.). The interviewee emphasized that these hurdles caused issues in the grant realization process, adding that organizational responsiveness also depended on the individual(s) coordinating the relevant matter, with some being more open to interpretations than others (ibid.). Indeed, individual perspectives and behavior are significant to aid flexibility, as they are to predictability, as demonstrated in the previous section in the case of the CAH.

Nevertheless, the issues with responsiveness and bureaucratic hurdles are also related to the organizational structure of the assistance. The Global Fund aims to ensure the recipients' ownership over grants by delegating the process of its realization to stakeholders present in the recipient country. The absence of field offices also intends to ensure organizational neutrality (IO Partner 4). Remaining in continuous communication with recipients, the Global Fund bases its judgments and decisions on the information provided by (inter)national stakeholders about the achievements and issues. Nevertheless, this concentration of decision-making in one place contrasts with decentralization and autonomy, contributing to the responsiveness of Swiss aid.

Furthermore, the flexibility of grants, similar to Swiss aid, is contingent upon external factors. Following the allegations of fraud in grants to multiple countries, the flexibility of projects decreased. The Global Fund introduced new regulations requiring all grant recipients to submit their training plans for approval, and it proved difficult to make any changes to these plans during the implementation process (Benjamin, 2011). There were also issues with adjusting activities to inflation in the country (see Ancker & Rechel, 2015a). Overall, the changes introduced after the fraud allegations intended to demonstrate the organizational ability to control finances, pointing to the organization's accountability before its funders. Once again, the increased control over finances seems to counterpoise aid flexibility.

Overall, Switzerland and the Global Fund emphasize and provide flexibility in their assistance. Both highlight recipients' ownership by adjusting the activities to changes occurring throughout the project realization process. Nevertheless, the extent of possible adjustment without bureaucratic hurdles is associated with the organizational structure of development partners. Thus, decentralization and a certain level of autonomy of field officers in Swiss aid contrast with concentrated decision-making in the Global Fund. However, both development partners struggle with balancing accountability before funders and recipients of their aid.

## 4.3   Capacity

Capacity in this book primarily refers to the abilities of organizations to fulfill their functions and set and achieve the stated objectives before them (Chap. 2). Accordingly, this section discusses the capacities of civil society organizations and state institutions addressed in the two case studies.

First, approaching the operationalization of civil society organizations in a broader sense, this section discusses the capacities of community-based organizations (CBOs) involved in the CAH and nongovernmental organizations (NGOs) participating in the Global Fund grants.

The CBOs established within the framework of this project include the Village Health Committees, *Rayon* Health Committees, and the Association of Village Health Committees.

Village Health Committees (VHCs) carry out preventive and health promotion activities among their communities in areas identified by community members as pressing and those targeted by national health care development programs (Chap. 5). These areas include hypertension, alcoholism, iodine and iron deficiency, influenza, brucellosis, and others (see Isabekova, 2021). It should, however, be noted that although they measure the blood pressure of their fellow villagers or the level of iodine in the salt sold in the local shop, VHCs do not provide medical services. Instead, the organizations serve as mediators between health care institutions and the population by noting health care issues and encouraging their villagers to refer to medical organizations and get timely treatment (AVHC, 2020).

VHCs are present in all seven regions, and most have official registration. Recent estimates suggest that there are 1606 VHCs in the country (AVHC, 2018, pp. 11–12). The organizations are composed of volunteers who come mostly from the villages in which they conduct their activities. The VHC members meet regularly, on average from 1–2 times per week to 2–3 times a month (Kickbusch, 2003, p. 18). Interviewees note that although generally proportional to the size of the relevant village, the number of volunteers fluctuated throughout the CAH from an initial 20–30 (CSO 4) down to 5–10 (CSO 7; State Partner 1). Subsequently, the total number of VHC members ranged from 10,215 in

2010 to 15,566 in 2014 and to 13,267 in 2016 (PIL Research Company, 2017, n.p.).

The *Rayon* Health Committees (RHCs) are composed of the leaders of the VHCs. Registered as nonprofit organizations (Schüth et al., 2014), they serve as a platform for VHCs to meet and discuss the work conducted and activities omitted and decide on the work plan for the next quarter (AVHC, 2018, pp. 5–6). This platform also serves two other purposes. First, it is used to pass on information from the association to the VHCs. Furthermore, at the end of quarterly meetings, RHCs report to the Association of VHCs by sending the following documents: a list of participants, a work plan, a meeting agenda, and a table of the VHCs' self-initiatives (ibid.). Since the RHCs have no office space of their own, their meetings take place on the premises of Family Medicine Centers or in the offices of the local self-government at the district level, where the authorities are able or willing to offer space for VHC meetings (CSO 4).

In addition to connecting and reporting functions, RHCs aim to solve health care issues at the village and district levels and coordinate capacity-building activities for VHCs (AVHC, n.d.). They are also expected to support Family Medicine Centers in their health care activities and coordinate the annual assessment of VHCs (ibid., pp. 8–9). According to recent estimates, there are 58 organizations in total in the country (AVHC, 2020, p. 3), with the number of RHC members proportional to the size of the population in the relevant district (*rayon*) (CSO 4).

The Association of Village Health Committees (hereinafter AVHC or the Association of VHCs) coordinates and represents *Rayon* and Village Health Committees before the state and donor organizations. Established in 2010 as a voluntary association of RHCs, it aims to promote health and improve sanitation and hygiene and the living circumstances of the rural population in Kyrgyzstan (AVHC, 2020, p. 3). The executive body of the AVHC has two permanent staff members and four staff on short-term contracts, although the number of staff at any given time largely depends on the workload since the staff on short-term contracts are taken on at times of increased workload (CSO 4). In contrast to the Village and *Rayon* Health Committees, the staff members of the executive body of the AVHC members receive a salary. Nevertheless, the AVHC is a nongovernmental and noncommercial organization whose primary responsibility is coordinating health committees and representing them before the state and donor organizations. Actors willing to work with health committees contact the Association of VHCs first (CSO 1). Examples of AVHCs' collaboration with other actors include disease prevention and health promotion activities conducted within the framework of programs funded by the World Bank, the German Corporation for International Cooperation (*die Deutsche Gesellschaft für Internationale Zusammenarbeit*—GIZ), the United States Agency for International Development (USAID), SDC, and others. With the Ministry of Health, the AVHC mainly works via the Republican Center for Health Promotion and Mass Communication under the Ministry of Health (hereinafter the Republican Center), but it also sought collaboration with the Mandatory Health Insurance Fund on the assessment of medical services (see Development Policy Institute, 2017).

Notably, the Association of VHCs has essential developmental and supervisory functions. It seeks collaboration with other actors primarily to support and strengthen the capacities of RHCs and VHCs. Therefore, in addition to organizations approaching the organization themselves, it also looks for potential donors and projects that could support the community-based organizations by financial means or by training and other forms of technical assistance (CSO 4). The AVHC also pursues supervisory activities by collecting the information provided by RHCs during their quarterly meetings to assess the VHCs' ongoing work and initiate supportive measures.

Organizational capacity closely relates to the composition and rotation of staff members. It should be noted that most CBO members (approximately 90%) are women (Tobias Schüth, 2011a). The CAH initiated several studies on the role of gender in VHCs and their activities (Development Planning Unit, 2010; Walker, 2013), but external evaluators did not find any conclusive impact of gender on CBOs. The project has also attempted to encourage male participation in the project, particularly in the areas of brucellosis, alcoholism, and tobacco abuse (SDC, 2014). These attempts also included organizing competitions, such as *"Ülgülüü Ata"* (a model Dad in Kyrgyz), where men had to complete certain tasks and answer questions relating to the VHCs' work to receive valuable prizes (PIL Research Company, 2017). The project, however, did not achieve its intended 70/30% gender representation since male participation dropped over time (Gotsadze & Murzalieva, 2017). Civil society representatives interviewed for this research similarly suggested that just two or three men participated in their activities, though these men were not members: one worked as a veterinary, one for the local authorities, and one in social services (CSOs 2 and 5). The majority of VHC members were women (ibid.).

Extensive labor migration and conventional gender roles in the Kyrgyz society contribute to the prevalence of women in community-based organizations. These roles, for instance, include the assumption that a household's health is viewed as a woman's "responsibility" and that women (in contrast to men) are not associated with a role of breadwinner. Men in rural areas leave for the cities, go abroad to work in construction, or go to the mountains to look after the livestock (CSO 5). As the men leave, the women stay at home to take care of the household. The VHCs' outreach activities target the villagers who are at home, and these are mainly women (Development Planning Unit, 2010). Traditional roles in the Kyrgyz society also view men as "breadwinners" and women as "caregivers" in households. My interviewee emphasized the fact that health continues to be seen as "the responsibility of women" (CSO 1). Men declined to participate in the CAH due to the unpaid nature of the work and the inconvenience of discussing health issues, such as female reproductive health (Development Planning Unit, 2010). Overall, the prevalence of women, however, was not limited to community-based organizations but mirrored the general tendency in civil society organizations in Kyrgyzstan (see the following section).

At the same time, not all women join community-based organizations. Depending on their age, women enjoy different statuses in society and in their families. Young women are expected to look after their children and in-laws and are under the strict supervision of their husbands and in-laws. Older women, however, have a higher status in society and in their families, fewer household responsibilities, and, therefore more time and freedom. My interviewee stressed the inability of younger women to participate in the VHCs despite their willingness due to resistance from their husbands and in-laws (CSO 7). For this reason, the VHC members are mainly women aged 40–50, who are unwilling to leave their positions in community-based organizations (ibid.). The majority of these women are housewives (CSO 5) and have just secondary school-level education (CSO 2). For this reason, training courses for the VHCs, for instance, were adjusted accordingly and used simplified terminology (CSO 1). Previous research on the VHCs was ambiguous about the social status and profession of VHC candidates (e.g., Kickbusch, 2003). Although this research cannot generalize the findings gained from a limited number of interviews, it nevertheless provides an important insight into the profile of the VHCs.

Remarkably, there was no issue with high staff turnover in the community-based organizations. On the contrary, my interviewees pointed to the opposite problem, the difficulty of cadre renewal. There are regulations in the statute of the VHCs or the statute of RHCs regarding the length of service of committee members (CSO 4), but some members are unwilling to step down or to delegate their authorities to younger counterparts (CSO 4; IO Partner 5). This concern was also expressed in external evaluations of the project, which stated that the majority of the VHC members who were interviewed had worked there for 10–13 years (T. Gotsadze & Murzalieva, 2017, p. 18). In other words, they were not newcomers to the committee. However, in addition to their unwillingness to leave, local culture was also a contributing factor in the age profile of the VHC leaders. The VHCs I interviewed stated that they were trying to attract younger volunteers, but young women cannot participate if their husbands or in-laws are against it (CSOs 5 and 7). These findings confirm the problem of recruitment of new volunteers but contradict the issue of high staff turnover pointed to in the literature on community-based organizations (e.g., Ajayi et al., 2012; Sebotsa et al., 2007; Walsh et al., 2012).

Consequently, VHCs as organizations depend on single leaders. Earle et al. (2004, pp. 31–32) point out that community-based organizations in Central Asia are built around a "charismatic, strong leader," and with fifteen members registered, only one can actually be active. Similarly, my interviewees, closely working with the VHCs, noted that the organizations largely depend on the leader (CSO 4). If a leader left without an "equally strong" successor, the VHC started "losing its positions" since the work and initiatives depended on one or two people, with others "passively" following them (CSO 1). In the long-term perspective, this dependence on one single leader jeopardizes the capacity of the community-based organizations.

Overall, community-based organizations demonstrate exceptional capacities due to their members' motivation to bring positive changes to their communities. Interestingly, the social and economic factors in the country, along with the conventional gender roles, contributed to the prevalence of women among community-based organizations. Nevertheless, this section emphasizes the necessity for differentiating the statuses of women in society (e.g., age), which finds its reflection in the CBOs' composition. Interestingly, in contrast to the literature on community volunteers in health, findings from Kyrgyzstan suggest that organizations struggle more with recruiting new members than with attrition. However, difficulties with recruitment further aggravate organizational dependence on single leaders and increase the vulnerability of organizational survival from a long-term perspective.

In contrast to the CAH, the Global Fund grants involve NGOs working on TB and HIV/AIDS. Overall, the political course Kyrgyzstan took during the initial years after gaining its independence provided a favorable environment for civil society development. The global agenda toward civil society engagement, in combination with the inflow of donor funding, nourished this situation further, encouraging the establishment and development of nongovernmental organizations. Once called the "land of NGOs" (see a quote by Edil Baisalov in Pétric, 2015, p. 49), the country had, in 2007 alone, more than 14,000 officially registered NGOs (Ancker et al., 2013). This increase also owes to the Global Fund grants to the country. Several respondents interviewed for this research noted that some NGOs were deliberately established to "siphon off" Global Fund grant money (IO Partner 21), implement the grants, and close right after the grant completion (IO Partner 3). However, the number of these "pocket" NGOs seems to have decreased with time under the pressure of other civil society organizations (ibid.) and their "collective action" (Spicer et al., 2011b, p. 1752).

Local NGOs in Kyrgyzstan have relatively good capacity. Organizations are known to have a "strong workforce" compared to their counterparts in state institutions or even civil society organizations in Central Asia

(G. Murzalieva et al., 2009, pp. 64–69). Like community-based organizations, the NGO sector is dominated by female members (Development Planning Unit, 2010). Despite some attrition, particularly at the level of outreach workers (Harmer et al., 2013), core staff members remain in their positions, contributing to the advantage of NGOs also *vis-à-vis* state organizations discussed in the following subsection. However, problems with organizational skills were reported in areas such as legal protection for the participants of harm reduction programs (Wolfe, 2005), data collection, and some staff members' limited understanding of the end goals of their activities (Murzalieva et al., 2009). Reported among the social and outreach workers, the last problem could also be the outcome of frequent staff rotation at this level.

The Global Fund contributed to the NGO's capacity. The grants advanced the managerial and administrative capacities and "professionalization" of the NGOs (Harmer et al., 2013, pp. 302–304) by organizing multiple training activities for the organizations implementing the grants (see UNDP, 2014). The grants also facilitated the recruitment of additional staff members (Spicer et al., 2011a) and the introduction of new positions, such as "social workers" and "outreach workers" (G. Murzalieva et al., 2009, p. 58). In addition, the limits on personnel costs in the grants resulted in low salary levels and significant staff rotation (ibid.). The NGOs solved these issues by decreasing the number of outreach workers and increasing the workload of the existing staff members (ibid.).

Capacities vary across NGOs depending on the area in which they are working. Smaller organizations have limited resources, fewer skills, and less knowledge than larger NGOs (Spicer et al., 2011b). Moreover, the organizations established earlier and those working with multiple partners have gathered sufficient experience and networks to rely on. These NGOs, as a rule, are less dependent on single funding sources than those working with few partners (Chap. 9). In addition to size, organizational abilities seem to vary across areas. Several interviewees noted the higher capacities of organizations working in HIV/AIDS (State Partner 4) than those in TB (CSO 6). This difference results from varying opportunities and emphasis in the two areas.

Tuberculosis was seen as a state realm, with detection and treatment provided mainly by state medical institutions. HIV, in contrast, involved

NGOs in detecting persons affected by this infection and persuading them to commence and continue their treatment. This difference in attitudes precipitated the incentives and opportunities for civil society organizations. Underdeveloped in tuberculosis, NGOs flourished in the area of HIV. In 2007, for instance, 200 organizations focused on this area (Ancker & Rechel, 2015b). Although insignificant in relation to the total number of registered NGOs, this number is still impressive in the context of population size and the burden of disease. NGOs working in HIV also have a network of organizations with considerable advocacy and community-mobilization skills, which contributes to their participation in HIV policy and decision-making processes (Foundation for AIDS Research, 2015). One interviewee emphasized that there were continuous training, roundtable, and meeting opportunities in HIV, with analogous activities for the organizations working in tuberculosis having commenced rather recently (CSO 6). Following the growing emphasis on TB and the decreasing share of external aid for HIV, civil society organizations seem to have reshaped their profiles and worked accordingly (CSO 8).

Overall, local NGOs, similar to community-based organizations, have relatively high capacities. Organizations are dominated by female members and face a certain level of attrition, although not necessarily among the core staff members. The Global Fund has contributed to the increase in the number of NGO staff and the growth of this sector in the country in general. Indeed, the organizations also developed in response to the emphasis and consequent funding and opportunities, which found its reflection in differences among the organizations working in TB and HIV/AIDS. As the emphasis is changing, the organizations seem to reshape their focus accordingly, which may also change the capacities and number of organizations working in tuberculosis.

Furthermore, similar to other developing countries, Kyrgyzstan faces the problem of human resources in state organizations, which affects their abilities to perform their functions. There is a general problem with high staff turnover (Majtieva et al., 2015), political instability (Ancker et al., 2013), and the low human resources capacity in government organizations (Spicer et al., 2011b).

The Ministry of Health is a natural choice for a state partner for health care projects, but this book focuses on the organizations subordinate to the Ministry and directly involved in the selected health care projects. The Ministry of Health is the major state actor in health, which is responsible for defining and implementing the national policy in this area, ensuring access to and the quality of health care, and coordinating all actors in this area (see Government of KR, 2009). Although critical to health care programs and policies at the national level, the Ministry rarely participates in health care programs directly, instead via agencies representing it. For this reason, this section focuses on the capacities of relevant agencies and not the Ministry itself.[1]

The CAH closely collaborated with the Republican Center for Health Promotion and Mass Communication under the Ministry of Health (hereinafter the Republican Center) and its subunits at district and regional levels, also known as Health Promotion Units. The Republican Center (2022) is responsible for health promotion and disease prevention. Although recently renamed, it was established as early as 2001 to separate health promotion and protection services traditionally provided by the Department of State Sanitary-Epidemiological Surveillance under the Ministry of Health and its branches (Meimanaliev et al., 2005). The Republican Center has branches in Bishkek and Osh, as well as at regional and district levels. The Health Promotion Units (HPUs) at district levels were piloted and supported within the framework of the CAH (ibid.). HPUs are part of primary health care (Family Medicine Centers) but report directly to the Republican Center (Tobias Schüth, 2011a). There is approximately one HPU per 10 villages or 20,000 people (ibid.). HPUs support the organizational development of the Village Health Committees by providing training and monitoring their health care activities (Schüth, 2011b). As of 2017, there were approximately 130 HPUs in the country (Gotsadze & Murzalieva, 2017).

HPUs are critical to the activities and development of community-based organizations, but low salaries and extensive workloads jeopardize HPUs' ability to perform their functions. HPUs have firsthand

---

[1] For more information on the issues with the Ministry of Health of the Kyrgyz Republic capacities, see Isabekova and Pleines (2021).

experience working with Village Health Committees by supporting organizations in their awareness-raising activities and conducting training
areas targeted in national health care programs. HPUs also collaborate
with *Rayon* Health Committees by jointly conducting awareness-raising
activities and evaluating VHCs' capacities. At the end of the CAH, many
trainers who had previously worked with the SRC moved to jobs in the
HPUs, which contributed to the continuity of knowledge and experience
of the project (Gotsadze & Murzalieva, 2017). However, one interviewee
emphasized that some positions were unfilled since trainers were unwilling to work for a monthly salary of 6000 KGS (about €64).[2] In these
cases, the responsibilities were reassigned to existing medical personnel
already tasked with receiving patients and home visits and would therefore have little time to engage with community-based organizations
(State Partner 4). Combining the functions of the HPU with another job
certainly affects the HPU's abilities to work with the VHCs.

Accordingly, the actual work of HPUs with community-based organizations is contingent upon the motivation and willingness of individual
HPU members. It also depends on Family Medicine Centers (FMC)
employing the HPUs. The interviewee noted that the HPUs collaborated
closely with the VHCs in cases in which the heads of the FMCs were
committed to working with community-based organizations (State
Partner 4). Indeed, support from FMCs is also critical to the evaluation
of VHCs because medical organizations provide transportation and per
diem costs for HPUs to conduct evaluation activities.

Furthermore, local self-government bodies are pivotal to community-
based organizations and their activities. Regulated at the national level by
the Cabinet of Ministers, these are elected (representative) and appointed
(executive) at the local level but are accountable to the President and the
Cabinet of Ministers (Government of KR, 2021).[3] The sizes of local self-
governance bodies differ. The executive bodies are set by the Cabinet of
Ministers, whereas the representative bodies are proportional to the sizes
of related constituencies (ibid.). The local self-governance bodies have

---

[2] The exchange rate, as of March 17, 2023, was applied throughout this book.
[3] On October 12, 2021, the President of the Kyrgyz Republic dissolved the Cabinet of Ministers
(see Gunkel, 2021).

critical responsibilities in their domain. The organizations are responsible for drafting, approving, and implementing the local budgets and for social and economic development of their constituencies, including issues with access to potable water, sanitation, waste disposal, and other matters (ibid.). Major sources of funding for these purposes come from public finances received from higher levels and finances obtained from local taxes (Tobias Schüth, 2011a).

Nevertheless, the financial and administrative capacities of local self-government bodies are case dependent. One interviewee reported that the financial capacity of the local self-governments varies throughout the country, and yet, most are subsidized by the national government (State Partner 9). In addition to the budget deficit, administrative capacities are further hindered by the unstable political situation in the country, causing rapid turnover among local authority officials. In this regard, several interviewees aptly noticed that representatives of local self-governments change as if "one is changing dresses" (CSO 2). As soon as the village health committee starts collaborating with a state official, (s)he is replaced by a new one (CSO 7). This high turnover of state officials has a negative impact on collaboration with community-based organizations (ibid.). Indeed, the financial and administrative capacities are case-dependent, and a more general overview of this matter requires a comprehensive analysis of rotations in local self-government bodies throughout the country. However, frequent changes of state officials at the national level support the assumptions made by the interviewees.

CAH also involved representatives of family group practices and *feldsher-midwife (akusher)* points, which are the first points of contact with the health care system in rural areas (Meimanaliev et al., 2005). These organizations supported the VHCs during the initial stages of the project, also in terms of the analysis of population health (Tobias Schüth, 2011a). This collaboration has also continued beyond the project duration, vividly demonstrated by joint activities on infectious and noncommunicable diseases (AVHC, 2022). Their capacities are reviewed in the following subsection, as the role of medical professionals is equally significant to the Global Fund grants.

In contrast to CAH, the Global Fund grants essentially collaborate with the agencies responsible for tuberculosis and HIV/AIDS services.

These are the National Center of Phthisiology (NCPh) and the Republican AIDS Center. Both organizations represent a broader network of vertical services focusing on and responsible for preventing and treating related diseases.

Tuberculosis services in the country include NCPh at the national and tertiary levels, regional and city tuberculosis clinics and centers at secondary levels, and tuberculosis cabinets in family medicine centers at primary care levels (Ministry of Health of KR, 2013). NCPh is responsible for the diagnosis, treatment, research, and coordination related to tuberculosis services throughout the country (Government of KR, 2014). The organization dates back to the Kyrgyz Scientific Institute for Tuberculosis Research, established in 1957 (NCPh, 2022).

HIV services in Kyrgyzstan include the Republican AIDS Center (2021a), its regional units, and the center in the capital Bishkek. HIV testing is provided by 34 labs, including 7 in the regional AIDS centers, 24 in district and city hospitals, and 3 in the medical organizations at the republic level (ibid.). Treatment is available in AIDS centers and family medicine centers in all seven regions of the country (ibid.). The AIDS centers were established in 1989 following the first cases of HIV in the country (Republican AIDS Center, 2021b). The Republican AIDS Center is responsible for coordinating HIV-related services, including detection and treatment, as well as monitoring the HIV situation in the country (ibid.).

Despite the broader outreach, multiple factors, including political instability, staff rotation, and excessive workload, limit the state institutions' capacity. Frequent changes in decision-makers (Majtieva et al., 2015) and staff rotation have paralyzed state agencies and ministries, affecting their ability to carry out their functions (Spicer et al., 2011b). Furthermore, the Global Fund grants increased the number of staff members of Sub-Recipient NGOs, but the number of employees in state agencies involved in the grants remained the same (Center for Health System Development et al., n.d., p. 19). In this way, the tasks related to the grants were distributed among the existing staff members of the Republican AIDS Center and the National Center of Phthisiology. However, the limited capacities of NCPh and the Republican AIDS Center also prevented them from remaining Primary Recipients of the

Global Fund grants. Misappropriation and mismanagement of grants lead to the transfer of the Primary Recipient functions to the United Nations Development Programme (UNDP) (see Chap. 8).

Limited evaluation of training efforts and the broader structural issues in the country jeopardize the outcome of capacity-building activities. Donor organizations are criticized for neglecting the capacity problem in state institutions (UNAIDS, 2005b). However, multiple organizations, including the Soros Foundation Kyrgyzstan, the United Kingdom's Department for International Development (DFID),[4] U.S. Agency for International Development (USAID), the World Health Organization (WHO) (Manukyan & Burrows, 2010), and the Global Fund (UNDP, 2015a), provided training to state officials.

Yet, the coverage and intensity of training remain unclear, as there is no system tracking the number of seminars and their attendees (Murzalieva et al., 2009). The capacity-building activities are also jeopardized by staff rotation at the ministries and agencies. As a respondent interviewed for this book noted, capacity-building presupposed having people in relevant positions. However, this was difficult because of high staff turnover, the brain drain from state agencies to donor organizations, and the appointment of relatives and friends instead of candidates with the necessary qualifications (IO Partner 4). This way, capacity-building activities seem to be trapped in a vicious cycle that can be broken only after solving broader issues related to political instability, staff appointment procedures, and low salaries.

In addition to the managerial level, the capacity of state organizations closely relates to the availability of health workers. Primary health care (PHC) workers are critical to health promotion and disease prevention activities. The reforms in the health care system since the early independence toward strengthening PHC and reducing the capacities of secondary and tertiary care levels also reemphasized its broader significance. Commencing in tuberculosis earlier, the tendency toward moving away from the vertical service provision toward its integration into PHC is also growing in HIV.

---

[4] In 2020, it was replaced by the Foreign, Commonwealth and Development Office.

Nevertheless, the capacities of both PHC workers and those working in specialized services are uneven due to staff attrition and geographic inequity in distribution. There are a sufficient number of medical graduates in the country, but most prefer specialization over general practice. There are 700 family medicine centers in the country employing approximately 2000 family doctors, although at least 3000 are needed for the growing size of the population (Bengard, 2021). Most PHC workers are of retirement age, and there are problems attracting new cadres (ibid.). Low salary levels and limited incentives at the PHC level are among the few reasons. Furthermore, the distribution of PHC workers in urban and rural areas remains unequal. These issues affect TB and HIV, particularly given the risks of nosocomial infections associated with these services. The availability of medical professionals is also affected by larger issues in the country and beyond, including extensive internal and external migration and limited incentives for attracting new and retaining existing health care workers.

Despite government efforts, the salaries of medical workers remain low. According to the Republican AIDS Center, the salaries of a nurse and a doctor were 4000 and 7500 KGS monthly (approximately €43 and €80), respectively (Government of KR, 2017a). The situation was similar in tuberculosis. According to an interviewee, the base rate salaries were approximately 2500 for nurses and 6000 KGS for doctors (approximately €27 and €64 Euro) (Health Worker 1). The final salaries in tuberculosis also depended on the number of successfully treated patients who added to additional payments for nurses and doctors. However, an interviewee noted that with all bonuses, salaries amounted to 10,000 KGS (€107) (ibid.). In this way, the suggested salaries of medical workers in tuberculosis and HIV were lower than the average salary at the country level for 2017 and 2018 (National Statistical Committee of KR, 2023). The government initiated a number of reforms to offer additional incentives in PHC and the health care sector in general. In 2018, it introduced a payment system stipulating additional payments to doctors' and nurses' base rate salaries depending on their work experience and work performance (Kudrâvceva, 2018). In 2022, the government initiated an increase in the base salary levels, as a result of which monthly salaries of medical personnel increased to 9000–15,000 KGS (€97–161) (Today.kg, 2022). Despite these increases, the monthly salaries of medical workers remain below the national average (see National Statistical Committee of KR, 2023).

Overall, the capacities of state organizations involved in the "Community Action for Health" and the Global Fund grants are significantly affected by the general economic and political situation in the country. In contrast to civil society organizations, state institutions are particularly disadvantaged by frequent rotation and unequal distribution of staff members. Development organizations have attempted to support capacities by organizing training activities. However, their outcomes remain unclear. Similarly, salary rates remain below the national level despite government efforts. All these factors result in capacity issues that continue to prevent state organizations from exercising their functions to the full extent.

## 4.4    Aid Dependency

Aid dependency in this book refers to the abilities of organizations to perform their functions and achieve their objectives in the absence of external aid (Chap. 2). This book focuses on the provision of services, be it health promotion, disease prevention, or treatment, by examining dependency in relation to technical (e.g., expertise) and financial assistance.

First, in terms of state organizations, aid dependency varied across the two cases studied in this book. The objectives and activities of the CAH echoed the ideas enunciated in the national programs. Community involvement and strengthening PHC were in the foreground of both "Manas" (1996–2000) and "Manas Taalimi" (2006–2010) (Government of KR, 2006; WHO/Europe and UNDP, 1997). Although commencing as a pilot project in selected districts, the CAH demonstrated the ability of the rural population to take responsibility for its own health, which was essential to the abovementioned reform programs.

The governmental commitment to learning was also evident in the division of health promotion from public health. The Ministry transferred the relevant responsibilities to the newly established Republic Center for Health Promotion and, in so doing, moved away from the system inherited from the Soviet Union, in which the Sanitary Epidemiological Service was responsible for both (Schüth, 2011a). The *Semashko* health care system generally emphasized treatment over prevention, with limited efforts targeted at health promotion. The newly established institution had no prior experience working with communities. Therefore, the Republican Center for Health promotion closely collaborated with the CAH, supporting its capacity for health promotion and working with communities. Notably, HPUs had limited prior experience in these areas. Specifically established by the Ministry of Health for the expansion of the CAH initiative, they were intended to strengthen the abilities of PHC workers to cooperate with communities (ibid.). HPUs received extensive training within the framework of the CAH (ibid.).

Equivalently, primary health care workers had neither prior knowledge nor experience in engaging with community-based organizations. The paternalistic health care system inherited from the Soviet Union precluded citizen participation (Ferge, 1998) and treated patients as passive service recipients (Field, 1988). Working with community organizations was never a part of the PHC activities before the CAH (Schüth, 2011a). Preventive activities were conventionally limited to individual consultations with patients during their visits or home visits of medical workers to specific groups of the population, such as pregnant women, those with newborns, and those with chronic diseases (ibid.). Therefore, the PHC workers equally received extensive training in the project.

It should be noted that the compelling expertise in community capacity-building the CAH offered was further strengthened with the project collaboration with primary health care and public health initiatives. The Vaccine Alliance (Gavi), United Nations Children's Fund (UNICEF), USAID, Global Fund, World Bank, and Aga Khan Foundation are among the organizations with which the SRC collaborated within the framework of the CAH. Among others, cooperation with Gavi supported the immunization program in the country (Akkazieva et al., 2009), and work with UNICEF (2016) and other partners launched the *Gulazyk* program for the distribution of micronutrient sprinkles. USAID was critical to the CAH in multiple aspects (also countrywide expansion) within the framework of its programs implemented between 1994 and 2009 on reforming and strengthening primary health care in Central Asia (see Abt Associates Inc., 2015). The Global Fund supported disease prevention, and the World Bank (n.d.), in turn, provided access to potable water and sanitation systems in rural areas. However, the closest in design was the community-based health care initiative in fifty villages by the Aga Khan Foundation, which adjusted its activities to match the CAH (Schüth, 2011b, p. 31). This coordination benefited community capacity-building efforts in health by reducing project activity duplications and contradictions. Integration with other projects has also strengthened the position of the CAH.

Overall, limited prior experience and knowledge in health promotion and community engagement left the Ministry of Health and its institutions dependent on the knowledge and skills the project offered. The CAH demonstrated the very outcomes of communities taking responsibility for their own health, which the state organizations were interested in. Thus, although not necessarily dependent on financial terms, the recipient state depended on the donor's technical expertise.

In contrast to CAH, the aid dependency in the Global Fund grants is related mainly to financing. Donor organizations finance a large share of tuberculosis and HIV/AIDS programs in Kyrgyzstan. At its peak, in 2007, donors provided 94% (297.8 million KGS or €3,193,395) and the state approximately 6% (20.3 million KGS or €217,683) of total expenditures on HIV/AIDS services (G. Murzalieva et al., 2009, p. 18). The share of donor contributions decreased with time, but it still represents more than half of HIV-related funding.[5] Multiple donor organizations participate in TB and HIV/AIDS programs in Kyrgyzstan. The German Development Bank (*die Kreditanstalt für Wiederaufbau*—KfW) finances laboratory construction, and GIZ provides technical assistance in the area of reproductive health. The International Committee of the Red Cross and Doctors Without Borders cover TB services in prison. The President's Emergency Plan for AIDS Relief (PEPFAR) and USAID

---

[5] The data on the share of external financing is inconsistent: the UNDP (2015a, p. 56) suggests that international financing to HIV/AIDS was 62% in 2012, 66% in 2013, and approximately 56% in 2014. A state representative, however, in her presentation during the SWAp, notes that external financing to health care was 71% in 2012, 76% in 2013, and 57% in 2014 (Majtieva et al., 2015, p. 20).

finance TB and HIV programs in the civilian sector, along with an HIV grant from the Russian Federation.

Despite the multiplicity of donors, the Global Fund remained the leading financier of TB and HIV/AIDS programs in the country. In 2004–2006, it covered 69% of all HIV/AIDS-related services, with other donors and the government providing the remainder of the financing (Gulgun Murzalieva et al., 2007, p. 31). Representing over half of the external assistance, the Global Fund finances HIV treatment and nearly all HIV prevention programs among the key groups (e.g., men who have sex with men, commercial sex workers, persons who inject drugs, and others) (Majtieva et al., 2015). Similarly, in the area of TB, the Global Fund covered medications against drug-resistant forms of TB, laboratory supplies, co-payments to health care workers, and other expenses (State Partner 9). In this way, the Global Fund remained the principal financier of TB and HIV services in the country.

In contrast, multiple organizations provide technical assistance in TB and HIV/AIDS. The interviewees specifically emphasized the Joint United Nations Programme on HIV/AIDS (UNAIDS), World Health Organization (WHO), KfW, World Bank, and USAID's contributions to the development of regulatory documents, management of health care, and building the capacity of state organizations (State Partner 10 and Academic Partner 2; IO Partner 3). Similarly, Global Fund grants stipulate training and capacity-building activities for medical personnel involved in TB and HIV/AIDS services. Therefore, the state officials interviewed for this research suggested that technical assistance was among the "most significant" benefits development organizations offered (State Partner 3) and that without it, the country would end up establishing ineffective and cumbersome systems (State Partner 6). Studies on health aid to Kyrgyzstan similarly highlight donors' contributions to strengthening laboratory services, establishing sentinel surveillance systems (Wolfe et al., 2008), and revising HIV/AIDS-related legislation (Ancker & Rechel, 2015b).

Despite the significance of all development partners, one could specifically highlight the role of the two United Nations agencies, namely, the WHO and UNAIDS, as *primus inter pares* in health. Their recommendations are equally followed by the state, civil society, and donor organizations. The Global Fund itself complies with the WHO standards (e.g., Global Fund, 2009) and the UNAIDS (2005a) suggestions. In this regard, Kaasch (2015) notes that although insignificant in terms of financing, the WHO has established itself as a standard setter and a leading actor in the area of health. Similar conclusions could also be made regarding UNAIDS, which specifically maintained its expertise in the area of HIV.

Overall, the Ministry of Health and state agencies on TB and HIV/AIDS collaborate with multiple donors, but they still heavily rely on the financing provided by the Global Fund. However, in technical assistance, the Global Fund, like other donors, conforms to the standards and regulations of other partners that established themselves as standard and norm setters for TB and HIV/AIDS.

As noted in the previous section on capacity, civil society organizations refer to CBOs in CAH and NGOs in the Global Fund grants. The CAH initiated community engagement in health care and facilitated the mobilization of community members to join the VHCs. However, newly established, these organizations had neither the experience nor the resources to pursue their objectives. The literature on grassroots organizations suggests that illiteracy (Jana et al., 2004), gender-related biases (WHO, 2008), political situation, and poverty (Morgan, 2001) all make communities dependent on external aid. According to UNESCO (2023) estimates, over 99% of the population in Kyrgyzstan is literate.[6] Moreover, members of the Village and *Rayon* Health Committees faced and overcame multiple issues, including gender-biased treatment from their communities and local authorities, frequent rotation of local self-governments, and resource mobilization hardships (Chaps. 5 and 6).

Nevertheless, similar to the recipient state, communities did not have prior knowledge or skills to participate in the health care system. Through technical and financial assistance, the project intended to build the capacities of community-based organizations throughout the project, but this extensive support has unintentionally contributed to the dependence of community-based organizations on the donor. The CAH was the only project providing comprehensive coverage of the Village and *Rayon*

---

[6] As of March 2023, the relevant data is available until 2019.

Health Committees throughout the country. Other donors just engaged with VHCs from specific regions and in certain areas that were compliant with their project objectives. Although continued donor assistance was not the only factor relevant to community capacity-building, the end of the CAH in 2017 exposed community-based organizations to a certain level of uncertainty about their future (Chap. 6).

At the same time, although providing significant technical and financial support at an organizational level, the CAH offered only minor financial incentives to the community volunteers. The project may have covered travel and per diem costs related to health promotion activities. However, the members of community-based organizations did not receive salaries from the SRC. One interviewee pointed out that the CAH would have been able to pay salaries, and initially, volunteers did request payment for their work (CSO 4). However, no salaries were paid to support the continuity of organizations and activities beyond the project, which was also explained to community members and accepted by them (ibid.).

Not everyone stayed, but those who remained were not driven by financial gains but by the willingness to bring positive changes to their communities. Several interviewees noted that those who joined community-based organizations for financial reasons soon resigned (CSO 5). As a result, very few VHC members sought financial gains or declined to conduct certain activities because they were volunteers (CSO 1). Instead, as unpaid volunteers, the community volunteers implemented the project-related activities because of their willingness to bring changes to their communities. Seeing the outcomes of activities generated enthusiasm among the volunteers and a belief that they could "bring something good" to their villages through their work (CSO 5). In this way, the incentives offered by the SRC supported but did not define the VHCs' willingness to carry out their activities.

As unpaid volunteers, the VHC members could discontinue their activities without any financial consequences to themselves. This financial independence at an individual level evened out the organizations' dependence on the project because the SRC also depended on the VHC members' willingness to work. As volunteers, the community members were able to decide whether to continue their work or not. As one of the interviewees noted, no one could point to them, saying, "you should work," and only those with "initiatives in their hearts" continued (CSO 4). In this way, the SRC (implementing) and the SDC (financing the CAH) depended on the community members' willingness to work. The lack of financial incentives limited the leverage of development partners over community volunteers.

Civil society organizations' aid-dependency dynamics differed in the Global Fund grants. NGOs working in TB and HIV/AIDS areas depend on external aid. This dependency is evident in the interruption of services during disruptions in external funding. The delays in Global Fund financing affected NGOs' service delivery (Harmer et al., 2013). To address the short-term breaks, the organizations involved volunteers; however, the long-term interruptions in 2007–2008 caused the termination of activities and staff turnover due to the disruption of salary payments (Spicer et al., 2011a). Some activities, such as diagnostic and treatment services to commercial sex workers, resumed only after the Global Fund restarted its financing (Murzalieva et al., 2009). Several donor organizations committed their resources to cover the financial gap and address the issue of NGOs' service interruptions. The Soros Foundation Kyrgyzstan and the UNDP provided "emergency coverage" during funding disruptions in 2004 (Wolfe, 2005, pp. 23–24). The UNDP has also used its own resources for staff recruitment, procurement of condoms, methadone, and the "emergency stock" of antiretroviral medications during the delays in the HIV grant of the Global Fund (2011–2016) (Grant Performance. Report External Print Version. Kyrgyzstan KGZ-910-G07-T, 2016, p. 31). In doing so, the UNDP ensured a continuous supply of medications (ibid.) and provision of other services stipulated in the grants. However, in contrast to the UNDP, local NGOs do not have sufficient financing to cover these costs, even temporarily, during financial interruptions.

However, financial dependency varies across organizations. Those with multiple sources of financing were less affected by the delays (Murzalieva et al., 2009) compared with those solely dependent on Global Fund grants. Accordingly, the perspective of the NGOs interviewed for this research on the continuity of their activities beyond the Global Fund grants varied. While some were optimistic about their continuance (CSO 6), others acknowledged the inability to implement the initiatives on their own (CSO 8) and that the breadth of their activities depended on donors (CSO 9).

In terms of technical assistance, NGOs received multiple but inconsistent training opportunities from donors. The Soros Foundation, UN agencies (Godinho et al., 2005), USAID, Global Fund, and other actors offered technical assistance to NGOs. The Global Fund financed the seminars on social support, strategic planning and fundraising, accounting and document management (see UNDP, 2015b, 2015c), and other areas. However, assessing dependency in terms of technical assistance is challenging, as neither donors, recipient states, nor civil society organizations have a broad understanding of all training activities conducted in the areas of TB and HIV/AIDS. Accordingly, the impact, selection criteria, and compliance of training with the needs of targeted groups are unclear (G. Murzalieva et al., 2009). An NGO representative interviewed for this research suggests that the selection criteria for participants are guided by their rotation and not the NGOs' specialization. However, the rotation does not contribute to the institutional memory of organizations, which would be enhanced by the more consistent and continuous support of fewer organizations for a longer period (CSO 8).

Overall, both community-based organizations and NGOs depend on external assistance. However, in the case of the former, this dependency was evened out because the community volunteers were unpaid by the project and could halt the activities at any time without any financial losses. In this situation, the donor depended on the willingness of community members to continue their activities. In the case of the latter, the dependency remained. On its own, financial benefits are natural to economic interaction. However, in the context of development assistance, they may unintentionally strengthen the conventional "gift-giving" and "gift-receiving" dynamics between stakeholders (see Hinton & Groves, 2004).

## 4.5 Summary

This chapter explored the structural factors relevant to both interactions between stakeholders and the sustainability of health projects. Focusing on the actors relevant to the selected cases, it examined the predictability and flexibility of aid on the sides of donors, as well as capacities and aid dependencies on the sides of recipients.

1. The chapter has vividly demonstrated that despite the acknowledgments of and commitments to ensuring predictable aid, Switzerland and the Global Fund varied considerably in their achievements. Although performing better than the average bilateral partner or a vertical health care program, the two actors nevertheless provided varying predictability due to their development cooperation structure. Indeed, Switzerland has performed better by offering a longer duration of assistance than does the Global Fund.
2. The comparative overview of the two actors has equally demonstrated their commitment to providing flexibility. Both put a great emphasis on the recipient's ownership. Accordingly, the Global Fund finances the proposals developed by applicants and is in accordance with their national strategies. Switzerland, in turn, places great value in defining the objectives in collaboration with partner countries. Correspondingly, the decentralization and autonomy of field offices provide the space for adjusting the activities to local needs and priorities. Driven by similar objectives, the Global Fund focused on concentrated decision-making with the rest of the activities performed in recipient countries. However, this concentration hinders the flexibility of assistance by affecting the responsiveness to changes occurring in the applicant countries. Another hindering factor is accountability before financiers, which equally affected the Global Fund and Swiss aid, having to balance accountability toward the funders and recipients of aid.
3. The capacities of state organizations involved in both health initiatives are extensively constrained by the political and economic instability in the country, causing frequent staff rotation, limited institutionalization, and a limited range of incentives available to state employees.

Notably, these factors seem to have an equally annihilatory impact at both the decision-making and service-provision levels. Interestingly, their effect on civil society organizations was less and uneven. The economic situation in the country facilitated migration, which, combined with conventional gender roles in households and society, contributed to the recruitment and retention of women in community-based organizations. Nevertheless, not all women join organizations that seem to depend on single leaders. Similarly, dominated by women, the NGO sector has also demonstrated a relatively strong capacity, particularly in comparison to state institutions. However, the capacities greatly vary across the organizations and sectors, with the organizations in HIV performing better than those in TB.

4. The broader political and economic instability in the country has contributed equally to the dependency of national actors on external aid. At the same time, the dependency on technical assistance observed in the CAH contrasts with the dependency on financial support offered in the Global Fund grants. Certainly, the organization provides substantial technical support to all stakeholders involved in its grants. However, the significance of this support in relation to financing seems smaller, particularly in the context of other technical partners. Having limited funding to offer, these have established themselves as arbiters of standards and norms equally followed by other development partners and state and civil society organizations. In the case of civil society, nonpayment of community volunteers evened out the dependencies of CBOs on donors, which did not occur in NGOs, whose activities remain financially dependent on donors.

# References

Abt Associates Inc. (2015). *Anatomy of health care transformation: USAID's legacy in health systems strengthening in Central Asia: 1994–2015* (pp. 1–72). Retrieved March 3, 2023, from https://2017-2020.usaid.gov/sites/default/files/documents/1861/USAID_Central%20Asia_Healthcare_20-year-Legacy-document_ENG.pdf

Ajayi, I. O., Jegede, A. S., & Falade, C. O. (2012). Sustainability of intervention for home Management of Malaria: The Nigerian experience. *Journal of Community Medicine and Health Education*, 1–8. https://doi.org/10.4172/2161-0711.1000175

Akkazieva, B., Samiev, A., & Temirov, A. (2009). *Kyrgyzstan case study: The global alliance for vaccination and immunization health systems strengthening tracking study* (pp. 1–74). Center for Health System Development; Gavi; InDevelop. Retrieved February 3, 2023, from http://hpac.kg/wp-content/uploads/2016/02/prp59_e.pdf

Ancker, S., & Rechel, B. (2015a). 'Donors are not interested in reality': The interplay between international donors and local NGOs in Kyrgyzstan's HIV/AIDS sector. *Central Asian Survey, 34*(4), 516–530. https://doi.org/10.1080/02634937.2015.1091682

Ancker, S., & Rechel, B. (2015b). HIV/AIDS policy-making in Kyrgyzstan: A stakeholder analysis. *Health Policy and Planning, 30*, 8–18. https://doi.org/10.1093/heapol/czt092

Ancker, S., Rechel, B., McKee, M., & Spicer, N. (2013). Kyrgyzstan: Still a regional 'pioneer' in HIV/AIDS or living on its reputation? *Central Asian Survey, 32*(1), 66–84. https://doi.org/10.1080/02634937.2013.771965

AVHC. (2018). *"Kyrgyzstan ayyldyk den sooluk komitetteri" Assotsiatsiyacy [The Association of Village Health Committees]: "Assotsiatsiyanyn 2018-jyldyn ishmerdüülügünün" otchetu [Report of activities for 2018]* (pp. 1–19).

AVHC. (2020). *Godovoj otčet 2019goda [Annual report 2019]: Associaciâ «Kyrgyzstan ajyldyk den sooluk komitetteri» Deâtel'nost' Associacii KADK v ramkah proekta USAID «Vylečit' tuberkulez» s 1 ânvarâ po 30 sentâbrâ 2020 g. [Association of Village Health Committees. The work of the Association within the framework of the USAID "Defeat TB" project from 01 Jan to 30 Sep 2020]* (pp. 1–24). n.p.

AVHC. (2022). *Official Facebook page of the association of village committees.* Retrieved December, 2022, from https://www.facebook.com/associationkadk/?ref=page_internal

AVHC. (n.d.). *Rayonduk Den Sooluk Komiteti Tuuraluu Jobo [Regulations on Rayon Health Committee]:* n.p.

Bengard, A. (2021). *Okolo tysâči semejnyh vračej ne hvataet v Kyrgyzstane [About a thousand family doctors are missing in Kyrgyzstan].* Retrieved February 3, 2023, from https://24.kg/obschestvo/193745_okolo_tyisyachi_semeynyih_vrachey_nehvataet_vkyirgyizstane/

Benjamin, H. (2011). *Examining the impact of global fund reforms on implementation: Results of the global fund implementers survey.* Open Society Foundations.

Retrieved February 3, 2023, from https://www.opensocietyfoundations.org/sites/default/files/global-fund-implementers-20120305_0.pdf

Brown, J. C., & Griekspoor, W. (2013). Fraud at the Global Fund? A viewpoint. *The International Journal of Health Planning and Management, 28*(1), 138–143. https://doi.org/10.1002/hpm.2152

Center for Health System Development, American University of Central Asia, & London School of Hygiene & Tropical Medicine. (n.d.). *Policy brief № 19 tracking global HIV/AIDS initiatives and their impact on the health system: the Experience of the Kyrgyz Republic: (Interim Report)* (pp. 1–2). Retrieved February 3, 2023, from http://hpac.kg/wp-content/uploads/2016/02/PB19HIVAIDS_E.pdf

Development Planning Unit. (2010). *Gender analysis of the Kyrgyz-Swiss-Swedish health project (KSSHP) Phase V* (pp. 1–33). University College London.

Development Policy Institute. (2017). Mehanizm peredači informacii o kačestve uslug sistemy zdravoohraneniâ snizu vverh: ot mestnyh soobŝestv na nacional'nyj uroven' [The bottom-up mechanism for transferring information on the quality of health care services: from local communities to the national level]. *Municipalitet, 12*(134). Retrieved March 3, 2023, from http://www.municipalitet.kg/ru/article/full/1715.html

Earle, L., Fozilhujaev, B., Tashbaeva, C., & Djamankulova, K. (2004). Community development in Kazakhstan, Kyrgyzstan and Uzbekistan: Lessons learnt from recent experience. *Occasional Papers Series, 40*, 1–63.

Embassy of Switzerland in KR. (2013). *List of projects in Kyrgyzstan supported by Swiss Government.* Retrieved March 1, 2021, from https://www.eda.admin.ch/dam/countries/countries-content/uzbekistan/en/resource_en_225065.pdf

Federal Department of Foreign Affairs, & State Secretariat for Economic Affairs. (2020). *Switzerland's international cooperation strategy 2021–24.* Retrieved March 3, 2023, from https://www.eda.admin.ch/eda/en/fdfa/fdfa/publikationen.html/content/publikationen/en/deza/diverse-publikationen/broschuere-iza-2021-24

Ferge, Z. (1998). Social policy challenges and dilemmas in ex-socialist systems. In J. M. Nelson, C. Tilly, & L. Walker (Eds.), *Transforming post-communist political economies* (pp. 299–321). National Academy Press.

Field, M. G. (1988). The position of the soviet physician: The bureaucratic professional. *The Milbank Quarterly, 66*(Suppl 2), 182–201.

Foundation for AIDS Research. (2015). *Harm reduction and the global HIV epidemic. Interventions to prevent and treat HIV among people who inject drugs* (pp. 1–22). Foundation for AIDS Research Public Policy Office. Retrieved March 27, 2019, from https://www.amfar.org/uploadedFiles/_amfarorg/vArticles/On_The_Hill/2015/DC-PWID-Policy-Report_08-31-15v205.pdf

Global Fund. (2009). *Amended and restated grant agreement for the rolling continuation channel ('RCC') program.* Retrieved May 10, 2020, from https://data.theglobalfund.org/investments/grant/ARM-202-G06-H-00/2

Global Fund. (2013). *The global fund's new funding model* (pp. 1–12). Global Fund. Retrieved February 3, 2023, from https://www.theglobalfund.org/media/1467/replenishment_2013newfundingmodel_report_en.pdf?u=63648680736000000

Global Fund. (2023a). *Government and public donors.* Retrieved February 3, 2023, from https://www.theglobalfund.org/en/government/

Global Fund. (2023b). *Allocation funding.* Retrieved February 3, 2023, from https://www.theglobalfund.org/en/applying-for-funding/sources-of-funding/allocation-funding/

Global Fund. (2023c). *Replenishment.* Retrieved February 3, 2023, from https://www.theglobalfund.org/en/replenishment/

Global Fund. (n.d.-a). *Eligibility list 2022.* Retrieved February 3, 2023, from https://www.theglobalfund.org/media/11712/core_eligiblecountries2022_list_en.pdf

Global Fund. (n.d.-b). *A strategy for the Global Fund. Accelerating the effort to save lives* (pp. 1–48). n.p.: Global Fund. Retrieved February 3, 2023, from https://www.theglobalfund.org/media/2525/core_globalfundstrategy2006_strategy_en.pdf?u=636486807020000000

Global Fund. (n.d.-c). *Introduction to the 2017–2019 funding cycle and the differentiated funding application process.* Retrieved February 3, 2023, from http://www.stoptb.org/assets/documents/global/fund/Differentiated%20Approaches%20for%20Countries%20to%20Access%20Funding_Panel.pdf

Global Partnership for Effective Development Co-operation. (n.d.). *Monitoring dashboard: Partner comparison.* Retrieved February 3, 2023, from https://dashboard.effectivecooperation.org/partner

Godinho, J., Renton, A., Vinogradov, V., Novotny, T., & Rivers, M.-J. (2005). Reversing the tide: Priorities for HIV/AIDS prevention in Central Asia. *World Bank Working Paper, 54,* 1–186.

Gotsadze, T., & Murzalieva, G. (2017). *Impact evaluation of the Community Action for Health (CAH) project in Kyrgyzstan: Phase I–VII* (April 2002–March 2017) Report (pp. 1–44). n.p. Retrieved March 3, 2023, from https://www.newsd.admin.ch/newsd/NSBExterneStudien/880/attachment/en/3725.pdf

Government of KR. (2006). Nacional'naâ programma reformy zdravoohraneniâ Kyrgyzskoj Respubliki "Manas taalimi" na 2006–2010 gody [National Health Care Reform Program "Manas Taalimi" for 2006–2010]: Utverždena postanovleniem Pravitel'stva Kyrgyzskoj Respubliki ot 16 fevralâ 2006 goda № 100 [Approved by the Decree of the Government of the Kyrgyz Republic dated February 16, 2006 No. 100]. Retrieved March 3, 2023, from http://cbd.minjust.gov.kg/act/view/ru-ru/57155

Government of KR. (2009). Položenie o Ministerstve zdravoohraneniâ Kyrgyzskoj Respubliki (utverždeno postanovleniem Pravitel'stva KR ot 4 dekabrâ 2009 goda №730) [Regulations on the Ministry of Health of the Kyrgyz Republic (approved by the Decree of the Government of the Kyrgyz Republic dated December 4, 2009 No. 730)]. Retrieved February 3, 2023, from http://cbd.minjust.gov.kg/act/view/ru-ru/90385

Government of KR. (2014). Postanovlenie Pravitel'stva Kyrgyzskoj Respubliki ot 26 iûnâ 2014 goda № 352 O Koordinacionnom sovete po obŝestvennomu zdravoohraneniû pri Pravitel'stve Kyrgyzskoj Respubliki [Decree of the Government of the Kyrgyz Republic dated June 26, 2014 No. 352 On the Coordinating Council for Public Health under the Government of the Kyrgyz Republic]. Retrieved February 3, 2023, from http://cbd.minjust.gov.kg/act/view/ru-ru/96604?cl=ru-ru

Government of KR. (2017a). Programma Pravitel'stva Kyrgyzskoj Respubliki po preodoleniû VIČ-infekcii v Kyrgyzskoj Respublike na 2017–2021 gody [The Government of the Kyrgyz Republic Program on Overcoming HIV Infection in the Kyrgyz Republic for 2017–2021]: Priloženie 1 Utverždeno postanovleniem Pravitel'stva Kyrgyzskoj Respubliki ot 30 dekabrâ 2017 goda № 852 [Annex 1 approved by the Decree of the Government of the Kyrgyz Republic dated December 30, 2017 No. 852]. Retrieved February 17, 2023, from http://cbd.minjust.gov.kg/act/view/ru-ru/11590

Government of KR. (2017b). O Programme Pravitel'stva Kyrgyzskoj Respubliki po preodoleniû VIČ-infekcii v Kyrgyzskoj Respublike na 2017–2021 gody [The program of the Government of the Kyrgyz Republic sight to overcome HIV infection in the Kyrgyz Republic for 2017–2021]: Postanovlenie Pravitel'stva Kyrgyzskoj Respubliki ot 30 dekabrâ 2017 goda № 852 [Decree of the Government of the Kyrgyz Republic dated December 30, 2017 No. 852]. Retrieved February 3, 2023, from http://cbd.minjust.gov.kg/act/view/ru-ru/11589

Government of KR. (2021). Zakon Kyrgyzskoj Respubliki ot 20 oktâbrâ 2021 goda № 123 O mestnoj gosudarstvennoj administracii i organah mestnogo samoupravleniâ [Law of the Kyrgyz Republic of October 20, 2021 No. 123

"The Law on Local State Administration and Local Self Government Bodies"]. Retrieved March 2, 2023, from http://cbd.minjust.gov.kg/act/view/ru-ru/112302

Grant Performance. Report External Print Version. Kyrgyzstan KGZ-910-G07-T. (2016). Retrieved March 3, 2023, from (pp. 1–30). http://docs.theglobalfund.org/program-documents/GF_PD_003_e99065eb-b1c1-409a-a5f5-e0db338541f2.pdf

Gunkel, E. (2021). Prezident Kyrgyzstana raspustil kabinet ministrov [The president of Kyrgyzstan dismissed the cabinet of ministers]. Retrieved March 4, 2023, from https://www.dw.com/ru/prezident-kyrgyzstana-sadyr-zhaparov-raspustil-kabmin/a-59483342

Harmer, A., Spicer, N., Aleshkina, J., Bogdan, D., Chkhatarashvili, K., Murzalieva, G., et al. (2013). Has global fund support for civil society advocacy in the former Soviet Union established meaningful engagement or "a lot of jabber about nothing"? *Health Policy and Planning, 28*, 299–308. https://doi.org/10.1093/heapol/czs060

Hinton, R., & Groves, L. (2004). The complexity of inclusive aid. In L. C. Groves & R. B. Hinton (Eds.), *Inclusive aid: Changing power and relationships in international development / edited by Leslie Christine Groves and Rachel Barbara Hinton* (pp. 3–20). Earthscan.

Isabekova, G. (2021). Mutual learning on the local level: The Swiss Red Cross and the village health committees in the Kyrgyz Republic. *Global Social Policy, 21*(1), 117–137. https://doi.org/10.1177/1468018120950032

Isabekova, G., & Pleines, H. (2021). Integrating development aid into social policy: Lessons on cooperation and its challenges learned from the example of health care in Kyrgyzstan. *Social Policy & Administration, 55*(6), 1082–1097. https://doi.org/10.1111/spol.12669

Jana, S., Basu, I., Rotheram-Borus, M. J., & Newman, P. A. (2004). The Sonagachi project: A sustainable community intervention program. *AIDS Education and Prevention, 16*(5), 405–414. https://doi.org/10.1521/aeap.16.5.405.48734

Kaasch, A. (2015). *Shaping global health policy: Global social policy actors and ideas about health care systems.* Palgrave Macmillan.

Kickbusch, I. (2003). External review of the Kyrgyz-Swiss health reform support project (KSHRSP) Phase II. In *Components strengthening primary health care and public health* (pp. 1–37). n.p.

Kudrâvceva, T. (2018). *Novye zarplaty medikov Kyrgyzstana. Vse, čto sleduet o nih znat' [New salaries of medical professionals in Kyrgyzstan. Everything one needs to know about it].* Retrieved February 3, 2023, from https://24.kg/ekonomika/97704_novyie_zarplatyi_medikov_kyirgyizstana_vse_chto_sle-duet_onih_znat/

Majtieva, V. S., Čokmorova, U. Zh., Ismailova, A. D., Asybalieva, N. A., Ânbuhtina, L. F., Sarybayeva, M. E., et al. (2015). *Stranovoj otčet o dostignutom progresse v osûŝestvlenii global'nyh mer v otvet na vič-infekciû za 2014 god [Kyrgyzskaâ Respublika] [2014 Country Progress Report on the Global Response to HIV [Kyrgyz Republic]]* (pp. 1–29). Ministry of Health, Republican AIDS Center, UNAIDS, WHO, UNICEF. Retrieved February 3, 2023, from http://www.unaids.org/sites/default/files/country/documents/KGZ_narra-tive_report_2015.pdf

Manukyan, A., & Burrows, D. (2010). *Country-level partnership case study—Kyrgyzstan. For the global fund to fight AIDS, TB and Malaria* (pp. 1–27). AIDS Projects Management Group. Retrieved November 10, 2019, from http://apmglobalhealth.com/project/country-case-study-partnerships-kyrgyzstan

Meimanaliev, A.-S., Ibraimova, A., Elebesov, B., Rechel, B., & McKee, M. (2005). *Health care systems in transition: Kyrgyzstan* (pp. 1–116). WHO Regional Office for Europe on behalf of the European Observatory on Health Systems and Policies. Retrieved February 3, 2023, from https://www.euro.who.int/__data/assets/pdf_file/0006/95109/E86633.pdf

Ministry of Health of KR. (2013). *Položenie "O strukture protivotuberkuleznoj služby Kyrgyzskoj Respubliki". Utverždeno Prikazom Ministerstva zdravoohraneniâ Kyrgyzskoj Respubliki [Regulation "On the structure of the TB service of the Kyrgyz Republic" (approved by Decree of the Ministry of Health].* Retrieved December 23, 2022, from https://continent-online.com/Document/?doc_id=31531690#pos=0;100

Morgan, L. M. (2001). Community participation in health: Perpetual allure, persistent challenge. *Health Policy and Planning, 16*(3), 221–230.

Moszynski, P. (2011). Global fund suspends new projects until 2014 because of lack of funding. *BMJ, 343*, 1–2. https://doi.org/10.1136/bmj.d7755

Murzalieva, G., Aleshkina, J., Temirov, A., Samiev, A., Kartanbaeva, N., Jakab, M., Spicer, N. and Network, G.H. (2009). *Tracking global HIV/AIDS initiatives and their impact on the health system: The experience of the Kyrgyz Republic: Final REPORT* (pp. 1–89). Royal College of Surgeons in Ireland. Retrieved March 4, 2023, from https://repository.rcsi.com/articles/report/Tracking_Global_HIV_AIDS_Initiatives_and_their_Impact_on_the_Health_System_the_experience_of_the_Kyrgyz_Republic/10776524/1

Murzalieva, Gulgun, Kojokeev, K., Manjieva, E., Akkazieva, B., Samiev, A., Botoeva, G., Ablezova, M. and Jakab, M. (2007). *Tracking global HIV/AIDS initiatives and their impact on the health system: The experience of the Kyrgyz Republic: Context Report* (pp. 1–48). Center for Health System Development; American University of Central Asia. Retrieved March 3, 2023, from http://elibrary.auca.kg/bitstream/123456789/220/1/Tracking%20Global%20HIV-AIDS%20Initiatives_AUCA.pdf

National Statistical Committee of KR. (2023). *Srednemesâčnaâ zarabotnaâ plata (somov) [Average monthly salary (soms)]*. Retrieved February 3, 2023, from http://www.stat.kg/ru/opendata/category/112/

NCPh. (2022). *Stanovlenie i Razvitie Protivotuberkuleznoj Služby v Kyrgyzskoj Respublike [The making and development of the TB Service in the Kyrgyz Republic]*. Retrieved December 20, 2022, from http://tbcenter.kg/ru/info/about-us/

OECD. (2005). *Switzerland*. Retrieved February 15, 2023, from http://www.oecd.org/dac/peer-reviews/35297586.pdf

OECD. (2009). *Switzerland*. Development assistance committee (DAC) peer preview. Retrieved March 2, 2023, from http://www.oecd.org/dac/peer-reviews/44021195.pdf

OECD. (2014). *Switzerland 2013*. n.p.: OECD Publishing. Retrieved February 3, 2023, from http://www.oecd.org/dac/peer-reviews/Switzerland_PR_2013.pdf

OECD. (2019). *OECD development co-operation peer reviews: Switzerland 2019*. OECD Publishing. Retrieved February 3, 2023, from https://read.oecd-ilibrary.org/development/oecd-development-co-operation-peer-reviews-switzerland-2019_9789264312340-en#page3

Pétric, B.-M. (2015). *Where are all our sheep?: Kyrgyzstan, a global political arena / Boris Pétric* (Vol. 16, 1st ed.). Berghahn Books.

PIL Research Company. (2017). *Community action for health (CAH) project impact assessment report* (pp. 1–49). Kyrgyz Republic.

Republican AIDS Center. (2021a). *Ob Organizacii [About the organization]*. Retrieved December 23, 2022, from https://aidscenter.kg/ob-organizatsii/?lang=ru

Republican AIDS Center. (2021b). *Missiâ i celi [Mission and goals]*. Retrieved December 23, 2023, from https://aidscenter.kg/missiya-i-tseli/?lang=ru

Republican Center for Health Promotion. (2022). *About*. Retrieved December 30, 2022, from https://saksalamat.kg/o_nas/

Schüth, T. (2011b). *Appreciative principles and appreciative inquiry in the Community Action for Health Programme in Kyrgyzstan.* Tilburg University, n.p. https://pure.uvt.nl/ws/portalfiles/portal/1359087/Schueth_appreciative_07-11-2011.pdf

Schüth, T., Jamangulova, T., Aidaraliev, R., Aitmurzaeva, G., Iliyazova, A., & Toktogonova, V. (2014). Community action for health in the Kyrgyz Republic: Overview and results. *Sharing Experiences in International Cooperation.* Issue Paper on Health Series, (3a), 1–31.

Schüth, T. (2011a). Glava 9 Dejstviâ soobŝestv po voprosam zdorov'â v Kyrgyzstane [Chapter nine community action for health in Kyrgyzstan]. In G. Laverack (Ed.), *Ukreplenie zdorov'â i rasširenie vozmožnostej [Health promotion and empowerment]: Translated into Russian with financial assistance from the SDC and technical support from the SRC* (pp. 132–175). Bishkek.

Schüth, Tobias. (n.d.). *Community action for health in Kyrgyzstan.* n.p.

SDC. (2014). *Local communities engage actively in their village's health* (pp. 1–4). Retrieved February 3, 2023, from https://www.eda.admin.ch/dam/deza/en/documents/aktivitaeten-projekte/projekte/factsheet-kyrgyzstan-community-action-for-health_EN.pdf

Sebotsa, M. L. D., Dannhauser, A., Jooste, P. L., & Joubert, G. (2007). Assessment of the sustainability of the iodine-deficiency disorders control program in Lesotho. *Food and Nutrition Bulletin, 28*(3), 337–347. https://doi.org/10.1177/156482650702800310

Spicer, N., Bogdan, D., Brugha, R., Harmer, A., Murzalieva, G., & Semigina, T. (2011a). "It's risky to walk in the city with syringes": Understanding access to HIV/AIDS services for injecting drug users in the former Soviet Union countries of Ukraine and Kyrgyzstan. *Globalization and Health, 7*, 22. https://doi.org/10.1186/1744-8603-7-22

Spicer, N., Harmer, A., Aleshkina, J., Bogdan, D., Chkhatarashvili, K., Murzalieva, G., et al. (2011b). Circus monkeys or change agents? Civil society advocacy for HIV/AIDS in adverse policy environments. *Social Science & Medicine,  73*(12),  1748–1755.  https://doi.org/10.1016/j.socscimed.2011.08.024

The Lancet. (2011). Supporting the Global Fund to fight fraud. *Lancet, 377*(9764), 440. https://doi.org/10.1016/S0140-6736(11)60143-8

Today.kg. (2022). *S 1 aprelâ povysilas' zarplata medikov [Medical professionals' salaries increased as of April 1].* Retrieved March 3, 2023, from https://today.kg/news/632467/?utm_source=last&hl=ru

UNAIDS. (2005a). *The global task team on improving AIDS coordination among multilateral institutions and international donors* (pp. 1–34). UNAIDS.

Retrieved February 3, 2023, from https://www.theglobalfund.org/media/1393/replenishment_2005romegtt_report_en.pdf?u=6367279102000000000

UNAIDS. (2005b). *The "Three Ones" in action: where we are and where we go from here* (pp. 1–51). UNAIDS. Retrieved February 3, 2023, from http://data.unaids.org/publications/irc-pub06/jc935-3onesinaction_en.pdf

UNDP. (2014). *Annual report on the implementation of grants provided by the Global Fund to fight AIDS, Tuberculosis and Malaria in Kyrgyzstan—2013* (pp. 1–70). UNDP. Retrieved February 3, 2023, from https://www.kg.undp.org/content/kyrgyzstan/en/home/library/hiv_aids/annual-report-on-the-implementation-of-grants-provided-by-the-gl.html

UNDP. (2015a). *Annual report on the implementation of UNDP project in support of the Government of the Kyrgyz Republic, funded by The Global Fund to Fight AIDS, Tuberculosis and Malaria—2014* (pp. 1–108). UNDP. Retrieved February 3, 2023, from https://www.kg.undp.org/content/kyrgyzstan/en/home/library/hiv_aids/gfatmannualreport_eng.html

UNDP. (2015b). *Newsletter: Grants on HIV, TB and malaria.* April 2014 (pp. 1–10). Retrieved March 3, 2023, from https://www.kg.undp.org/content/kyrgyzstan/en/home/library/hiv_aids/april-2014-newsletter%2D%2Dgrants-on-hiv%2D%2Dtb-and-malaria.html

UNDP. (2015c). *Newsletter: Grants on HIV, TB and Malaria.* August 2014 (pp. 1–12). Retrieved March 2, 2023, from https://www.kg.undp.org/content/kyrgyzstan/en/home/library/hiv_aids/august-2014-newsletter%2D%2Dgrants-on-hiv%2D%2Dtb-and-malaria.html

UNESCO Institute for Statistics. (2023). *Kyrgyzstan.* Retrieved March 3, 2023, from https://uis.unesco.org/en/country/kg

UNICEF. (2016). *Analytical review of documents on the system of home visits in the Kyrgyz Republic.* UNICEF. Retrieved March 3, 2023, from https://www.unicef.org/kyrgyzstan/sites/unicef.org.kyrgyzstan/files/2018-01/ANALYTICAL%20REVIEW%20OF%20DOCUMENTS%20ON%20THE%20SYSTEM%20OF%20HOME%20VISITS%20IN%20THE%20KYRGYZ%20REPUBLIC%20eng.pdf

Vujicic, M., Weber, S. E., Nikolic, I. A., Atun, R., & Kumar, R. (2011). GAVI, the Global Fund and the World Bank support for human resources for health in developing countries. *HNP Discussion Paper,* 1–16.

Walker, J. (2013). Time poverty, gender and well-being: Lessons from the Kyrgyz Swiss Swedish health programme. *Development in Practice, 23*(1), 57–68. https://doi.org/10.1080/09614524.2013.751357

Walsh, A., Mulambia, C., Brugha, R., & Hanefeld, J. (2012). "The problem is ours, it is not CRAIDS'". Evaluating sustainability of community based organisations for HIV/AIDS in a rural district in Zambia. *Globalization and Health, 8*(1), 40. https://doi.org/10.1186/1744-8603-8-40

WHO. (2008). *Community involvement in tuberculosis care and prevention Towards partnerships for health: Guiding principles and recommendations based on a WHO review.* Retrieved February 28, 2023, from http://apps.who.int/iris/bitstream/10665/43842/1/9789241596404_eng.pdf

WHO/Europe, & UNDP. (1997). *Manas health care reform programme of Kyrgyzstan.* Retrieved March 3, 2023, from https://apps.who.int/iris/bitstream/handle/10665/108088/EUR_KGZ_CARE_07_01_11.pdf?sequence=1&isAllowed=y

Wolfe, D. (2005). *Pointing the way: Harm reduction in Kyrgyz Republic* (pp. 1–60). Harm Reduction Association of Kyrgyzstan. Retrieved February 17, 2023, from https://core.ac.uk/download/pdf/11872287.pdf

Wolfe, D., Elovich, R., Boltaev, A., & Pulatov, D. (2008). Chapter 25: HIV in Central Asia: Tajikistan, Uzbekistan and Kyrgyzstan. In D. D. Celentano & C. Beyrer (Eds.), *Public health aspects of HIV/AIDS in low and middle income countries* (pp. 557–581). Springer.

World Bank. (n.d.). *Second rural water supply and sanitation project* (pp. 1–27). https://documents1.worldbank.org/curated/en/644021468047365980/pdf/RP12020V130P1100Box385359B00PUBLIC0.pdf

# 5

# The "Community Action for Health": The Project Life Cycle

This chapter discusses the interaction among the principal actors over the life cycle of the "Community Action for Health" (CAH) project by grouping them into the following analytical categories.

First, the *recipient state* refers to the Ministry of Health, represented by the Republican Centre for Health Promotion and Mass Communication under the Ministry of Health (hereinafter the Republican Center) and its subunits, and primary health care workers who participated in the project and collaborated with community-based organizations. It also encompassed local self-governments at the village, city, and district levels, which are directly accountable to the President of the Kyrgyz Republic and the Cabinet of Ministers. Community-based organizations at the district level work with authorities at this level, but for those at the village level, the local self-governance bodies at the village level are of particular importance. These are local councils (*ayyl kengesh*) elected by local communities, with the size of these councils being proportional to the size of the related constituency (Government of KR, 2021). A structure of an executive body (*ayyl ökmötü*) is defined by the Cabinet of Ministers at the national level, but the head of the executive body at the district level appoints the head of *ayyl ökmötü* (ibid.).

© The Author(s) 2024

G. Isabekova, *Stakeholder Relationships And Sustainability*, Global Dynamics of Social Policy, https://doi.org/10.1007/978-3-031-31990-7_5

Second, *civil society organizations* are community-based organizations (CBOs) established within the framework of the CAH project. These include the Village Health Committees (VHCs) in a village, Rayon Health Committees in a district, and the Association of VHCs at the national level.

Third, *donors* denote the Swiss Agency for Development and Cooperation (SDC), which financed the CAH, and the Swiss Red Cross (SRC), which implemented it. I conceptualize each as a "donor" because the SRC was the key actor working with CBOs, and in so doing performed the role of the "donor" on the ground. However, a number of other development organizations supported CAH. The Swedish International Development Cooperation Agency (Sida) and the United States Agency for International Development (USAID) joined the project at a later stage and were essential to the expansion of the initiative throughout the country. The list of other international organizations contributing to the project includes the Liechtenstein Development Service (Schüth et al., 2014a), the German Corporation for International Cooperation (*die Deutsche Gesellschaft für Internationale Zusammenarbeit*—GIZ), the United Nations Children's Fund (UNICEF), the World Bank, the Soros Foundation, the Global Fund to Fight AIDS, Tuberculosis and Malaria (the Global Fund), the Interchurch Organization for Development Cooperation, the Asian Development Bank, and the World Health Organization (WHO), among others (www.cah.kg n.d.). Nevertheless, Sida and USAID remained the major donors (in addition to the SDC), since the contributions of other organizations were limited to specific project activities complying with the areas targeted by those organizations.

## 5.1    Initiation

The initiation of the "Community Action for Health" project coincided and corresponded with the country's transition from the Soviet-style *Semashko* health care system. The government aimed to optimize health care spending and emphasize citizens' responsibility for their health, as opposed to the idea of health care being a state responsibility, which was in the foreground of the previous system. First, as part of optimization reforms, the national health care reform program *"Manas"* (1996–2005) intended to address the majority of health care issues at the primary health care (PHC) level and decrease the number of referrals to secondary (or hospital) care (Government of KR 1995). Accordingly, the government aimed to increase public funding to PHC and cut the number of hospital beds per capita to decrease the maintenance and utility costs spent on health care facilities and secondary care. A state representative interviewed for this research estimated that in one district of the Naryn region, for instance, just two or three facilities were retained out of twenty, with the rest being demolished (State Partner 1). Second, the *"Manas"* program emphasized people's responsibility for their own health (Government of KR 1995). The government propagated the idea of citizens taking preventive measures to improve their health instead of depending on the health care system (State Partner 1). In so doing, it attempted to delegate at least part of its responsibility for health to the population.

The CAH project was in line with the health care reform agenda at that time, but it is not clear who initiated the project. According to Gotsadze and Murzalieva (2017, p. vi), the Ministry of Health approached SDC in the early 2000s to design a program for health promotion in rural areas. In other words, the Ministry was the one who initiated the project. However, the CAH also may have been the outcome of a donor initiative. Indeed, the national program *"Manas"* highlighted the responsibilities of the population for their own health (Government of KR 1995), but it did not stipulate any means for citizens to express their wishes and concerns about the reform process. The CAH, on the contrary, stressed the role of the local population in defining the issues to be targeted by the program. This emphasis on the involvement of groups targeted by health care programs in the decision-making process (SRC n.d.) and the empowerment of communities (SDC 2003) corresponded to the objectives of the organizations financing and implementing the project.

The CAH occurred in the second phase of the Kyrgyz–Swiss Health Reform Support Project. Following the request from the Ministry of Health, the first fifteen months of this initiative, from January 2000 to March 2001, were dedicated to the renovation of two remote hospitals in the Naryn region (Schüth 2011b). However, as the agreement between the SDC and the Ministry stipulated supporting health care reform in the Naryn region as a whole, the SRC planned to increase the scope of activities in the second phase of the project, which commenced in the summer of 2000 (ibid.). The organization invited the SRC for this purpose. This choice was not surprising since Swiss development agencies tend to provide a large part of the development assistance through Swiss nongovernmental organizations (OECD 2014). The increase in the scope of activities materialized through the involvement of communities in the planning, implementation, and evaluation phases of the project, corresponding to the principles of the SDC and SRC (Schüth 2011b). To identify the priorities of the population in health care reforms, the SRC invited a project coordinator, Dr. Tobias Schüth, to conduct a qualitative study among the communities.

The initiative on community involvement in health care reforms commenced with an appraisal of people's views on health care services and their priorities for reforms. The appraisal was conducted in the At-Bashi and Ak-Talaa districts of the Naryn region. The study covered district centers and three villages of various distances from the center (Schüth 2000), and used the Participatory Reflection and Action approach, formerly known as Participatory Rural Appraisal (PRA). PRA encompasses approaches and methods that "enable local (rural and urban) people to express, enhance, share and analyze their knowledge of life and conditions, to plan and to act" (Chambers 1994, p. 1253). By using this approach, the SRC intended to understand communities' perceptions through their analysis of problems and solutions. The SRC trained a study team of eight members, which also included representatives of different state departments (Schüth 2000). To cover various community groups, the organization engaged volunteers to conduct separate interviews with vulnerable groups, such as the poorest households, pregnant women, mothers with young children, and people with disabilities (ibid., pp. 16–19).

The communities were asked about the most pressing diseases, their priorities in health care, and their awareness of the "Manas" health care reform program. The most frequent diseases identified by people were brucellosis, anemia (mainly in women), high blood pressure, dental diseases, goiter, and liver disease (Schüth 2000). People's priorities in health reforms related to access and quality of health care. This included the availability of specialized health care services and ambulances in remote areas, higher salaries for medical staff, combatting bribery at district hospitals, and so forth (ibid., p. 8). The appraisal demonstrated uneven access to health care, dependent on the social status of a household. Traditional healers were the first point of contact for the villagers, though better-off households also used health facilities at the district level (ibid., p.7). In general, the villagers spent less on health care than the residents of districts did. Overall, the respondents "had heard" about the *"Manas"* health care reform program and were willing to learn more about family group practices and eligibility for the health insurance scheme (ibid., pp. 8–9, 47–48). People were even ready to pay a small amount of money for the health brochures (ibid.). In general, the initial study demonstrated the interest of communities in the health care reform program and their readiness to participate in health promotion.

Notably, the recipient state participated in the initial appraisal of the population's concerns and priorities (e.g., the PRA sessions), which was essential to state interaction with community members. The study team of eight members, trained by the SRC, included representatives of different state departments (Schüth 2000). The state actors worked with communities in defining their concerns. The participation of the recipient state in the appraisal was key to its interaction with communities. The SRC also encouraged community members to present the results of the initial study to the Ministry of Health in Bishkek, which was "well received" by the Ministry (Schüth 2011b, p. 24).

It should be noted that the Swiss actors (SDC and SRC) were the only donors involved in initiating the CAH, possibly due to the general division of labor among the donors in the country, a result of the sector-wide approach (SWAp) to health care in the country (see Chap. 1). Formalized in 2005, the SWAp has been in use in Kyrgyzstan since 1996 (see Isabekova and Pleines 2021). For this reason, the fact that the Swiss actors were the only donors working in the area of community engagement in health care reforms also may be the outcome of negotiations taking place in the SWAp.

Overall, the initiation phase suggests that the CAH may have equally been a donor initiative and an initiative of the recipient state. The initiation of this project coincided with the transition from the *Semashko* health care system. This transition was consistent with the interests of the recipient state, which, in the face of social and economic crises in the country, was willing to delegate part of its responsibilities to citizens. However, the project emphasized community engagement in the decision-making process, which was consonant with the principles of the SDC and SRC. Although the source of the initiative is ambiguous, the CAH nevertheless addressed pressing issues of the local population, which also was reflected in their interest and readiness to collaborate with the project.

## 5.2   Design

The design of the project was developed in collaboration with community members. The CAH commenced in the Jumgal district of the Naryn region. Selection of this region complied with the renovation of hospitals (IO Partner 11), which took place in the same area. Another reason for the selection of this region was poverty. My interviewees note that the project commenced at a time of extreme impoverishment (CSO 7), and the Naryn region was among the poorest in the country (IO Partner 5). The CAH pursued two overarching goals, namely, supporting the communities in taking action for their health and building the partnership between the state health care system and communities (Schüth n.d.). These goals were further divided into smaller objectives and project activities, jointly identified by the SRC and communities in the PRA sessions.

The PRA sessions followed the principle of "nondominance." In a nutshell, this principle meant respectful behavior, which aimed to provide a space for the actors to express themselves and be heard by another party. This respectful behavior intended to overcome conventionally unequal roles between the providers and recipients of aid by emphasizing the fundamental equality of all stakeholders involved in development assistance (Schüth 2011b). The sessions stressed the expertise of local people and noninterference in the discussions. The emphasis was on local people as the ones "who know" and the project team being the ones "who learn from the people" (ibid., pp. 23–24). The SRC and the primary health care staff aimed to encourage the discussions without "guiding" them. Noninterference in the discussions meant "accepting people's views without judging them as right or wrong" (ibid.).

Dr. Tobias Schüth, a project coordinator invited by the SRC, stressed the role of nondominance in relationships among the actors throughout the project cycle. Both the SRC and state representatives engaged in the project complied with this principle. The project recruited staff members who "were good with people, behaved in a good way, and were quick to pick up things" (IO Partner 11). The SRC also trained and involved the local primary health care staff in the PRA. A former state official interviewed for this study emphasized the collaboration of the Ministry of Health and the SRC in forming health committees in the Jumgal district (State Partner 1). The involvement of state institutions was critical for the further nationwide rollout of the program because the recipient state, and not the SRC, conducted the PRA sessions beyond the pilot districts. No other donor organization participated in the initial design of the CAH, as USAID and Sida joined the project at later stages.

The PRA sessions were intended to define those diseases that were of pressing concern to communities and to the community perspective on how to stay healthy (IO Partner 11). The sessions took place in every village and involved approximately 50–80% of households (Schüth 2011a, p. 147). A PHC representative gathered approximately ten people from a neighborhood and supported them during their analysis using the PRA approach (ibid.). Since most of the PRA participants were women (as they were the ones at home), separate sessions were organized for men to consider their opinions (ibid.). The outcomes of the survey varied across the regions but generally included goiter, alcohol consumption, anemia, hypertension, brucellosis, and so forth (see Isabekova 2021). In addition to listing problems, the PRA participants also brainstormed and listed their ideas on "what do you need to stay healthy in this village?" (Schüth 2011b, p. 32). They compiled a list of determinants of health, which included broader issues, such as the lack of public baths or access to potable water. The facilitator (e.g., the SRC or primary health care staff) compared this list to the elements of primary health care outlined in the

Alma-Ata Declaration (1978) (see WHO/Europe n.d.), which encouraged the participants since their list often contained most of or even went beyond the elements outlined in the declaration (Schüth 2011b).

In addition to defining the problems and potential solutions, the PRA sessions were used to mobilize community members. The participants were asked to nominate trustworthy, "active and community-minded" people from their neighborhoods (Schüth et al., 2014a, pp. 5–6) to become members of the VHCs, which intended to take action on the problems and determinants of health. My interviewee noted that the project, in a way, identified "people respected and influential in villages" (CSO 2). Nomination and election to the VHCs by village residents contributed to the recognition of candidates by the local population, which was essential to the subsequent implementation of the project. Importantly, the selection of the VHC members took place via secret voting of PRA participants to ensure the election of persons willing to work and not merely influential in their communities (Tobias Schüth 2011a, p. 151). During a public vote, people were often willing "to be seen" to vote for persons influential in their communities (ibid.).

The CAH was built around close collaboration with local communities. During the initial stages, project staff members lived in the local communities (Schüth 2011a). This has allowed continuous interaction with community volunteers. The interviewees noted that the project members incorporated the perspectives of local communities into ideas by asking for feedback from community members and adjusting these ideas accordingly (IO Partner 11). Thus, the decision could have been made in the morning and changed in the evening if the initial idea did not work out (IO Partner 5). This interaction allowed further adjustments of activities to the lives of community volunteers. As one interviewee noted, while present on site, the project workers did not limit themselves to the "usual" working hours but to the time the community members could spare between their daily responsibilities. The interviewee highlighted that this flexibility and immersion into the context discerned the differences with other projects following the "usual" working hours and visiting community members on an occasional basis (IO Partner 5).

Overall, this section demonstrated a close collaboration among community representatives, the SRC, and PHC workers commissioned by the Ministry of Health to support the initiative. The following country-wide rollout of this project involved the health promotion units (HPUs) established by the Ministry of Health. The expansion also has involved the USAID representatives that funded the Jalal-Abad and Issy-Kul regions. The countrywide extension of the project is elaborated on in the following section.

## 5.3    Project Implementation

Multiple stakeholders participated in implementing the "Community Action for Health" in Kyrgyzstan. Nevertheless, close collaboration among the donor, state PHC, and community-based organizations was a distinctive characteristic of this project. The health-related activities in the project included three components: essential research, awareness-raising, and data collection for monitoring and further research (Schüth 2011a).

First, the essential research conducted and analyzed by the VHCs was intended to provide deeper insights into community problems and further encourage the CBO members to work with them (Schüth 2011a, p. 151). A participant interviewed for this research notes that following the PRA seminar, its participants surveyed the local population by visiting "every second house" (CSO 2). In addition, they attempted to organize general meetings by gathering people "from every street." However, the participant admits that convincing people to attend these meetings was "difficult." The interviewee notes that surveying the local population and disseminating the information about the CAH in a way demonstrated the abilities of those nominated to become VHC members to reach out to the local population (ibid.).

Second, the awareness-raising was conducted within the project, mainly by providing information materials, although at times individual consultations and explanations aimed at behavioral and lifestyle changes (Schüth 2011a). The VHCs targeted a broad spectrum of health care issues (see Isabekova 2021). These included decreasing alcohol consumption, controlling brucellosis, anemia, tuberculosis (TB), smoking, hypertension, sexual–reproductive health (Schueth 2009), promoting "safe nutrition" (iodized salt, fortified flour, meat consumption), and increasing awareness of childhood diseases such as diarrhea, influenza, acute respiratory infection, and others (PIL Research Company 2017). Most of these issues were identified by the village population in the surveys conducted by the VHCs during the project design. In addition to survey results, the VHCs also targeted priority areas highlighted in the national health care program.

Following the focus of this research on TB and HIV/AIDS, I will describe the VHCs' activities in regard to these diseases. TB was not one of the priority areas defined by the population (Schüth et al., 2014a, p. 19), but it was among the issues targeted by the VHCs, also due to the problem of drug-resistant tuberculosis in the country. Kyrgyzstan, similar to other countries in the post-Soviet region, has a high prevalence of the multidrug-resistant form of tuberculosis, particularly among previously treated patients (Isabekova, 2019b). The absence of tuberculosis among the issues prioritized by the communities may relate to its prevalence in urban, rather than rural, areas (ibid.). The *"Manas"* (1996–2006), *"Manas taalimi"* (2006–2012), and *"Den Sooluk"* (2012–2018) health care reform programs listed TB among their priority areas (Government of KR 2006, 2012; WHO/Europe and UNDP 1997). Therefore, the inclusion of TB in the areas targeted by the VHCs made their work compliant with national health care policy. The VHCs received leaflets on the importance of treatment continuity and its completion, as well as nondiscrimination against patients with TB (Schüth et al., 2014a, p. 20). First piloted in Chui and Issyk-Kul regions, these dissemination campaigns were expanded to the country as a whole in 2013 (ibid.).

In contrast, HIV/AIDS was, in a way, among the issues prioritized by the villagers and the national health care programs. Reproductive tract infections were among the priorities listed by people in all *oblasts* (Schüth et al., 2014a, p. 19). HIV/AIDS also was among the priority areas listed in the national health care reform programs *"Manas"* (1996–2006), *"Manas taalimi"* (2006–2012), and *"Den Sooluk"* (2012–2018) (Government of KR 2006, 2012; WHO/Europe and UNDP 1997). Correspondingly, the VHCs implemented campaigns to raise awareness of sexually transmitted infections, including HIV/AIDs, in collaboration with the SRC, other donors, and local actors. Working with school parliaments (a body composed of pupils elected by pupils to represent their interests before the school administration) and teachers, the VHCs circulated an educational course called "The road to safety" for students of the 9th–11th grades. This course used DVDs on sexual and reproductive health, developed in the framework of CAH's collaboration with GIZ (Schüth et al., 2014a, p. 19). To target the working-age population, the VHCs visited local businesses (CSO 2) and conducted seminars with potential labor migrants—the youth—due to a large amount of labor migration to Russia and Kazakhstan. For instance, in the city of Osh, in the south of the country, the VHCs informed migrant workers about TB, HIV/AIDS, and treatment possibilities as part of CAH's collaboration with a global nongovernmental organization—the Interchurch Organization for Development Cooperation (Schüth et al., 2014a, p. 25). Thus, in contrast to the VHCs' activities for TB, the awareness-raising campaigns for HIV/AIDS complied with the priorities of both the local population and the national health care program.

Third, data collection and monitoring took place at a district level, based on the essential research conducted by the VHCs during their work with target groups or selected research (Schüth 2011a). The data compiled at the district level were further sent to the Republican Center and supporting health care projects at regional and national levels (ibid.). The information exchange also was intended to inform both state and donor organizations about the VHCs' findings and to compare the coherence of priorities with those identified at the community level.

It should be noted that the SRC supported the VHCs in their activities by providing technical and financial assistance for dissemination campaigns, organizational capacity, and resource mobilization. The SRC offered training courses in a number of areas, but I focused only on those indicated in the project-related documents and mentioned by my interviewees.

First, the VHCs learned how to work with the population and organize seminars. During the dissemination campaigns, the VHC members gathered the villagers to inform them about preventive measures and health promotion. In this regard, the VHCs followed the principle of nondominance promoted by the SRC. A VHC representative interviewed for this research noted that training pertained to building relationships with others and identifying issues. According to her, becoming a VHC member implied "understanding the work" and finding "a common language with people." Therefore, "giving orders to others" by pointing at the information they "should learn" about the diseases relevant to them was "not right" (CSO 5). The VHCs used the principle of nondominance during the seminars to build a dialogue between medical workers and the population groups affected by the various diseases. They also followed this principle in relation to each other, irrespective of their position in the VHC, be it a head or a member of the organization.

Through their close work with communities, the VHCs, unlike the state health care workers, were familiar with the health issues of specific households. By offering blood pressure checks, for example, the CBOs were aware of members of the community who had hypertension (CSOs 2 and 5). The VHCs prepared coffee breaks and gathered local health care workers and people affected by the different diseases (CSO 5) to increase awareness of danger signs, symptoms, and preventive measures against specific diseases, such as hypertension, anemia, diabetes, and others.

It should be noted that medical personnel were not always supportive of the VHCs' work. There were occasions when health care workers did not perceive community-based organizations as equals or even competed with them. However, this attitude changed due to the support the VHCs provided to primary health care professionals in outreaching the local population and the joint implementation of health promotion campaigns. This change also is demonstrated by medical and community-based organizations congratulating each other on their professional days, namely, September 9 for the VHCs and July 2 for medical professionals (AVHC 2022).

Similarly, the attitude of local self-government bodies has transformed from an initial disinterest to cooperation. My interviewees recalled the initial detachment of local authorities toward the VHCs and their activities (CSO 5) and questions of why VHC members "needed this" (CSO 1). According to one, there also were remarks hinting at a superior position of authorities over community-based organizations, such as "some five women are running around, are those the VHCs?" (CSO 4). However, this attitude changed during the joint implementation of activities. The CAH forethoughtfully offered small grants to which VHCs could apply jointly with local authorities. This collaboration strengthened further within the framework of the project implemented by the Development Policy Institute, which sought to enhance the partnership between the state and VHCs through their joint realization of initiatives (AVHC 2017a). The cooperation also has continued beyond donor assistance. The VHCs I interviewed in one of the northern regions participated in the meetings and the joint committees of the local authorities on social issues, for example, working with poor households (CSO 5). According to the local authority representative in this region, this collaboration had been going on for 4–5 years, and the authority had provided a Certificate of Merit to the VHC member in appreciation of her work (State Partner 12).

In addition to the joint implementation, the attitude of local self-governments toward community-based organizations changed as the authorities realized the potential of community-based organizations (CSO 1). The VHCs work closely with the local population and are aware of their concerns and their living circumstances (CSO 4). This contributes to the expertise of community-based organizations, which is valuable to the local authorities. One VHC representative from another region I visited noted that not a single activity organized by the local authorities took place without the VHC. The interviewee noted that in recent years, authorities often asked for support in mobilizing the local population on the grounds that people's attitude toward the VHCs was "positive," in contrast to their attitude toward the authorities (CSO 2). Engaging with the VHCs is essential for the work of local authorities since the VHCs have not only the capacity for dissemination activities but also a certain status in their communities.

Secondly, during the first two or three years after their formation, the community-based organizations received training on bookkeeping and budgeting, and were given office equipment, which intended to improve their organizational capacity (Schueth 2009). The CBOs learned essential budgeting skills to calculate the current financial balance of their organization, and plan their activities accordingly. The SRC also explained how to write appeals to local self-government and enclose the relevant attachments (CSO 5). The VHCs obtained their office spaces from local authorities or medical organizations (CSOs 2 and 5); however, maintenance of these offices and the relevant equipment were provided by the SRC. During my fieldwork, the VHCs presented me their books, receipts for activity-related expenses, as well as the equipment and furniture provided by the SRC, including table, chairs, PCs, printers, and so on (CSO 5).

Thirdly, the CBOs received training on how to write grant applications, and financing to mobilize their resources. My interviewee stressed that the SRC provided not only guidance on how to write proposals, but also the opportunity to work on relevant issues. Another community member interviewed for this research noted that members were unaware of how to write project applications, but trainers elaborated on the writing process. She added: "they explained to us [the application process]… taught us like children. Other projects do not do that" (CSO 7).

In addition, the SRC offered small grants and materials for the VHCs to top up their organizational budget. The VHCs applied for these grants to address the problems highlighted by communities in the initial survey. These grants were used to build public baths, *feldsher-midwife (akusher) points* (primary health care facility in rural areas), repair water pipes (Health Worker 3; State Partner 1), and support vulnerable households. Poor families received chickens, roosters, chicken feed (CSO 7), and chicken coops built by the VHCs (Schüth et al., 2014a, p. 25). The VHCs used their small grants to build public baths and establish social enterprises, such as sewing workshops and hairdressers, which contributed to the organizational funds of these community-based organizations.[1] Overall, mobilization of resources was emphasized throughout the CAH. The SRC provided project-related materials, such as gloves to prevent brucellosis, quality seeds to plant beetroot, carrots, tomatoes, and so forth to combat anemia, that were sold by the VHCs to the local population (CSO 4). At the end of the CAH, the SRC announced another round of small grants, namely 25,000 Kyrgyzstani som (KGS) (around €268)[2] to be provided to the VHCs based on their project applications (CSO 2). These grants were intended to ensure an additional financial

---

[1] For more details, see the section on income-generation in the chapter on sustainability of the CAH.

[2] The exchange rate, as of March 17, 2023, was applied throughout this book.

basis for the VHCs to continue their activities beyond the end of the project (CSO 4). In general, the SRC's technical and financial support was essential for the VHCs' organizational capacity. Yet this assistance complemented, rather than dominated, the project implementation, because it targeted the issues identified by communities themselves.

It should be noted that the donor did not conduct the training activities alone. The Ministry of Health supported the VHCs after it became acquainted with the VHC members and their work. During the pilot phase of the CAH in fifteen villages in the Jumgal district of the Naryn region, the VHCs organized a campaign against goiter, where they promoted the usage of iodized salt, and checked iodine in the salt sold by local retailers (see Isabekova 2021). This campaign caught the Ministry's attention and contributed to its acknowledgment of the initiative (Schüth 2011b). The VHC member I interviewed notes that the Ministry's support was dependent on the "success" of the project. If the initiative "worked out," the Ministry wanted to retain the VHCs to disseminate the information among the population; if it didn't, the community-based organizations (CBOs) would be discontinued (CSO 2). According to project-related documents, this acquaintance was decisive, since "no amount of explanation can be as convincing as an hour spent with a VHC" (Schueth 2009, p. 47; Schüth 2011b, p. 49). Equally significant was the support of individual persons, including the Minister of Health at that time, Tilek Meimanaliev, who supported community engagement, despite the relatively modest attention to this matter in the national health care program (Schüth 2011a).

The recipient state actively participated in training activities, particularly after the countrywide expansion of the CAH. The Ministry of Health included the CAH in the national health care program and requested its countrywide extension. The SRC, in turn, asked the Ministry to provide health care staff for this purpose and offered calculations on the number of staff needed. The Ministry agreed and promised to establish HPUs in regions in which donors funded the expansion of the "Jumgal model" (IO Partner 11). Notably, the HPUs are part of the health care system and are accountable to the Republic Center for Health Promotion under the Ministry of Health. The HPUs received extensive training on how to work with communities from the SRC before taking over the training of PHC workers on the PRA approach. They equally took over training the VHCs on how to work as an independent civil society organization and conduct health-related activities (Schüth 2011a).

The HPUs were selected and worked in compliance with the principle of nondominance. People with a "bossy attitude" were "avoided" during the selection process (Schüth 2011b, p. 48). The SRC trained the HPUs on the PRA tools (Schueth 2009, p. 22) and in the principle of nondominance. The HPU representative interviewed for this research emphasized that medical professionals should not "give orders to common people," and instead of acting as "teachers," they should be "equal" to people referring to them (CSO 5). The interviewee noted that the HPUs had already learned about the nondominance principle at the beginning of the project (ibid.). This timely training contributed to the HPUs' roles as "facilitators" of the PRA sessions and training activities that support but do not overlook the community initiative.

Following the endorsement by the Minister for Health, USAID and Sida joined the project implementation to support its national rollout. The Ministry of Health's inclusion of the "Jumgal model" in the national health care program (Schüth 2011b, p. 26) and a promise to provide the HPUs for the countrywide extension of the program encouraged other donors to support the initiative (IO Partner 11). Two organizations were critical to this expansion. First, the USAID covered Jalal-Abad and Issyk-Kul regions as part of its ongoing "Zdravplus" (2000–2005) and "ZdravPlus II" (2005–2009) projects (Dominis et al. 2018), which aimed to improve the quality of health care services in Kyrgyzstan, Kazakhstan, Uzbekistan, Tajikistan, and Turkmenistan (Abt Associates 2023). Health promotion by community members corresponded to community and population health—one of the four major components of these projects (Cleland et al. 2008). Second, Sida financed the SRC to include the Batken, Osh, and Chui regions (Schüth 2011b). Between 2006 and 2011, Sida was among the core financiers of the Sector-Wide Approach to health care (Sida 2008). Because of the joint financing from Sida and the SDC, the project changed its name in 2006 from the Kyrgyz Swiss Health Project to the Kyrgyz, Swiss, Swedish Health Project (Development Planning Unit 2010). With the Swiss organizations (i.e., SRC and SDC) taking over the expansion in the Naryn and Talas regions (IO Partner 11), the organizations ensured the countrywide extension of the program.

Despite the differences in engagement, both USAID and Sida followed the leadership of the SRC. USAID implemented the extension itself as part of its ongoing project, while Sida cofinanced the SRC. However, compliance with the Swiss model (IO Partner 5), or the SRC approach in the Jumgal district, was "part of the deal" (IO Partner 11). This was ensured throughout the extension process. The SRC trainers accompanied USAID and trained its staff on project implementation and monitoring (ibid.). In the case of Sida, no issues arose in terms of the differences in approaches, since it simply transferred finances without any direct involvement in the project implementation. As my interviewee noted, one "did not even notice that there was different money" (ibid.). The SRC reported on how the funds were used, and Sida visited the project sites. However, although it was cofinancing, Sida basically accepted the Swiss actors' approach to project implementation and monitoring (ibid.)· In this way, despite their differences in engagement in the CAH, both donors, USAID and Sida, followed the Swiss actors' approach to project implementation.

Overall, the project implementation phase shows that participation and support of the Ministry of Health intensified further as the project recommended itself as the "Jumgal model." It also allowed the countrywide expansion of the project, encouraging other donors to commit themselves. Notably, both Sida and USAID followed the SRC's approach in the CAH.

## 5.4    Project Evaluation

The CAH, similar to other development projects, went through a number of evaluations by external parties (e.g., Gotsadze and Murzalieva 2017; Kickbusch 2003). Both USAID and Sida also conducted an external evaluation of their contribution to the expansion of the Jumgal model (e.g., by hiring consultants). USAID conducted an external evaluation of its activities within the framework of the "ZdravPlus" program. Similarly, Sida assessed the use of financing by the SRC.

However, in addition to external assessments, the project developed annual evaluations of its activities by the project participants themselves. For this purpose, the project coordinator adapted Labonte and Laverack's (2001a, 2001b) framework for community capacity-building. This framework stresses participation, leadership, organizational structure, problem assessment, resource mobilization, "asking why," links with others, the role of outside agents, and program management (all categories listed verbatim) (Labonte and Laverack 2001a, p. 117). The original framework was adapted into 25 indicators (IO Partner 11) and further elaborated into clarifying questions, including those related to organizational abilities and essential accounting, conflict resolution, and sources of regular income (Schüth 2011a, p. 163). These indicators and questions aimed to ensure the evaluation of the CBO activities by the CBOs themselves and the organizations working with them.

The project evaluation emphasizes the roles of the state and community representatives in the assessment. The annual evaluation commences with the VHC members' reflection on the abovementioned indicators. The CBO members additionally fill out their "happiness" and "workload" indices. Following this "internal" self-assessment, the HPUs and Rayon Health Committees conduct the "external" evaluation of CBOs (IO Partner 11). Both assessments matter to the validity of evaluation outcomes. The "internal" evaluation demonstrates the VHC members' perception of and satisfaction with their work. The "external" assessment, in turn, shows the perspectives of organizations having firsthand experience with the VHCs. HPUs provide the training necessary for organizational development and connect CBOs to the national health care system. They are the ones having continuous contact with the CBOs and are aware of their organizational issues. Additional involvement of the Rayon Health Committees, composed of the VHC leaders, contributes to the validity of the CBO assessment by both state and civil society representatives. As one interviewee noted, one could claim many achievements on paper. However, during the actual visits to organizations, the VHC leaders witness the outcomes of the organizational work (CSO 2).

Both "internal" and "external" assessments are based on the same set of indicators. These include organizational membership, VHCs' abilities for collective decision-making and conducting activities, documentation quality, and attracting new members (AVHC 2018). The indicators also stipulate conducting formal events and essential accounting according to the VHCs' regulations (adopted by the Association of VHCs), engagement, and connections to authorities and other associations and organizations at a local level (ibid.). Another indicator signified and regularly monitored by the Association of VHCs is self-initiatives that, in addition to VHC funds, also can be conducted at the expense of local authorities and third-party funding sources (AVHC 2017a). Self-initiatives may include fundraising for health funds, support to the poor, community care, improving the environment in villages, organizational development, activities related to health, and participating in improving the village infrastructure (AVHC 2018). Overall, this similarity of assessment criteria ensures the consistency of internal and external evaluations (IO Partner 11).

One should specifically emphasize the roles of the Association of VHCs and Rayon Health Committees (RHCs) in the evaluation process, particularly after the end of the CAH. The evaluation of RHCs closely relates to their support for Village Health Committees. The organizations are expected to conduct at least four regional meetings funded by the organizations themselves, four self-initiatives on improving health determinants at a district level, monitoring health funds, and monitoring activities targeted at VHC development (AVHC 2018). The RHCs also are integral to the supervisory functions of the Association of VHCs. By the end of the quarterly meetings at a regional level, RHCs report to the AVHC a list of participants, meeting protocol, working plan, and a complete table with self-initiatives (ibid., pp. 15–16). This reporting is critical for the AVHCs' overview of the organizations and their activities. Delayed reporting because of nonparticipation of VHCs at regional meetings or the inability of RHCs to report the activities on time distorted the assessment of the actual situation (AVHC 2018). Therefore, as a corrective measure, the Association of VHCs asked the Rayon Health Committees to fill the tables on VHCs' activities right after the meeting and send the data to the AVHC immediately after the meeting via email or WhatsApp (ibid.).

Indeed, there have been multiple issues with evaluation, particularly since the end of the CAH. There were cases of HPUs not conducting the evaluation due to a lack of funding for transportation and *per diem* costs, although at large, the Family Medicine Centers provided the necessary funding (AVHC 2017b). The Association of VHCs discusses these issues directly with the Republican Center (ibid.), and it also intends to improve the mechanisms for collecting and streamlining HPU reports (AVHC 2017a). The attrition of medical professionals additionally challenges the evaluation process. However, foresightedly, the Association, in collaboration with donors, developed a training film for RHCs and HPUs on the assessment of VHCs (ibid.). This was intended to ensure the awareness of evaluation criteria and approaches irrespective of rotation in personnel. However, in the long run, the evaluation criteria are likely to evolve further. There also was a discussion on changing the self-assessment indicators as the organizations and their activities evolved further (AVHC 2018). These are only a few of the issues the Association of VHCs and organizations and members in the network face.

Nevertheless, the "Community Action for Health" project was remarkable in the sense that, in addition to the evaluation of project activities by external parties, it stipulated an opportunity for both state and community representatives to participate in the evaluation process. Although the SRC adapted the assessment criteria based on the academic analytical framework, these were the very HPUs, VHCs, and Rayon Health Committees that assessed the work and organizational capacity of the community-based organizations. This has changed the roles of the VHCs and HPUs from mere "subjects" of evaluation to actors assessing their own performance. It also laid down the basis for the Association of VHCs and its network members to continue evaluating their activities beyond the duration of the CAH.

# References

Abt Associates. (2023). *Improving the quality of health services in central Asia.* Retrieved February 2, 2023, from https://www.abtassociates.com/projects/improving-the-quality-of-health-services-in-central-asia

AVHC. (2017a). *Associaciâ "Kyrgyzstan ajyldyk den sooluk komitetteri" [The Association of Village Health Committees]: Otčet "Deâtel'nost' Associacii za 2016 g." [Report of activities for 2016]* (pp. 1–20).

AVHC. (2017b). *Associaciâ «Kyrgyzstan ajyldyk den sooluk komitetteri» [Association of Village Health Committees]: Otčet "Deâtel'nost' Associacii KADK za 1ânvarâ—15 maâ 2017 g." [Report "Activities of the Association from 01 January to 15 May 2017"].*

AVHC. (2018). *"Kyrgyzstan ayyldyk den sooluk komitetteri" Assotsiatsiyacy [The Association of Village Health Committees]: "Assotsiatsiyanyn 2018-jyldyn ish-merdüülügünün" otchetu [Report of activities for 2018]* (pp. 1–19).

AVHC. (2022). *Official Facebook page of the association of village committees.* Retrieved December 25, 2022, from https://www.facebook.com/associationkadk/?ref=page_internal

Chambers, R. (1994). Participatory rural appraisal (PRA): Analysis of experience. *World Development, 22*(9), 1253–1268.

Cleland, C., Boezwinkle, J., Cavanaugh, K., Duncan, F., & Heiby, J. (2008). *Mid-term evaluation of USAID/CAR project quality public health and primary health care in the central Asian republics "ZdravPlus II"* (pp. 1–38). n.p. Retrieved February 2, 2023, from https://pdf.usaid.gov/pdf_docs/Pdacl913.pdf

Development Planning Unit. (2010). *Gender Analysis of the Kyrgyz-Swiss-Swedish Health Project (KSSHP) Phase V* (pp. 1–33). University College London.

Dominis, S., Yazbeck, A. S., & Hartel, L. A. (2018). Keys to health system strengthening success: Lessons from 25 years of health system reforms and external technical support in Central Asia. *Health Systems & Reform, 4*(2), 160–169. https://doi.org/10.1080/23288604.2018.1440348

Gotsadze, T., & Murzalieva, G. (2017). *Impact evaluation of the community action for health (CAH) project in Kyrgyzstan: Phase I–VII* (April 2002–March 2017) Report (pp. 1–44). n.p. Retrieved March 3, 2023, from https://www.newsd.admin.ch/newsd/NSBExterneStudien/880/attachment/en/3725.pdf

Government of KR. (1995). *Nacional'naâ Programma Kyrgyzskoj Respubliki "Tuberkulez" na 1996–2000 gody [National program Kyrgyz Republic "tuberculosis" for 1996–2000].* Retrieved February 3, 2023, from http://cbd.minjust.gov.kg/act/view/ru-ru/36659

Government of KR. (2006). *Nacional'naâ programma reformy zdravoohraneniâ Kyrgyzskoj Respubliki "Manas taalimi" na 2006–2010 gody [National Health Care Reform Program "Manas Taalimi" for 2006–2010]: Utverždena postanovleniem Pravitel'stva Kyrgyzskoj Respubliki ot 16 fevralâ 2006 goda № 100 [Approved by the Decree of the Government of the Kyrgyz Republic dated February 16, 2006 No. 100].* Retrieved March 3, 2023, from http://cbd.minjust.gov.kg/act/view/ru-ru/57155

Government of KR. (2012). *Postanovlenie ot 24 maya 2012 goda № 309 O Natsional'noy programme reformirovaniya zdravookhraneniya Kyrgyzskoy Respubliki "Den sooluk" na 2012–2016 gody [Decree dated May 24, 2012 No. 309 On the National Healthcare Reform Program of the Kyrgyz Republic "Den Sooluk" for 2012–2016].* Retrieved February 2, 2023, from http://cbd.minjust.gov.kg/act/view/ru-ru/93628?cl=ru-ru

Government of KR. (2021). *Zakon Kyrgyzskoj Respubliki ot 20 oktâbrâ 2021 goda № 123 O mestnoj gosudarstvennoj administracii i organah mestnogo samoupravleniâ [Law of the Kyrgyz Republic of October 20, 2021 No. 123 "The law on local state administration and local self government bodies"].* Retrieved March 2, 2023, from http://cbd.minjust.gov.kg/act/view/ru-ru/112302

Isabekova, G. (2019b). The contribution of vulnerability of labour migrants to drug resistance in the region: Overview and suggestions. *The European Journal of Development Research, 31*(3), 620–642. https://doi.org/10.1057/s41287-018-0172-1

Isabekova, G. (2021). Mutual learning on the local level: The Swiss Red Cross and the village health committees in the Kyrgyz Republic. *Global Social Policy, 21*(1), 117–137. https://doi.org/10.1177/1468018120950032

Isabekova, G., & Pleines, H. (2021). Integrating development aid into social policy: Lessons on cooperation and its challenges learned from the example of health care in Kyrgyzstan. *Social Policy & Administration, 55*(6), 1082–1097. https://doi.org/10.1111/spol.12669

Kickbusch, I. (2003). External review of the Kyrgyz-Swiss Health reform support project (KSHRSP) Phase II. In *Components strengthening primary health care and public health* (pp. 1–37). n.p.

Labonte, R., & Laverack, G. (2001a). Capacity building in health promotion, part 1: For whom? And for what purpose? *Critical Public Health, 11*(2), 111–127. https://doi.org/10.1080/09581590110039838

Labonte, R., & Laverack, G. (2001b). Capacity building in health promotion, part 2: Whose use? And with what measurement? *Critical Public Health, 11*(2), 129–138. https://doi.org/10.1080/09581590110039847

OECD. (2014). *Switzerland 2013.* n.p.: OECD Publishing. Retrieved February 3, 2023, from http://www.oecd.org/dac/peer-reviews/Switzerland_PR_2013.pdf

PIL Research Company. (2017). *Community action for health (CAH) project impact assessment report* (pp. 1–49). Kyrgyz Republic.

Schueth, T. (2009). *Community Action for Health in Kyrgyzstan. A partnership between Village Health Committees and the governmental health system* (pp. 1–64). n.p.

Schüth, T. (2000). *If we were the minister of health… people's perspectives on health care: PRA study in two rayons of Naryn oblast (At Bashi and Ak-Tala)* (pp. 1–57). Swiss Red Cross.

Schüth, T. (2011b). *Appreciative principles and appreciative inquiry in the community action for health programme in Kyrgyzstan*. Tilburg University, n.p. https://pure.uvt.nl/ws/portalfiles/portal/1359087/Schueth_appreciative_07-11-2011.pdf

Schüth, T. (n.d.). *Community action for health in Kyrgyzstan*. n.p.

Schüth, T., Jamangulova, T., Aidaraliev, R., Aitmurzaeva, G., Iliyazova, A., & Toktogonova, V. (2014a). Community action for health in the Kyrgyz Republic: Overview and results. *Sharing Experiences in International Cooperation*. Issue Paper on Health Series, (3a), 1–31.

Schüth, T. (2011a). Glava 9 Dejstviâ soobŝestv po voprosam zdorov'â v Kyrgyzstane [Chapter nine community action for health in Kyrgyzstan]. In G. Laverack (Ed.), *Ukreplenie zdorov'â i rasširenie vozmožnostej [Health promotion and empowerment]: Translated into Russian with financial assistance from the SDC and technical support from the SRC* (pp. 132–175). V.R.S. Company llc.

SDC. (2003). *SDC health policy 2003–2010*. Swiss Agency for Development and Cooperation, Federal Department of Foreign Affairs.

Sida. (2008). *Sida country report 2007*. Retrieved September 2, 2016, from https://www.sida.se/en/publications/kyrgyzstan

SRC. (n.d.). *International cooperation health policy* (pp. 1–11). Swiss Red Cross.

WHO/Europe. (n.d.). *Declaration of Alma-Ata*. Retrieved February 3, 2023, from https://www.unicef.org/media/85611/file/Alma-Ata-conference-1978-report.pdf

WHO/Europe, & UNDP. (1997). *Manas health care reform programme of Kyrgyzstan*. Retrieved March 3, 2023, from https://apps.who.int/iris/bitstream/handle/10665/108088/EUR_KGZ_CARE_07_01_11.pdf?sequence=1&isAllowed=y

www.cah.kg. (n.d.). *The community action for health programme in Kyrgyzstan* (pp. 1–2).

# 6

# Sustainability of the "Community Action for Health" Project

This chapter introduces the "Community Action for Health" project in Kyrgyzstan and discusses the sustainability of this project. It commences with an overview of the project and its objectives. The following sections focus on the analysis of project sustainability as the long-term continuity of project activities, maintenance of benefits, and community capacity-building once the project has officially ended (Shediac-Rizkallah & Bone, 1998). The chapter also examines how factors relevant to the sustainability of health care interventions, principally funding, and account for the influence of general conditions, including political, economic, sociocultural, and organizational factors, unfold in this project.

## 6.1    Project Description

The "Community Action for Health" lasted for almost 17 years and had an overall budget of 24,500,000 Swiss francs (around €24,736,236[1]) (Gotsadze & Murzalieva, 2017, vi). The project was implemented in

---

[1] The exchange rate, as of March 17, 2023, was applied throughout this book.

© The Author(s) 2024
G. Isabekova, *Stakeholder Relationships And Sustainability*, Global Dynamics of Social Policy, https://doi.org/10.1007/978-3-031-31990-7_6

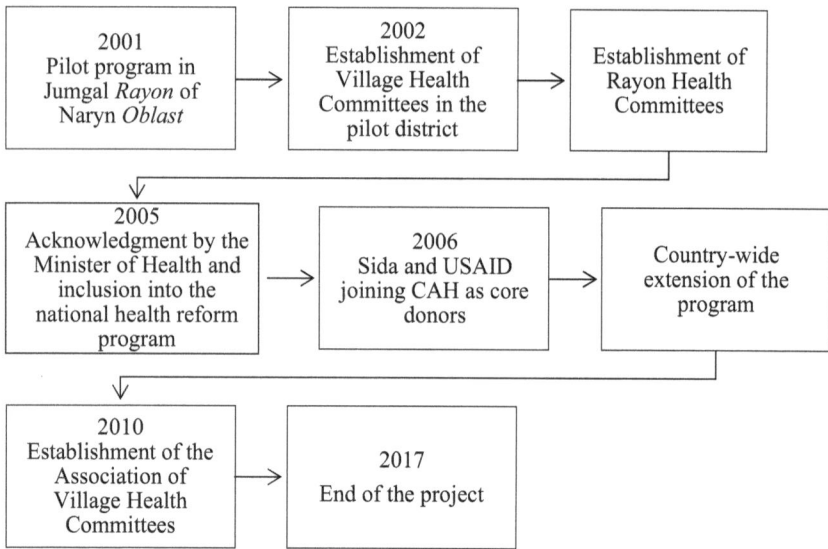

**Diagram 6.1** Chronology of the project (Source: Adapted from Schueth (2009, p. 11) and complemented with information from project-related documents)

seven phases (ibid.). It started off with a pilot project, which covered 16 villages in the Jumgal *rayon* of Naryn *oblast* (Schueth, 2009, p. 10). After its acknowledgment by the Minister of Health, who also referred to community engagement with the health care system as the "Jumgal model" (IO Partner 11), the project was included in the national health care reform program, *"Manas Taalimi"* (2006–2010) (Government of KR, 2006). The project was then expanded throughout the country, which was also made possible with assistance from the United States Agency for International Development (USAID) and the Swedish International Development Cooperation Agency (Sida) (see Diagram 6.1).

The "Community Action for Health" Project (CAH) was set up to empower[2] Kyrgyz communities through their engagement in health care. Previously known as the Kyrgyz-Swiss Health Reform Support Project (Schüth, 2000, p. 7), and the Kyrgyz-Swiss-Swedish Health Project (Jamangulova et al., n.d.), the project commenced by renovating five

---

[2] Here empower meaning "enabling communities to increase control over their lives" (WHO, 2023).

hospitals in Kyrgyzstan, before beginning its work in the community (SDC, 2008, pp. 1–2). As part of its collaboration with local communities, it had two goals: to "enable rural communities to act on their own for the improvement of their health" and to support the state health care system "to work in partnership with communities for improving health" (Schüth, n.d., n.p.). First, "acting on their own" means the emancipation of communities, which, following Kessler and Renggli's definition (2011), implies the participation of local communities in the health care system by defining the services those communities need and taking responsibility for their own health care. Second, the project was also intended to facilitate the collaboration of communities with Kyrgyz state institutions in this regard.

The emphasis of the project on communities and their engagement with health care provision echoes the ideas laid out in the Ottawa Charter for Health Promotion (1986) and the Alma-Ata Declaration (1978). The Charter stressed health as being the responsibility of individuals and communities, rather than just being the responsibility of the health sector alone (WHO/Europe, 1986). The Declaration went further by calling for the eradication of health inequalities, both between countries, and within countries, through the participation of individuals and communities in health care (WHO/Europe, n.d.-a). These international documents—the Ottawa Charter and the Alma-Ata Declaration—were referred to specifically in several of the documents setting out the CAH project (see Kickbusch, 2003; Schüth et al., 2005). Thus, in addition to increasing communities' control over their health (WHO, 2023), "empowerment" in the project also implied overcoming health inequalities within the country, a goal which was also reflected in the project's focus on rural areas.

The "Community Action for Health" project has been positively evaluated by academics and practitioners for its achievements in disease prevention and health promotion. According to the Swiss Tropical and Public Health Institute, preventive activities pursued by the project saved about US $3 million in patient travel and treatment costs, and around US $1.5 million in loss of income by patients who would otherwise have been unable to work (Schüth et al., 2014, p. 11). These preventive activities, together with health promotion, are believed to have contributed to improved public (Gotsadze & Murzalieva, 2017) and hypertension

(WHO/Europe, n.d.-b), as well as a decrease in the incidence of brucellosis (Schüth et al., 2014) and goiter (Schueth, 2009), decreased infant and maternal mortality, and decreased mortality from cardiovascular diseases (Gotsadze & Murzalieva, 2017). The project has been positively evaluated by representatives of the Government of Kyrgyzstan, local Kyrgyz communities, and external experts (see Ibraimova et al., 2011; Kickbusch, 2003; Maier & Martin-Moreno, 2011). Moreover, the CAH is referred to as a "good example" of collaboration between rural communities and their state health care system (Kessler & Renggli, 2011, p. 24), with "good practices" and "documented knowledge" of this project being beneficial to countries willing to adopt a similar model (Gotsadze & Murzalieva, 2017, p. 5).

## 6.2   Continuity of Project Activities

The "Community Action for Health" targeted wider community concerns including, but not limited to, tuberculosis (TB) and human immunodeficiency virus infection and acquired immune deficiency syndrome (HIV/AIDS). However, to ensure comparability with another case selected for this book, sustainability refers to the continuity of project activities and benefits related to TB and HIV/AIDS and community capacity-building. This section discusses the continuity of project activities by elaborating on the types of activities ("what") and the extent of their continuity ("to what extent") (Scheirer & Dearing, 2011, p. 2062). As previously noted, the Village Health Committees (VHCs) largely provided awareness-raising and health promotion in their villages but no medical services. The discussion also incorporates the factors critical to the continuity of activities, such as the sociocultural, economic, and political context in the country.

First of all, regarding the types of activities ("what"), my fieldwork in Kyrgyzstan in 2018 substantiated the continuity of TB-related services. The interviewees reported that there had been a continuity of awareness-raising activities for TB (State Partner 12; CSO 5), including dissemination campaigns in streets or schools (CSO 7). Each VHC decided on the timing of the activities by themselves. While some of the VHCs had a specific day for their campaign (CSO 5), others defined a longer time period and suggested that campaigns to fight TB should not be limited to one day, but should rather last up to a month. Overall, the campaigns had broad involvement, including representatives of local self-government, health care workers, school pupils, local residents, and others.

TB activities pursued multiple objectives. Community-based organizations intended to raise awareness among the population about the transmission of the disease and tackle discrimination and stigmatization against people with TB. A VHC representative noted widespread discrimination against persons with TB, driven by a misconception that this disease was not treatable. The VHCs informed the population that it was an airborne disease and not transmitted through the shared use of towels and dishes, as many believed (CSO 2). The organizations also aimed to prevent TB by raising population awareness of the symptoms of the disease and the need to refer cases to a health care facility. By stressing that no one is safe from TB and that it can be treated, the VHCs also intended to overcome the discrimination TB patients have to suffer from their family members and neighbors (ibid.).

Still, the awareness of TB and its treatment, as well as discrimination against persons affected by it, continued to be relevant in 2022. The Association of VHCs (AVHC) emphasizes the importance of treatment and that the treatment is provided for free (AVHC, 2022). The VHCs continue their awareness-raising activities. For instance, in Arkalyk village of the Jalal-Abad region, a VHC member used the break cotton pickers take after harvesting to bring information related to TB, social and behavior change, and other matters out into the open (ibid.). A VHC in Suusamyr village in the Chui region conducted a campaign among pupils of grades 5–11 by providing books, notebooks, and pens as rewards for active participation (ibid.). Similarly, discrimination and stigmatization of TB patients remained relevant, also reflected by a seminar that included representatives of local self-governments and police workers of two villages in the Chatkal district of the Jalal-Abad region (ibid.). On March 24, VHCs annually celebrate the World TB Day.

Similarly, the fieldwork demonstrated continuity of activities targeting HIV/AIDS (CSO 5). The VHCs organized seminars in schools, roundtables, and community walks to raise awareness (ibid.). As with the activities targeting TB, some VHCs chose a specific day for these activities (CSO 7), while others stipulated a longer duration, with the campaigns being conducted over the period of a week or even a month (VHC activities related to HIV/AIDS in 2019. Reports from Batken, Chui, Issyk-Kul, Naryn, Osh, n.d.). Similar to TB, these activities had a broad coverage, including representatives from local self-government and from the health care facilities in the village and at district levels (CSO 5; State Partner 12), as well as the local population, school pupils, and so on.

The main goal of these activities was to increase awareness among the population about HIV prevention and discrimination against persons living with HIV (PLHIV). However, the attitude of Kyrgyz society toward sexually transmitted diseases remains conservative and moralistic. In this regard, one of the project-related documents reports a case of a man coming forward during the PRA seminar to ask for information about syphilis. However, as elderly residents of the same village approached, the man fell silent (Schüth, 2000, p. 25).

The population's awareness of HIV transmission avenues, preventive measures, and nondiscrimination of persons living with HIV remained relevant. VHCs annually commemorate World AIDS Day (December 01). In 2019, the organizations organized a number of activities in collaboration with Rayon Health Committees, Health Promotion Units, primary health care workers, local self-governments, mass media, and other representatives. The awareness-raising activities included contests at schools, seminars by medical professionals (also for school teachers, pupils, and their parents), roundtables, processions of pupils, school performances, flashmobs, Q&A sessions, and essay-writing contests (VHC activities related to HIV/AIDS in 2019. Reports from Batken, Chui, Issyk-Kul, Naryn, Osh, n.d.). With support from other actors, the VHCs also organized walking campaigns and hung posters on HIV/AIDS in public spaces, public transportation, and bus stops (ibid.). In addition to increasing the awareness of the epidemiological situation via various media outlets, the activities touched upon themes such as "HIV is not transmitted through friendship" or "say no to drugs" (ibid., n.p.).

Similar to the smaller Village Health Committees, Rayon Health Committees have continued their awareness-raising activities in TB and HIV/AIDS. They have also continued supporting the VHCs in their organizational development and activities (AVHC, 2017b). This continuity of activities both at the district and at the village level suggests that the TB and HIV/AIDS campaigns did not cease immediately after the end of the project in 2017. Still, the simple fact of the continuity of activities does not tell us much about their extent. This topic will be discussed in the following subsection.

Furthermore, regarding the extent of their continuity ("to what extent"), the number of general activities implemented by the VHCs fluctuated throughout the project. One development partner noted that the VHCs had around sixteen different areas of activity at one particular moment in time, which caused the "burn-out" of VHC members (IO Partner 5). In this regard, the decision was taken to highlight certain key areas, leaving other areas up to the VHCs' discretion (ibid.). This was intended to ease the workload of volunteers. The community-based organizations also used this momentum to reshuffle their objectives. One VHC member who was interviewed stated that since 2013 the organization had started to discontinue campaigns that had achieved their goals. These included activities targeting alcohol abuse, iodine deficiency, and brucellosis. As of 2018, the interviewee stressed that the organization was currently focusing on five or six activities (including TB and HIV/AIDS), but had the relevant material to revive the discontinued campaigns, if necessary (CSO 5). This availability of multiple brochures on various health care issues was also evident during my visit to the offices of other VHCs.

Still, a certain level of letup in activities seems to have accompanied the end of the project. For example, in response to my question about the changes since the end of the CAH, several interviewees pointed to a general "slowing down" in the VHC's work. One emphasized that the workload decreased without funding, and with the decrease in the frequency of meetings, some VHC members wondered if they were "unemployed" now (CSO 4). Indeed, some volunteers seem to have perceived their work in the CAH as employment. Therefore, the end of the project brought a sense of uncertainty about the future of their activities.

One important indicator was the number of meetings between VHC members. One interviewee noted that these meetings, also taking place through tea gatherings, were crucial to bonding between members and their discussions of ongoing issues and future plans (CSO 4). It should be noted that the number of meetings among Rayon Health Committees (RHCs) also fell from 393 in 2014 to 275 in 2018 (AVHC, 2018). Devoted to a specific topic suggested by the Association of VHCs or Health Promotion Units (HPUs), these meetings are also used to discuss the outcomes, opportunities, and issues in the VHCs' work (AVHC, 2017b). They were also used to discuss the yearly report, the work plan, and activities targeted at the VHCs' organizational development (AVHC, 2022).

Still, the community-based organizations continued their activities. As of 2021–2022, the VHCs were implementing campaigns on multiple issues, including noncommunicable diseases, healthy nutrition, non-smoking, physical activity, clean water, handwashing, and awareness of breast cancer symptoms (AVHC, 2022). Similarly, despite the decrease in the number of meetings, the RHCs increased the share of meetings funded entirely on their own (without third-party funding). Thus, if in 2014, 74 out of 393 meetings were self-funded, in 2018, 209 out of 275 meetings were financed entirely by the RHCs (AVHC, 2018). As one interviewee acknowledged, the range of activities may not have been as extensive as before, and there was an overall "slowing down," but the community-based organizations continued their work (CSO 7).

The collaboration with donor organizations supported the continuity of some, but not all areas. The CAH coordinated its campaigns with other donor organizations throughout the entire duration of the project. Some examples thereof are awareness-raising activities conducted in collaboration with the German Corporation for International Cooperation (*die Deutsche Gesellschaft für Internationale Zusammenarbeit*—GIZ) and the Interchurch Organisation for Development Cooperation or complementarity ensured through the USAID funding the RHC meetings (Gotsadze & Murzalieva, 2017). After 2017, the Association of VHCs continued working with the World Bank, USAID, GIZ, SDC, and the United Nations (UN) agencies. Though beneficial to the capacities of the Village and Rayon Health Committees, the projects implemented by these organizations did not necessarily target infectious diseases. For example, the GIZ project pursued the incorporation of community priorities in its socioeconomic development plans (Development Policy

Institute, 2016). The World Bank initiative aimed to strengthen the capacities of Village Health Committees to collaborate with local self-governments (Independent Auditor's Report, 2018). The joint project of four UN agencies (the UN Women, the Food and Agricultural Organization, the World Food Programme, and the International Fund for Agricultural Development) focused on providing economic opportunities for women in rural areas (AVHC, 2022). The SDC initiative, in its turn, focused on the management and prevention of noncommunicable diseases (SDC, the Federal Department of Foreign Affairs, n.d.-a).

In the Kyrgyz Republic, USAID is among the few organizations, except for the Global Fund to Fight AIDS, Tuberculosis and Malaria (discussed in this book), focusing on TB and HIV/AIDS. The AVHC has collaborated with several of its initiatives, including "Defeat Tuberculosis" (2014–2019) and "Cure Tuberculosis" (2019–2024). The former involved RHCs and VHCs in selected regions, with main activities targeted at raising the population's awareness of ambulatory treatment, non-discrimination, and protecting the interests of TB patients (AVHC, 2018). The latter closely involves the AVHC and the Rayon and Village Health Committees in five regions, and the Kara-Suu district of the Osh region. Activities range from raising the population's awareness of TB treatment opportunities and the nondiscrimination of persons affected by this disease (AVHC, 2022), to fundraising and advocacy, to financially assisting TB patients from vulnerable groups (JSI Research & Training Institute, 2020).

Yet the activities also continued beyond the donor-funded areas. These include, for instance, the VHCs' annual countrywide awareness-raising campaigns dedicated to International TB Day (March 24) (AVHC, 2022). The scale of activities varies. Yet, delineating the campaigns supported by a development partner or conducted at the expense of the VHC is tricky, especially since the "Cure TB" project expanded its campaigns throughout the country in 2021. The activities are inextricably related, complementing each other. There are, for instance, cases in which the development partner provided the leaflets, but the VHCs organized and conducted the walking campaigns or seminars at their own expense. Notably, the VHCs conduct TB activities also at their own organizational expense (AVHC, 2017b) but mostly in collaboration with other actors, such as representatives of local self-governments, HPUs, RHCs, medical workers, school administrations, and others.

In contrast to TB, USAID's involvement in HIV/AIDS activities is somewhat limited. The organization instead collaborates with local NGOs having access to and working with targeted groups, such as PLHIV and intravenous drug users (see USAID, 2019). Due to stigma, discrimination, and anonymity concerns, these groups are closed to the state health care system and presumably to community-based organizations working on broader issues. This may explain the financier's inclination toward NGOs specializing in and closely working with persons affected by HIV/AIDS. In this way, the community-based organizations continued the awareness-raising activities mainly at their own expense. The lack of donor support is also reflected in the limited availability of information materials in different languages during the awareness-raising campaigns in 2019 (VHC activities related to HIV/AIDS in 2019. Reports from Batken, Chui, Issyk-Kul, Naryn, Osh, n.d.).

In these circumstances, state support proved critical to the continuity of HIV/AIDS-related activities. Decrees of the Ministry of Health and local state administrations on HIV/AIDS prevention were the basis for medical professionals and representatives of local self-governments to organize and support the campaigns (VHC activities related to HIV/AIDS in 2019. Reports from Batken, Chui, Issyk-Kul, Naryn, Osh, n.d.). As of 2019, the scale of activities varied across the country, ranging from small seminars to large-scale campaigns involving up to 900 participants (ibid.). The extensive involvement of actors allowed a broad range of activities, including printing articles in local newspapers and broadcasting videos (ibid.). Thus, similar to TB, VHCs organized HIV/AIDS-related activities in collaboration with a wide range of actors, including Rayon Health Committees, the Republic Center, HPUs, regional AIDS centers, mass media representatives, medical professionals, pupils, religious leaders, and others (ibid.).

Overall, the community-based organizations continued their activities despite the end of the project. The presence of a donor organization in a relevant field, such as USAID in TB, did surely strengthen the campaigns by providing additional resources. However, the awareness-raising in the areas not covered by donors, such as HIV/AIDS, has continued mainly at the expense of community-based organizations. Notably, state support proved to be critical to these campaigns. However, state support is also changeable, which impacts the maintenance of benefits (discussed in the following subsection).

## 6.3  Maintaining Benefits

The VHCs' activities in relation to TB and HIV/AIDS largely relate to dissemination campaigns. Therefore, the benefits maintained refer to the information received by the communities in regard to these two diseases. This assumes, however, not just the existence of the information activities but also the quality of the information.

One of the ways to look at the quality of the information provided by the VHCs is to look at the external evaluation of project activities both during the project and at the end of it.

The external assessment of the VHC activities on TB is inconclusive. Becoming a "tradition" among the VHCs (PIL Research Company, 2017), the information activities for TB improved the population's awareness of the disease. Randomized cluster surveys show greater awareness of TB indicators in the areas with VHCs than in the areas without (Schüth et al., 2014), which has also been confirmed by the external evaluation of the project (Gotsadze & Murzalieva, 2017). At the same time, another assessment found that increased awareness of the disease and its symptoms did not necessarily influence people's knowledge of TB treatment, and discriminatory attitudes toward people with TB remained (PIL Research Company, 2017). Though contributing to improved population awareness about the disease, the activities seem to have had limited effect in regard to treatment of the disease and discrimination against patients with TB.

The VHCs' activities also contributed to public awareness of HIV/AIDS. The VHCs surveyed school pupils from the 9th grade from the districts of Naryn, Talas, Chui, Batken, and Osh regions (five schools per district were covered), before and after the training course on HIV/AIDS. The surveys demonstrated increased awareness of HIV prevention among the pupils as a result of the training course conducted by the

VHCs (Schüth et al., 2014, p. 19). Similarly, the external evaluations of the project pointed to increased population awareness of HIV transmission, preventive measures (Gotsadze & Murzalieva, 2017), and increased awareness of sexually transmitted diseases (PIL Research Company, 2017). However, the impact of activities on stigma and discrimination against PLHIV is unclear.

Overall, the evaluations conducted both within the project and by external actors point to the contribution of the VHCs' information activities to increasing population awareness of the two diseases. Yet these assessments alone are not sufficient to evaluate the quality of information. I propose looking at the training received by the VHCs as another way to estimate the maintenance of benefits. VHC members are volunteers, and the majority of them have no medical education. For this reason, the quality of information they provide closely relates to their training.

The primary source of training for the VHCs was the Swiss Red Cross (SRC), which gradually transferred this function to the Health Promotion Units (HPUs). This transfer took place during the rollout of the program from the initial pilot districts to the country as a whole. The HPUs are essential to the VHCs' training. A development partner closely working with the community-based organizations supported this assumption, suggesting that the quality of, or problems with, HPUs inevitably reflected upon VHCs (IO Partner 11). Indeed, the HPUs continuously train the VHCs by visiting them in the villages. Therefore, they have firsthand knowledge of the issues faced by community-based organizations, as well as the opportunities to address them.

However, the frequency of HPU visits and the scope of training areas have decreased over time. During the period of operation of the CAH, the SRC covered the relevant travel costs for HPU staff to travel to the villages. After the project ended, the Ministry of Health took over the financing but decreased the frequency of visits. Previously monthly visits changed to quarterly (CSO 4). The Ministry of Health also limited the scope of training to four areas prioritized by the national health care program "*Den Sooluk*" (2012–2018), namely hypertension, HIV/AIDS, tuberculosis, and mother and child care (IO Partner 5). Explicit prioritization of TB and HIV/AIDS was beneficial to the continuity of training by state-funded HPUs. Albeit with decreased frequency, these activities nevertheless contributed to the uniformity of the information received and provided by the VHCs, and their compliance with the state health care program.

Still, the quality of training largely depends on staff availability and motivation. One interviewee noted that the organizational decline of VHCs was, to a certain extent, expected without the CAH but also dependent on HPUs and broader issues, such as the availability of qualified medical professionals in the country (IO Partner 11). By the end of the CAH, some trainers continued their work with community-based organizations in HPU roles. However, as elsewhere, low salaries and limited motivation contribute to the high rotation of medical professionals and inequity between urban and rural areas.

Furthermore, with the adoption of the "Healthy Person—Prosperous Country" program (2019–2030), Kyrgyzstan's priorities changed toward a systemic and away from its previous area-specific approach to health care. In contrast to "*Den Sooluk*" (2012–2018), which, along with other activities, targeted the four areas mentioned above, this program pursues a systemic approach to the health care system and reforms instead. Priority directions of the new program are public health, further strengthening primary health care, improving and rationalizing the hospital sector, developing emergency medical care and lab services, and improving the regulations of and access to medicines and medical devices (Government of KR, 2018a). It also intends to ensure strategic management of the health care system, target the problems with human resources in this sector, develop E-Health and health financing, and ensure the successful realization of stated objectives (ibid.). In contrast to "*Den Sooluk*," it does not explicitly prioritize TB or HIV/AIDS, but rather integrates them into the public health and primary health care areas of the program (ibid.).

Despite the aforementioned changes, community-based organizations are still central to the Kyrgyz health care system. The systemic approach of the "Healthy Person—Prosperous Country" (2019–2030) program envisions a broad engagement of stakeholders and community-based organizations. In addition to emphasizing citizens' responsibilities for their own health, the new program also intends to increase awareness of the right to quality health care and modernize the planning and organization of health care according to the population's needs (Government of KR, 2018a). VHCs are indispensable to achieving these objectives. Not explicitly prioritizing TB and HIV/AIDS, the program still offers distance learning modules on organizational development and public health and training activities on population needs assessment for health care (ibid.). However, the actual implementation of training activities largely depends on the availability of funding.

Furthermore, along with state institutions, donor organizations provide training to VHCs within the scope of their activities. For example, the World Bank-funded project (2014–2017) implemented by the Development Policy Institute aimed to build the capacities of VHCs and AVHC in identifying social determinants of health and working with local authorities to solve them (Development Policy Institute, 2014). It also allowed VHCs to expand their activities in unexplored areas, such as participating in the formation of local budgets at a village level. The emphasis on the role of the PRA in defining social determinants of health has also allowed the VHCs and the AVHC to then use this approach later to assess health care quality (see Development Policy Institute, 2017). Despite the wide range of benefits offered by this initiative, its coverage was limited to 30 pilot villages (Development Policy Institute, 2014). In addition to geographic coverage, the scope of activities may also be related to specific areas. The SDC-funded project on the "Effective Management and Prevention of Non-communicable Diseases" targeted Chui, Naryn, Issyk-Kul, and Talas regions in the first phase (2017–2022), and Batken, Osh, and Jalalabad regions and two cities, Bishkek and Osh, in the second phase (2022–2026) (SDC, the Federal Department of Foreign Affairs, n.d.-a, n.d.-b).

As noted above, USAID's "Cure TB" program is among the few projects with countrywide coverage and a focus on TB. In collaboration with the AVHC and the Republican Health Promotion Center, this project offered a series of trainings for HPUs, which, in turn, conducted seminars for VHCs to increase awareness of TB, reduce stigma and discrimination, and support adherence to treatment (JSI Research & Training Institute, 2021). Initially, the project covered only Talas, Naryn Chui, and Jalal-Abad, but in 2021 it expanded to the Batken region and the Kara-Suu district of the Osh region (ibid.). To date, USAID's "Cure TB" program seems to be the main source of training for HPUs and VHCs in the area of TB.

Despite the fluctuations in development assistance, the AVHC serves as a stabilizing factor by coordinating training activities. Through its coordinating role and direct engagement in initiatives, the Association of VHCs keeps an overview of development assistance provided to VHCs, including a record of organizations covered and excluded from aid. This perspective is essential to quality assurance and equity among community-based organizations, as the AVHC uses health projects to support and expand the training offered to VHCs. For instance, during the Development Policy Institute, the VHCs outside the piloted areas also expressed their interest in learning more about collaboration with local self-government bodies (AVHC, 2017b). In response, the Association developed a strategy for sharing experiences within the network. The Rayon and Village Health Committees discussed this strategy further, along with funding options and mechanisms for methodological support, during the RHC meetings (ibid.). Based on these discussions, the AVHC stipulated funding for experience-sharing within the network depending on the willingness of RHCs and VHCs and their financial capacities (ibid.). As a result, the coverage of training activities expanded beyond those piloted in the project. The VHCs from piloted areas conducted 1–2 seminars in areas not covered by aid, the organizers taking over small tea and coffee breaks, and the visiting CBOs covering commuting costs (ibid.).

Overall, both state and donor support are critical to maintaining benefits. However, the Association of VHCs and its network organizations and members demonstrated a remarkable initiative in extending training programs beyond their initial scope. In so doing, they contributed to the equality of awareness-raising activities in regions not covered by aid. Certainly, the shift in government priorities toward a systemic approach affects TB and HIV/AIDS, which had been explicitly prioritized in the previous health care program. Still, the maintenance of benefits also depends on the availability of training material, as discussed below.

In addition to training, dissemination campaigns presume the availability of relevant leaflets and other supporting material, which had previously been ensured by the CAH. VHCs interviewed for this research used the leaflets they accumulated during their work with the SRC and other international organizations (CSO 2). However, replenishment of these stocks is uneven.

Indeed, training material and handouts provided within the framework of the "Cure TB" project ensure access to updated information on TB, also in the context of the COVID-19 pandemic (see JSI Research & Training Institute, 2021). The project supported the preparation of information in Kyrgyz and Russian, online and in the form of postcards and videos (AVHC, 2020). There has been an increased use of social networks, such as Facebook, *Odnoklassniki*, and Instagram, among the AVHC and the VHC members that, for instance, follow the relevant pages of the Association (JSI Research & Training Institute, 2020). The project also supported the development of methodological handouts for conducting seminars (on- and offline formats) and booklets for volunteers providing extensive and brief information on TB and its prevention (AVHC, 2020). This support ensures access to updated information across the VHCs, which contributes to the uniformity of the information provided.

The situation with HIV/AIDS is different. Due to the lack of an ongoing project with countrywide coverage, the information provided by community-based organizations is limited to content from previous projects. The AVHC aims to increase public awareness of HIV via its social media posts. Yet, a more systematic approach to and the broader availability of information on treatment options, preventive measures, and risks of HIV in the context of the global pandemic would be desirable, certainly benefiting the efforts of community-based organizations.

Overall, a closer look at training provided within the areas of TB and HIV/AIDS vividly demonstrates the changing agenda and differing stakeholder involvement, which also contributes to inequity in terms of access to training and supporting materials.

## 6.4   Community Capacity-Building

Survival of civil society organizations (CSOs) beyond the end of a donor-financed project is a key indicator of community capacity-building. Therefore, in this section, I examine the extent to which community-based organizations set up under the CAH continue to exist beyond the end of the project, and I look at their leadership and mobilization of resources (see Labonte & Laverack, 2001a, 2001b).

In 2018, the Association of VHCs conducted a "mapping" of the VHCs and RHCs to identify the number of VHCs still operating and those who discontinued their work or needed additional support. In so doing, the Association intended to support "quality" over "quantity" of community-based organizations (CSO 4). The mapping showed some attrition, but most of the VHCs, and all of the RHCs continued their work. As of 2020, the AVHC (2020, p. 3) reported that there were 58 RHCs and over 1500 VHCs in the country. Following these results, the AVHC organized a general meeting of its members to discuss the VHCs' self-evaluation outcomes and strengthen the VHCs in need of assistance (ibid.). Participants divided themselves into groups and worked on own initiatives, support to the poor, reanimation of organizations, and VHCs' connections to other actors as part of the VHCs' and RHCs' work plans (ibid.). The processes and issues encountered during this activity would require research on their own.

As noted above, there has been some attrition of members and organizations. However, most VHCs have continued to exist after the end of the project. In response to my question about organizational performance, my interviewee, closely working with the VHCs, noted that the majority of "weak" organizations were in close proximity to the capital. The interviewee stressed that in contrast to their rural counterparts, members of these organizations had little time and did not have such close communication with local residents (CSO 4). This corresponds to the findings in the literature about the strength of social bonding (e.g., Agnitsch et al., 2006) and the persistence of community-based organizations in rural areas in contrast to urban settings (Gryboski et al., 2006).

Indeed, the survival of community-based organizations depends on several factors, including the leadership of its members (see Labonte & Laverack, 2001a, 2001b). My interviewees similarly stressed the importance of leadership of VHCs (IO Partner 5) and the ability of its members to express and formulate their concerns (IO Partner 11). In this section, I elaborate on the issues the VHCs faced in their work, the solutions they developed, and the strategies they used to overcome the structural inequalities.

During their work, the VHCs came across a number of issues, including mistrust from the local population and the local authorities. There were cases of people throwing away the health information brochures provided by the VHCs (CSO 2) and actually chasing the VHC members out of the seminars (CSO 5). There were also negative remarks toward the members, most of whom were women. There were claims that these women had "nothing else to do" but were "just fishwives running around the streets"[3] (CSO 2). A similar misunderstanding was common among representatives of local authorities. My interviewees recalled disinterest on the part of local authority officials (CSOs 4 and 5). At times, the remarks were also related to gender, with individual government representatives pointing to the VHC members to "go and look after husband and children" and not to "interfere" in matters that did not concern them (CSO 1).

However, not all VHCs continued their work under these circumstances. One interviewee noted that only people capable of saying, "no, you sit and listen to what I say," remained in the VHCs (CSO 1). The interviewee noted that those remaining had to be (using a Kyrgyz saying to describe it) "barking dogs"[4] in order to be resilient to the "attitude" of others (ibid.). It should be noted that the VHCs interviewed for this research were those who continued their work despite resistance from the local population and local authorities. These women continued to advocate for their ideas and developed their own strategies to overcome the structural barriers.

---

[3] Translated from Kyrgyz, "fishwives" is the closest expression in meaning to the original Kyrgyz expression "*dankyldagan ayaldar,*" which refers to noisy or shouting women.

[4] Translated by meaning, original is "*azhyldagan ayaldar.*"

First, the VHCs used "existing resources" for their dissemination campaigns. They targeted public gathering places and asked people for "five minutes" to share their information with them (CSO 2). In addition to visiting schools and local organizations, the VHC members also attended celebrations and visited communal grazing areas.[5] The VHC members used all available means to conduct their awareness-raising activities. One interviewee, for instance, told me that she could not find a place for the seminar after the representatives of the local self-governments ignored her request. However, on the way back, the VHC member saw a young woman hanging her laundry outside and paid attention to her yard as she approached and noticed that it was "large and clean." The interviewee asked for permission to host her event in the woman's yard, and was granted a permit to conduct a seminar for the local community on sanitary-hygiene issues there. The VHC member recalled that this seminar turned out to be even larger than expected as neighbors and other people from the street came in response to her and the woman's invitations to attend it (CSO 5).

Second, VHC members tried to "popularize" health care practices by following these practices themselves. Their adherence raised the interest of other people in the village. One VHC noted that as members started practicing what they called for, neighbors began to wonder why the person was "so obsessed" with a specific practice, for instance, cleaning the yard (CSO 5). This curiosity developed into interest, which was the exact objective of VHCs. But beyond this interest, the VHC practices also brought tangible results. The same interviewee emphasized that the CBOs contributed to halting the problem of alcohol abuse, which was a pressing problem in the 1990s. The VHC members persuaded people not to bring alcohol to funerals. They followed this practice and pointed out that alcohol consumption at funerals was inappropriate, also during their conversations with community members at tea gatherings after burials (ibid.). In this way, adherence to certain practices went hand in hand with information dissemination.

---

[5] In rural areas of Kyrgyzstan people take their cattle out to the mountains by giving their cattle to shepherds who carry out large-scale herding in the mountains in spring-summer period.

Third, the VHCs sought solutions for socially significant issues. Sometimes, these issues included those not initially anticipated in the work plan. For example, another VHC member interviewed for this research recalled a problem the members encountered during their dissemination campaign. One community they visited shared its concern with the dump on their street. The volunteers supported the local population in writing the relevant petition to the local self-government, which the local population had not considered before. As a result, the landfill was closed, and another one was opened elsewhere. The interviewee brightly concluded that though headed to a neighboring community for one reason, the VHCs were able to support it in solving a separate issue that was pressing to them (CSO 5).

Fourth, the VHCs pursued own initiatives on matters relevant to local communities. In contrast to socially significant issues, these initiatives were not limited to problems raised by local communities, but included support to its members and opportunities for community development. As part of the support to community members, the VHCs continued to assist vulnerable groups, including the poor and those facing catastrophic health expenditures (see Isabekova, 2021). The organizations also regularly commemorate Victory Day (May 9) and International Children's Day (June 1) by arranging presents and organizing events for war veterans and children, particularly those from vulnerable households (AVHC, 2022). In addition, the organizations sought further development of their villages in collaboration with other stakeholders. With small grants provided during the CAH, the VHCs cooperated with local self-government institutions and local sponsors to realize the projects. These included constructing a mini-football field with changing room and shower facilities, building a bus stop, maintaining bridges in emergency conditions, renovating and equipping a kindergarten, and renovating a local medical center (AVHC, 2017a).

By the end of the CAH, the AVHC intended to support the own initiatives that varied across organizations. In the case of VHCs, the number of initiatives fluctuated over the course of the project. Though growing between 2014 and 2016, the number fell by half in 2018 to 1254 (AVHC, 2018, p. 19). In contrast, own initiatives organized by Rayon Health Committees increased over time. Thus, if in 2017, only five organizations implemented over four initiatives, by 2018, twenty-seven organizations did (AVHC, 2018, p. 16). The number of organizations that did not implement own initiatives decreased from 14 to 1 over the same period (ibid.). It should, however, be noted that in the case of VHCs the low number was related to not only the actual work but also logistical issues. The organizations not participating in regional meetings failed to pass on the information at the RHC level, which in turn delayed reporting to the AVHC (2018, p. 19). As a result, some organizations were not included in the statistics of the Association of VHCs. As a corrective measure, the Association of VHCs asked the Rayon Health Committees to fill out tables on VHCs' activities and send this data to the AVHC immediately after the meeting via Google Forms or WhatsApp (ibid.).

The abovementioned are only a few examples of the strategies used by the VHCs to overcome structural barriers and gender-biased attitudes in their society. They used existing resources for their dissemination campaigns, popularized practices by following these themselves, sought solutions to local issues, and raised initiatives in regard to the matters relevant to the development of their communities.

It should be noted that the misunderstanding from the local population and disinterest on the part of the local authorities gradually changed into appreciation and the inclusion of the VHCs into decision-making processes (CSOs 2 and 5; State Partner 12). This appreciation is also reflected in the cases of individual VHC members receiving a medal for distinguished labor ("*emgek kaarmandygy*") from local organizations, or the broader fact that September 9 is now celebrated as the "Day of Village Health Committees," with organizations receiving congratulations from local authorities and medical institutions (AVHC, 2022). I do not make countrywide generalizations about the strategies the VHCs used to overcome the social barriers, as, in fact, not everyone did overcome them. However, the VHCs I interviewed in the north of the country demonstrated their leadership through their ability to define problems, suggest solutions, and develop various strategies to overcome gender-biased attitudes in their local society.

Finally, mobilization of resources via donor or state financing and fundraising is an essential component of sustainability, as it relates closely to the continuity of civil society organizations and their activities beyond the end of the donor-funded project. In this section, I examine resource mobilization through donors, fundraising, and income-generation activities conducted by community-based organizations and state support.

First of all, in terms of donor financing, donor organizations cover specific geographic locations or issues relevant to project objectives. For instance, the World Bank-financed "Sustainable Rural Water Supply and Sanitation Development Project" (2016–2025) operates in Osh, Chui, and Issyk-Kul *oblasts* and provides training of trainers to the VHCs on water quality, handwashing and hygiene, improvement of sanitation facilities, and food hygiene (World Bank, 2016). As part of the USAID-funded "SPRING Project" (2014–2016), the VHCs disseminated information on nutrition and hygiene among pregnant and lactating women and parents of children under two years of age in Jalalabad and Naryn regions (USAID, 2021). There are indeed countrywide initiatives, such as the USAID-funded "Cure TB" or the SDC-financed "Effective Management and Prevention of non-communicable Diseases" projects. The former expanded countrywide in 2021 (SDC, the Federal Department of Foreign Affairs, n.d.-a), while the latter focused on a select number of regions in each phase of the project to ensure countrywide coverage (SDC, the Federal Department of Foreign Affairs, n.d.-b, p. 2). Despite their countrywide coverage, both focused on areas relevant to project objectives (i.e., infectious or noncommunicable diseases). One interviewee similarly referred to the uneven coverage of donor assistance, with some providing training but not financing (CSO 4). None covered the broad spectrum of activities, including those identified by VHCs and not necessarily prioritized by the project, as the CAH had done.

What happens to activities or VHCs not covered by donors? They remain the sole responsibility of the community-based organizations. In 2018, I interviewed one VHC representative who stated that they had no collaboration with any donor organization at the time of our conversation (CSO 2). Yet the VHC continued its activities, and the interviewee stressed that other VHCs did the same and did not necessarily wait for development projects (ibid.). Although non-generalizable, this finding suggests some continuity of community-based organizations without donor support. One could specifically emphasize the role of the Association of VHCs, which serves as a stabilizing factor, also in coordinating the project implemented by donors and experience-sharing activities to VHCs not covered by donor activities. As seen from the mapping exercise, the AVHC has also been critical in identifying and organizing support for VHCs in need of assistance.

Second, fundraising is another option for the VHCs' resource mobilization. One of the VHC members showed me photos of the fundraising campaign the organization organized for a villager in need of surgery, namely a sports competition which raised around 27,000 Kyrgyzstani som (KGS, national currency of Kyrgyzstan), which is around €290 (CSO 2). The organizations also use other methods in addition to community fundraising for a specific purpose. Some organizations introduce membership fees to replenish their budget. The fees vary, but are relatively small, about 5–20 KGS (approximately 5–21 Euro cents) (CSO 4). However, poverty and unemployment in rural areas hinder the VHCs' fundraising possibilities. For this reason, during the CAH, the VHCs received small grants in the amount of 25,000 KGS (around €268) based on their project applications (CSO 2). A VHC noted that in 2017, a commission was formed among the representatives of the *rayon* administration (ibid.). Its aim was to evaluate the VHCs and redistribute the financing previously received from the SRC. The commission visited each village, checked the documentation, VHCs' activities, fundraising, and links to local institutions, and so on. As a result of this assessment, eleven organizations received small grants for two years, and the VHCs nominated for the first four places received additional rewards (ibid.).

Third, community-based organizations mobilize resources through social entrepreneurship and by using their organizational funds for income generation, though not all initiatives were "successful." During the CAH, the VHCs used the small grants provided by the SRC to solve community problems and establish small social enterprises. Public baths (*banya*) were built to address sanitation problems. VHCs provided free entrance to vulnerable groups in the population, including the elderly and people with disabilities. In addition to covering the maintenance costs, the entrance fees to the public baths supplemented the VHC's budget/fund (CSO 5). Community-based organizations also established sewing workshops and hairdressers. In addition to providing employment for the local population, these enterprises brought in 150 KGS (around

€1.6) to the VHC's budget on a quarterly basis (ibid.). However, not all of these initiatives were successful. The majority of public baths in the district I visited were in need of an overhaul, and the VHCs I interviewed were negotiating their transfer to the ownership of the local self-government, as the entrance fees for the public baths did not cover the amount needed for the overhaul (ibid.). In addition to social entrepreneurship, the VHCs also used their organizational funds to generate additional income. Some increased their funding by lending funds to VHC members at low interest rates (CSO 2). Others invested in cattle breeding, which was unsuccessful (CSO 7).

VHCs were not alone in their struggles. Rayon Health Committees similarly received stimulus grants within the framework of the CAH to generate additional income for organizational support. However, not all organizations benefited from this income, the amount of which also turned out to be less than expected. For instance, in 2014, fourteen RHCs received these grants, and nine of them managed to receive additional income in 2015; among twenty-nine organizations receiving grants in 2015, only six received extra income in 2016 (AVHC, 2017a, pp. 18–19). As a result, the total amount of revenue obtained through grants was considerably lower than expected due to internal as well as external factors. Internal factors were related to the organizations' abilities to maintain income-generating activities, decision-making in crises, and their skills in financial management, further investment, accountability, and taxation (ibid.). External factors included the low level of income, lack of marketing, and falling cattle prices, among others (ibid.).

As noted above, further capacity-building may be desirable for both VHCs and RHCs in the areas of social entrepreneurship and income generation. My interviewees noted that "good" leadership was critical to the size of the organizational budget (CSO 4), and yet the community-based organizations were "not ready for business," and despite their willingness to invest, they were unsure how to (CSO 7). Similarly, external evaluations of the CAH suggest that although the VHCs gained fundraising and strategic planning skills during the CAH, these may not be enough for them to work independently (PIL Research Company, 2017) beyond the end of the project. At this point, the Association of VHCs continues exploring investment opportunities at the national level to support community-based organizations at both village and district levels. However, further training in social entrepreneurship and income generation would undoubtedly benefit the organizations by allowing effective use of existing resources and more effective sourcing of further resources.

The fourth source of resource mobilization is state support. National authorities, including the Ministry of Health, the Republican Center for Health Promotion, and regional and district administrations, largely support the continuity of activities in the areas prioritized in the national health care program. Examples include HIV/AIDS and tuberculosis activities which benefited from the Ministry of Health and state administrations' decrees, in turn stipulating the organization of relevant activities by medical professionals and by extension supporting the community-based organizations. In addition, the Ministry and the Republican Center for Health Promotion were also critical to the continuous training of community-based organizations provided by Health Promotion Units, thus contributing to the maintenance of benefits (i.e., quality and uniformity of health promotion information provided by VHCs). The acknowledgment and support at the national level are also critical to the capacity of community-based organizations. Thus, inclusion in the national health care program allows the CBOs to develop additional skills. For instance, the ongoing "Healthy Person—Prosperous Country" (2019–2030) program stipulates the development and implementation of remote training modules for VHCs also in areas of organizational development (Government of KR, 2018b).

However, as part of resource mobilization, I focus on local self-government institutions that are somewhat unexplored and yet vital to the Village and Rayon Health Committees. The local self-government institutions provide administrative support to community-based organizations. My interviewees referred to meeting rooms the local authorities provided for the VHCs to gather and conduct their seminars and other awareness-raising activities (CSOs 4 and 5; State Partner 12). There are also cases of VHCs receiving office spaces from local authorities for their exceptional work and contribution to local development (e.g., AVHC, 2022). The VHCs often conduct maintenance work at the expense of their organizations. However, the availability of a fixed location for their activities and organizations indeed contributes to their capacity.

Community-based organizations may also receive financial assistance from local authorities, though the amounts in question are rather small due to budget deficits within state organizations. One interviewee estimated that the financial support to VHCs might range between 2000 and 5000 KGS (around €21–54) in the case of "poorer" and up to 10,000 KGS (about €107) in the case of "well-off" local authorities (CSO 2). A more accurate assessment of financial assistance would undoubtedly require access to local budgets and their countrywide comparison. However, another community-based organization representative similarly corroborated the small share of financial support. The interviewee emphasized that the eagerness of VHCs to work with donors also relates to the "little money" local self-governments had, which was not sufficient to meet the population's needs (CSO 7). In this way, although offering administrative support, the local authorities can provide only limited financing to the VHCs and their activities.

It should, however, be noted that the state support both at national and local levels is also contingent on the support of individual officials to community-based organizations and their work. The significance of state support and understanding was emphasized already during the CAH. The project-related documents reported changing attitudes of officials, also at the level of the Ministry of Health, after their acquaintance with the VHCs and their work (Schüth, 2011). Yet, awareness does not always equal support for the VHCs, as there were cases of individual candidates for political positions attempting to involve the VHCs in their election campaigns. Driven by their own agenda, they do not necessarily consider the interests or organizational development of CBOs. One of my interviewees noted that the presence of (former) VHC members among local authorities contributes to those authorities' understanding of and support for the VHCs' work (CSO 5).

Notably, the Association of VHCs endorses the political aspirations of its members. In 2016, for instance, it prepared guidelines for trainers and brochures to support the VHC members running for election to local councils (*jergiliktüü kengeshter*) (AVHC, 2017a, pp. 15–16). The AVHC conducted 12 seminars at the regional level throughout the country for candidates to enhance their capacity and ability to participate in political processes and advance their leadership skills (ibid.). Out of 325 VHC members trained, 45 obtained seats in local councils, with two-thirds of them being women (ibid.). It should be noted that in addition to strengthening the skills of the individual CBO members, these training activities were beneficial to the local self-governance institutions' understanding of health issues and their affinity toward the VHCs' work.

Still, the unstable economic and political situation in the country contributes to the frequent rotation of state officials. In these circumstances, relying on state officials or, in fact, also relying on the changing agenda of donor organizations does not seem sensible.

In addition to country-specific problems, the global COVID-19 pandemic constituted an unanticipated challenge, which, however, ended up demonstrating the relevance of community-based organizations in Kyrgyzstan. The volunteers had to halt their activities during the state of emergency declared in the country. Similarly, the HPUs had to cancel the training activities planned for the Village and Rayon Health Committees (AVHC, 2020). The pandemic has profoundly impacted community-based organizations and their work. Indeed, the seminars were renewed by online means in the second quarter of 2020, and the HPUs actively used social media (e.g., WhatsApp) to communicate with volunteers and share information (ibid., pp. 13–15). However, a better understanding of capacity-building, continuity of activities, and maintenance of benefits during (and, at some point, after) the pandemic require further research. Though challenging, the pandemic has also demonstrated the relevance of community-based organizations. In collaboration with local medical workers, Village and Rayon Health Committees organized awareness-raising activities on protective measures and campaigns calling for people to vaccinate against COVID-19 (AVHC, 2022). Similarly, the Association of VHCs continued sharing information on vaccination and the virus, also in relation to TB, on its social media pages (ibid.).

## 6.5  Summary

This chapter evaluated the sustainability of the "Community Action for Health" project by focusing on the continuity of activities after the end of the project, maintenance of benefits received by the targeted population, and community capacity-building.

First, it demonstrated that community-based organizations continued their activities, also in TB and HIV/AIDS, after the end of the project. However, there has been a general "slow-down," further amplified by the lack of donor assistance covering all aspects of VHCs' work throughout the country.

Second, in terms of maintenance of benefits, since the external assessment of the VHCs' activities and their impact on TB and HIV/AIDS was inconclusive, I used training as the assurance for the quality of the information disseminated by the VHCs. Most assistance focused and provided training on either specific issues or geographic areas and thus was incomparable to the CAH, which encompassed all initiatives and activities of health committees throughout the country. The community-based organizations continued receiving TB and HIV/AIDS-related training from the Health Promotion Units. However, the change in the national health care strategy from a disease-specific toward a systemic approach also affected the training offered to RHCs and VHCs. Still, the organizations continued their activities and demonstrated exceptional learning and training skills by sharing their experience and skills within the network. The coordinative and developmental roles of the Association of VHCs were crucial to this effort, though the leaflets and other materials used for the dissemination campaigns still primarily come from donors.

Third, as part of community capacity-building, despite some attrition, most of the organizations continued their survival beyond the end of the CAH. The organizations mobilize resources via donor funding, fundraising, social entrepreneurship, and state support, all of which are challenging in an environment of poverty, unemployment, and limited skills/training in the relevant area. Overall, the VHCs demonstrated leadership and continued their activities beyond the end of the development project. At the same time, there are a number of internal factors (mentioned here in the chapter) and external factors (the political, economic, and sociocultural situation in the country, the global pandemic) which challenge the sustainability of the community-based organizations, VHCs, and their activities in the area of TB, HIV/AIDS, and beyond.

Overall, the detailed analysis provided in this chapter demonstrated both issues and opportunities associated with the sustainability of health aid. Indeed, the CAH is a "success story," demonstrating the long-term sustainability of the activities and organizations initially supported by the project beyond the duration of donor funding. The resilience of community-based organizations and activities to (un)anticipated challenges, including the global COVID-19 pandemic, is extraordinary. This resilience also suggests that members of community-based organizations not only overcome obstacles, but also seek opportunities for their activities and places for their organizations in these very challenges. Still, this chapter also enlisted multiple issues in this regard, including the contingency of the long-term sustainability of health projects on the broader political, economic, and social situation in the country. In so doing, it showed that the sustainability of health aid is not a categorical "yes/no" matter, but a complex phenomenon requiring a fine-grained analysis of each of its three dimensions and related factors.

# References

Agnitsch, K., Flora, J., & Ryan, V. (2006). Bonding and bridging social capital: The interactive effects on community action. *Community Development, 37*(1), 36–51. https://doi.org/10.1080/15575330609490153

AVHC. (2017a). *Associaciâ "Kyrgyzstan ajyldyk den sooluk komitetteri" [The Association of Village Health Committees]: Otčet "Deâtel'nost' Associacii za 2016 g." [Report of activities for 2016]* (pp. 1–20).

AVHC. (2017b). *Associaciâ «Kyrgyzstan ajyldyk den sooluk komitetteri» [Association of Village Health Committees]: Otčet "Deâtel'nost' Associacii KADK za 1ânvarâ—15 maâ 2017 g." [Report "Activities of the Association from 01 January to 15 May 2017"].*

AVHC. (2018). *"Kyrgyzstan ayyldyk den sooluk komitetteri" Assotsiatsiyacy [The Association of Village Health Committees]: "Assotsiatsiyanyn 2018-jyldyn ish-merdüülügünün" otchetu [Report of activities for 2018]* (pp. 1–19).

AVHC. (2020). *Godovoj otčet 2019goda [Annual report 2019]: Associaciâ «Kyrgyzstan ajyldyk den sooluk komitetteri» Deâtel'nost' Associacii KADK v ramkah proekta USAID «Vylečit' tuberkulez» s 1 ânvarâ po 30 sentâbrâ 2020 g. [Association of Village Health Committees. The work of the Association within the framework of the USAID "Defeat TB" project from 01 Jan to 30 Sep 2020]* (pp. 1–24). n.p.

AVHC. (2022). Official Facebook page of the Association of Village Committees. Retrieved December 25, 2022, from https://www.facebook.com/association kadk/?ref=page_internal

Development Policy Institute. (2014). *Voice of Health Committees and social accountability of local self-government bodies on health determinants of rural communities of Kyrgyzstan (WB, 2014–2017)*. Retrieved March 3, 2023, from http://dpi.kg/en/activity/projects/full/0/86.html

Development Policy Institute. (2016). *Social'no-èkonomičeskoe razvitie ajylnyh ajmakov na osnove potrebnostej mestnogo obŝestva (GIZ, 2016) [Socio-economic development of aiyl aimaks based on the needs of the local community (GIZ 2016)]*. Retrieved February 3, 2023, from http://dpi.kg/ru/activity/projects/ full/2/77.html

Development Policy Institute. (2017). Mehanizm peredači informacii o kačestve uslug sistemy zdravoohraneniâ snizu vverh: ot mestnyh soobŝestv na nacional'nyj uroven' [The bottom-up mechanism for transferring informa-tion on the quality of health care services: from local communities to the national level]. *Municipalitet, 12*(134). Retrieved March 3, 2023, from http://www.municipalitet.kg/ru/article/full/1715.html

Gotsadze, T., & Murzalieva, G. (2017). *Impact evaluation of the Community Action for Health (CAH) project in Kyrgyzstan: phase I–VII (April 2002–March 2017) report* (pp. 1–44). n.p. Retrieved March 3, 2023, from https://www. newsd.admin.ch/newsd/NSBExterneStudien/880/attachment/en/3725.pdf

Government of KR. (2006). *Nacional'naâ programma reformy zdravoohraneniâ Kyrgyzskoj Respubliki "Manas taalimi" na 2006–2010 gody [National Health Care Reform Program "Manas Taalimi" for 2006–2010]: Utverždena postanov-leniem Pravitel'stva Kyrgyzskoj Respubliki ot 16 fevralâ 2006 goda № 100 [Approved by the Decree of the Government of the Kyrgyz Republic dated February 16, 2006 No. 100]*. Retrieved March 3, 2023, from http://cbd.minjust.gov. kg/act/view/ru-ru/57155

Government of KR. (2018a). *Programma Pravitel'stva Kyrgyzskoj Respubliki po ohrane zdorov'â naseleniâ i razvitiû sistemy zdravoohraneniâ na 2019–2030 gody "Zdorovyj čelovek—procvetaûŝaâ strana" [Program of the Government of the Kyrgyz Republic for the protection of public health and the development of the healthcare system for 2019–2030 "Healthy Person—Prosperous Country"]: Priloženie 1 (k postanovleniû Pravitel'stva Kyrgyzskoj Respubliki ot 20 dekabrâ 2018 goda № 600) [Annex 1 (to the Decree of the Government of the Kyrgyz Republic dated December 20, 2018 No. 600)]*. Retrieved February 2, 2023, from http://cbd.minjust.gov.kg/act/preview/ru-ru/12976/10?mode=tekst

Government of KR. (2018b). *O Programme Pravitel'stva Kyrgyzskoj Respubliki po ohrane zdorov'â naseleniâ i razvitiû sistemy zdravoohraneniâ na 2019–2030 gody "Zdorovyj čelovek—procvetaûŝaâ strana" [On the Program of the Government of the Kyrgyz Republic for the protection of public health and the development of the healthcare system for 2019–2030 "Healthy Person—Prosperous Country": Pravitel'stvo Kyrgyzskoj Respubliki Postanovldenie ot 20 dekabrâ 2018 goda № 600 [Decree of the Government date 20 December 2018, N 600]*. Retrieved March 3, 2023, from http://cbd.minjust.gov.kg/act/preview/ru-ru/12975/10?mode=tekst

Gryboski, K., Yinger, N., Dios, R., Worley, H., & Fikree, F. (2006). *Working with the community for improved health (No. 3)* (pp. 1–22). Population Reference Bureau. Retrieved March 3, 2023, from https://u.demog.berkeley.edu/~jrw/Biblio/Eprints/PRB/files/WorkingWithTheCommunity.pdf

Ibraimova, A., Akkazieva, B., Ibraimov, A., Manzhieva, E., & Rechel, B. (2011). Kyrgyzstan. Health system review. *Health Systems in Transition, 13*(3), 1–152.

Independent Auditor's Report. (2018). *Voice of Village Health Committees and social accountability of local self-government bodies on health determinants. GPSA Grant No. TF015846 Implemented by Development Policy Institute and the Union of Legal Entities "Association of Village Health Committees."* Special Purpose Financial Statements and Independent Auditor's Report For the Period from 29 January 2014 to 4 April 2018 (pp. 1–14). Bishkek, Kyrgyz Republic: BDO Armenia; Development Policy Institute. Retrieved February 2, 2023, from https://documents1.worldbank.org/curated/en/769221526448558713/pdf/Kyrgyz-Republic-Voice-of-Village-Health-Committees-and-social-accountability-of-local-self-government-bodies-on-health-determinants-Audit-Report-2014-2018.pdf

Isabekova, G. (2021). Mutual learning on the local level: The Swiss Red Cross and the village health committees in the Kyrgyz Republic. *Global Social Policy, 21*(1), 117–137. https://doi.org/10.1177/1468018120950032

Jamangulova, T., Iliyazova, A., Aidaraliev, R., Baktygul, T., Aitmurzaeva, Gu, & Schüth, T. (n.d.). *Community involvement to detect people with hypertension in Kyrgyzstan* (pp. 1–4). n.p.

JSI Research & Training Institute. (2020). *USAID Cure tuberculosis project: Year 1 annual report October 1, 2019–September 30, 2020* (pp. 1–106). USAID/Kyrgyz Republic. Retrieved March 3, 2023, from https://pdf.usaid.gov/pdf_docs/PA00X5RT.pdf

JSI Research & Training Institute. (2021). *USAID cure tuberculosis project: Year 2 annual report October 1, 2020–September 30, 2021* (pp. 1–97). USAID/

Kyrgyz Republic. Retrieved March 3, 2023, from https://pdf.usaid.gov/pdf_docs/PA00XWN7.pdf

Kessler, C., & Renggli, V. (2011). *Health promotion: Concepts and practices. A key issue paper focusing on the relevance for international cooperation* (pp. 1–36). SDC.

Kickbusch, I. (2003). External review of the Kyrgyz-Swiss health reform support project (KSHRSP) phase II. In *Components strengthening primary health care and public health* (pp. 1–37). n.p.

Labonte, R., & Laverack, G. (2001a). Capacity building in health promotion, Part 1: For whom? And for what purpose? Critical Public Health, 11(2), 111–127. https://doi.org/10.1080/09581590110039838

Labonte, R., & Laverack, G. (2001b). Capacity building in health promotion, Part 2: Whose use? And with what measurement? Critical Public Health, 11(2), 129–138. https://doi.org/10.1080/09581590110039847

Maier, C. B., & Martin-Moreno, J. M. (2011). Quo vadis SANEPID? A cross-country analysis of public health reforms in 10 post-soviet states. *Health Policy, 102*(1), 18–25. https://doi.org/10.1016/j.healthpol.2010.08.025

PIL Research Company. (2017). *Community action for health (CAH) project impact assessment report* (pp. 1–49).

Scheirer, M. A., & Dearing, J. W. (2011). An agenda for research on the sustainability of public health programs. *American Journal of Public Health, 101*(11), 2059–2067. https://doi.org/10.2105/AJPH.2011.300193

Schueth, T. (2009). *Community action for health in Kyrgyzstan. A partnership between village health committees and the governmental health system* (pp. 1–64). n.p.

Schüth, T. (n.d.). *Community action for health in Kyrgyzstan.* n.p.

Schüth, T. (2000). *If we were the minister of health… People's perspectives on health care: PRA study in two rayons of Naryn oblast (at Bashi and Ak-Tala)* (pp. 1–57). Swiss Red Cross.

Schüth, T. (2011). Glava 9 Dejstviâ soobŝestv po voprosam zdorov'â v Kyrgyzstane [chapter nine community action for health in Kyrgyzstan]. In G. Laverack (Ed.), *Ukreplenie zdorov'â i rasširenie vozmožnostej [health promotion and empowerment]: Translated into Russian with financial assistance from the SDC and technical support from the SRC* (pp. 132–175). Bishkek.

Schüth, T., Jamangulova, T., Janikeeva, S., & Tologonov, T. (2005). Power from below: Enabling communities to ensure the provision of iodated salt in Kyrgyzstan. *Food and Nutrition Bulletin, 26*(4), 366–375. https://doi.org/10.1177/156482650502600406

Schüth, T., Jamangulova, T., Aidaraliev, R., Aitmurzaeva, Gu., Iliyazova, A., & Toktogonova, V. (2014). *The Community Action for Health Program in the Kyrgyz Republic. Overview and Results* (pp. 1–50). Bishkek, Kyrgyz Republic: The Ministry of Health, Association of Village Health Committees, Swiss Agency for Development and Cooperation, Swiss Red Cross.

SDC. (2008). *Healthcare for remote regions—An SDC project in Kyrgyzstan sets the standard* (pp. 1–4). SDC.

SDC, the Federal Department of Foreign Affairs. (n.d.-a). *Effective management and prevention of non-communicable diseases*. Retrieved February 3, 2023, from https://www.eda.admin.ch/deza/en/home/countries/central-asia.html/content/dezaprojects/SDC/en/2017/7F09476/phase1.html?oldPagePath=

SDC, the Federal Department of Foreign Affairs. (n.d.-b). *Effective management and prevention of non-communicable diseases: Phase 2*. Retrieved March 3, 2023, from https://www.eda.admin.ch/countries/kyrgyzstan/en/home/international-cooperation/projects.html/content/dezaprojects/SDC/en/2017/7F09476/phase2?oldPagePath=/content/countries/kyrgyzstan/en/home/internationale-zusammenarbeit/projekte.html

Shediac-Rizkallah, M. C., & Bone, L. R. (1998). Planning for the sustainability of community- based health programs: conceptual frameworks and future directions for research, practice and policy. *Health Education Research, 13*(1), 87–108. https://doi.org/10.1093/her/13.1.87

USAID. (2019). *Improving tuberculosis prevention and care in central Asia. A story of 20 years of USAID commitment, partnership, and support 1997–2017* (pp. 1–67). Retrieved March 3, 2023, from https://pdf.usaid.gov/pdf_docs/PA00W4MZ.pdf

USAID. (2021). *Strengthening partnerships, results, and innovations in nutrition globally (SPRING) project*. Retrieved November 2, 2022, from https://www.usaid.gov/kyrgyz-republic/fact-sheets/strengthening-partnerships-results-and-innovations-nutrition-globally-0

VHC activities related to HIV/AIDS in 2019. Reports from Batken, Chui, Issyk-Kul, Naryn, Osh. (n.d.) (pp. 1–39).

WHO. (2023). *Health promotion. Track 1: Community empowerment*. Retrieved February 3, 2023, from https://www.who.int/teams/health-promotion/enhanced-wellbeing/seventh-global-conference/community-empowerment

WHO/Europe. (1986). *Ottawa charter for health promotion*. Retrieved March 3, 2023, from https://www.euro.who.int/__data/assets/pdf_file/0004/129532/Ottawa_Charter.pdf

WHO/Europe. (n.d.-a). *Declaration of Alma-Ata*. Retrieved February 3, 2023, from https://www.unicef.org/media/85611/file/Alma-Ata-conference-1978-report.pdf

WHO/Europe. (n.d.-b). *Good Practice Brief. Community action for health in Kyrgyzstan: an integrated approach of health promotion and an integrated approach of health promotion and primary health care provision in rural areas to scale up hypertension detection* (pp. 1–2). Retrieved February 3, 2023, from https://kyrgyzstan.un.org/sites/default/files/2019-12/Community%20 action%20for%20health%20in%20Kyrgyzstan%20-%20hypertension%20 detection%20%282016%29_ENG.pdf

World Bank. (2016). *International development association project appraisal document on a proposed credit in the amount of SDR 9.30 million (US$12.92 million equivalent) and a proposed grant in the amount of SDR 7.60 million (US$10.58 million Equivalent) to the Kyrgyz Republic for a sustainable rural water supply and sanitation development project* (pp. 1–80). n.p. Retrieved March 3, 2023, from https://documents1.worldbank.org/curated/en/573121475460021965/text/P154778-PAD-Board-Package-1392016-v2-09132016.txt

# 7

## Aid Relationships and Power Dynamics in the "Community Action for Health" Project

This chapter discusses the types of relationships between the actors in the "Community Action for Health" (CAH) project in Kyrgyzstan based on the findings from previous chapters. It builds around the findings regarding stakeholders' roles throughout the project realization process described in Chap. 5. It also considers the evolution of structural factors, including aid predictability and flexibility of providers, as well as the capacities and aid dependency on the recipients' sides, presented in Chap. 4. These two chapters constitute the basis for applying the analytical framework about power dynamics in relationships among stakeholders elaborated in Chap. 2. Informed by the findings and analytical frameworks laid down in these chapters, this chapter defines the following types of aid relationships (Table 7.1). It also elaborates on the impetuses these aggregated analytical categories of actors may have for pursuing the selected types.

© The Author(s) 2024
G. Isabekova, *Stakeholder Relationships And Sustainability*, Global Dynamics of Social Policy, https://doi.org/10.1007/978-3-031-31990-7_7

Table 7.1 Aid relationships between actors in the "Community Action for Health" project

| Actors | Reference | Type of relationships |
| --- | --- | --- |
| The Swiss actors—community-based organizations (CBOs) | Donor–civil society organization | "Empowerment" approach |
| Ministry of Health/Health Promotion Units/local authorities—CBOs | Recipient state–community | "Utilitarian" approach |
| The Swiss actors—the United States Agency for International Development (USAID) and the Swedish International Development Cooperation Agency (Sida) | Donor–donor | Unequal cooperation |
| The Swiss actors—Ministry of Health/Health Promotion Units | Donor–recipient state | (Contingent) Equal cooperation |

# 7.1   Donor–CSOs: The "Empowerment" Approach

I conceive of the relationships between the donor and civil society organizations (CSOs), which in the case of the CAH mainly refer to community-based organizations (CBOs), as an "empowerment" approach because of the equal participation of both actors throughout the project, structural factors favorable to this approach, and altered power dynamics between the provider and the recipient of the assistance.

First, in the CAH, both the "donor" and CBOs participated equally throughout the project. Ideally, the "empowerment" approach presumes the active role of CSOs throughout the period of the development assistance, but their role may vary in practice. The Swiss Red Cross (SRC) dominated the initiation phase of the project by suggesting the idea of community involvement in health care and establishing the Village Health Committees (VHCs) for this purpose. Furthermore, it was the SRC and not the participants themselves who suggested the Participatory Reflection and Action (PRA) approach and developed the assessment criteria used by the project participants. Yet, the SRC largely pursued a supportive role by offering relevant technical and financial support, following the needs and demands of the community-based organizations and the issues they encountered. It also initiated a process of annual self-assessment for the VHCs and the Health Promotion Units (HPUs) to emphasize their role in evaluating the project, something more commonly conducted by external consultants.

The community members played an equally significant role in the project. Their interest and agreement to participate in the initiative were essential to the advancement of the CAH beyond the pilot areas. Thus, in the following stages of the project—design and implementation—the VHCs took the leading role in the project by defining the issues of importance to them and implementing their solutions. These roles were consonant with the idea of empowerment as the "process of gaining influence over [the] conditions that matter to people" (Fawcett et al., 1995, p. 679), presuming the abilities of community-based organizations to express their concerns, set priorities, and participate in negotiations and decision-making process. By defining the issues and taking initiatives into their own hands, the CBOs were the source of initiative for the project and not merely its "passive" recipients (Rasschaert et al., 2014, p. 7).

However, the project initiation by an external actor, which also suggested the approaches followed by communities (e.g., PRA approach and assessment criteria), contrasts with the definition of the "empowerment" approach in ideal terms, which presumes the active role of civil society organizations throughout the development assistance. Yet, structural issues, such as illiteracy (Jana et al., 2004), gender-related biases (WHO, 2008), the political situation, and poverty (Fawcett et al., 1995), prevent civil society organizations from taking this active role throughout the assistance. Thus, the domination of the SRC in the initiation and evaluation phases points to structural issues hindering the ability of community members to initiate a project such as the CAH or suggest an assessment framework for their activities.

Yet, in the case of the CAH, these issues cannot be put down to illiteracy or economic hardships, as most of the population in the country (over 99%) is literate (UNESCO Institute for Statistics, 2023), and the organizations thrived in spite of the economic issues at the local and national levels. I argue that the structural issues stemmed from path dependency from the *Semashko* health care system inherited from the Soviet Union, which foresaw little space for public participation and initiative in health.

Despite the formal changes, both state and population continued living in practice within the old, paternalistic health care system, and therefore had limited perspectives on possible alternatives. My reasoning largely matches the analysis of health projects in Costa Rica by Morgan (1993, pp. 5, 15), who suggested that the "induced" or "sponsored" community participation could also be an outcome of a lack of citizen involvement in health projects.

Second, in addition to the equal involvement of VHCs throughout the project life cycle, the "empowerment" approach toward the CBOs was possible due to favorable structural factors. Remarkably, conventional gender roles in society contributed to the participation and retention of women in VHCs. These roles, for instance, include the assumption that a household's health is viewed as a woman's "responsibility" and that women (in contrast to men) are not associated with a role of breadwinner. The capacity of CBOs was further assured by the outstanding leadership of members who continued pursuing the organizational objectives amid misunderstandings from other community members or local authorities. Furthermore, the volunteer status of VHC members altered the hierarchy between the donor and CBOs by making the donor dependent on the willingness of community members to engage in the project, and, in so doing, evening out the aid dependency of CBOs on the donor. Certainly, the community-based organization members did receive minor incentives for taking part in the project, such as reimbursement of any project-related travel costs, training courses, seminars, and coffee breaks. However, these incentives were not the reason for community engagement—the reason was their willingness to work.

Equally, the flexibility and predictability of the Swiss aid assured the responsiveness and longevity of the project, providing a sense of security to stakeholders involved in this initiative and offering the time necessary to establish and build the capacities of community-based organizations. Flexibility and responsiveness were the foundation for the active roles of communities in the initiative. In total, the capacities of CBOs, the mutual dependence of stakeholders on each other's willingness to work, along with the predictability and flexibility of aid resulted in circumstances in which hierarchic relations between the provider and recipient of aid rendered themselves irrelevant.

Third, the altered power dynamics are another reason for defining the relationships of the "donor" with CBOs as an empowerment approach. Despite the dominance of the financier in specific phases, the relationships between the SRC and CBOs were characterized by the existence of the "power to," qualifying it as the "empowerment" approach. The "power to" manifested itself through a combination of the systems of thought and transformation of tacit knowledge into discursive, which empowered communities by attributing a decisive role to them, and a supportive one to the donor.

The systems of thought on the relationship with community members advanced by the SRC created the "power to" empower the CBOs. Following Haugaard (2003, pp. 107–108), the systemic biases and specific meanings "do not simply exist out there," but are rather supported by knowledge based on the "particular interpretative horizons." This way, stakeholders use and promote specific interpretations to create power for themselves or other actors. In the case of the CAH, the Village Health Committees benefited from the social consciousness the SRC and the project coordinator endorsed in relation to the role of communities in health, resulting in their decision-making and expert roles.

The SRC and the project coordinator advocated for the *decisive role of communities* in defining the issues targeted by project activities, which found its reflection in the active participation of community members in the initiation and design phases of the CAH. As demonstrated in Chap. 5, community members surveyed households and mobilized the local population to determine the pressing health care problems. The community members also brainstormed possible solutions to these problems. As a result, the issues targeted by the CAH were defined *by the communities themselves* and not induced by a donor. The SRC aimed to provide community members, who later joined the VHCs, the space to discuss the issues at hand and suggest possible solutions. This space presumed the altered roles: the donor and state representatives involved in the project were the ones listening, and the community members were the ones who spoke (Schüth, 2011b). This attitude, in combination with the nondominance approach, emboldened community members by placing them in the position of experts, those who knew the local needs and potential solutions.

By assigning the expertise to CBOs, the SRC altered the conventional perspective of donors, including their staff members and external consultants, as those who share their expertise with local people. In so doing, the staff members aimed to overcome the tradition of subordination of local expertise and knowledge to their international counterparts (Sending, 2015). This decisive role and the expertise of communities were supported by the transformation of tacit knowledge into its discursive form. Following Haugaard (2003, p. 108), stakeholders may be supportive of the existing social structure due to tacit knowledge rooted in practical consciousness, but changing this knowledge into a discursive form may "empower the powerless," who would use this knowledge to question the existing order. In the case of the CAH, this transformation of practical or tacit knowledge into discursive knowledge occurred throughout the project cycle.

During the initiation and design phases, the supportive role of facilitators, composed of both SRC and state representatives, included providing a space for discussions and encouraging community initiatives. Positioning the local communities as "those who know," the facilitators not only listened to, but actually encouraged the discussions. One of these encouragement tools was, for instance, comparing the community members' brainstorming on "how to stay healthy" to the Alma-Ata declaration on Primary Healthcare (1978) (Schüth, 2011b, p. 32). In addition to encouraging the community members, this comparison reaffirmed their position as "experts." This way, practical consciousness based on the tacit knowledge of the brainstorming exercise during the PRA sessions turned into discursive knowledge resulting from the community members' realization of their roles. The discursive knowledge complemented the formalized systems of thought, advancing the expertise of communities and their decisive roles in community health.

This realization about the roles of the local community and community-based organizations in health care continued during the implementation and evaluation phases. The VHC members used the means and knowledge obtained during the seminars to target the issues outlined by their communities. By applying this practical knowledge, the VHC members also realized their roles in targeting these problems and changing lives in their communities, which contributed to their willingness to continue their work. Despite the local self-governance representatives being adamant about the purpose of CBOs and their work, particularly at the beginning of the project, the community members proceeded with their activities (Chap. 5). I argue that it was the transformation of tacit knowledge into discursive that emboldened community members in their work. Self-reflection during evaluation, endorsed in the community capacity-building indicators adopted by the SRC, continued to emphasize the roles of VHCs also during the evaluation phase. Notably, the same tacit knowledge could have disempowered community members had it been used to support the existing hierarchies between the "donor" and CBOs.

This transformation of knowledge was supported by the systems of thought through which the SRC took the supportive, rather than leading, role in the project. As vividly demonstrated in the implementation phase, it provided necessary means and training to VHCs. Notably, the SRC could have also used these resources differently to increase its "power over" the community-based organizations, but it chose to advance the VHCs' position instead and create the "power to." This points to an important distinction between power and resources. The presence of resources does not automatically equal power, as power is about *using* resources. As Dahl (2005, pp. 273–276) noted, actors may use the same resources differently.

The SRC used the resources to highlight the nondomination principle, which was equally critical during the implementation process as it was during the design and initiation phases. My interviewee suggested that the project implementation involved and emphasized the importance of all participants and their contributions (IO Partner 5). The emphasis on non-dominance was particularly strong in the case of communities. The VHC representative endorsed the project coordinator's idea. The interviewee reflected that VHC members had differing levels of education, but were asked not to correct each other. Neither the SRC nor other VHC members corrected anyone who misspelled, for instance, while writing on the board. CBO members corrected the spellings later in their own notes, based on the protocols they received from training facilitators at the end of the seminar (CSO 5). This seemingly simple yet introspective idea nourished community participation and prevented possible building of hierarchy based on educational level.

Overall, a combination of a system of thought and transformation of tacit knowledge into its discursive form laid the foundation for changing the conventional power dynamics, characterized by "power over," to aid recipients' "power to." These altered power dynamics, in combination with the favorable structural factors and the equal engagement of an aid provider and recipient throughout the project, contributed to the formation of the "empowerment" approach of the Swiss actors toward community-based organizations in the CAH.

Why were the community members interested in cooperating with the SRC? I suggest two reasons for this, namely motivation for change and the opportunity for self-development. The interviewees noted that community members joining the VHCs were interested in "changing something," "not just existence" (CSO 1) but rather bringing "at least something good for the village" (CSO 5). This motivation to bring positive changes to their communities is the primary reason behind the VHCs' relationships with the SRC and with all other donors and state organizations.

Furthermore, the willingness to bring changes is related to another reason driving the community members, namely self-realization. Interviewees closely working with VHCs remarked/observed that having "some authority" motivated them to learn (CSO 1), and members often say that "instead of doing nothing, better work for free, everyone needs health" (CSO 4). The motivations of VHC members outlined in the testimony of the second interviewee suggest that women do not perceive their household work as labor. Still, outside their households, women in rural areas often have limited opportunities to participate in decision-making processes. A VHC representative opined that participation in the CAH offered opportunities for training and meetings at district, regional, and national levels for women that rarely left their village (CSO 5). These women were eager to take advantage of the knowledge and skills the project offered. According to another VHC representative, as a result of their work with the VHCs, many women were elected onto the local council (*kenesh*), got jobs in local government institutions (*aiyl okmotu*), or became nurses at primary health care facilities or cooks in schools (CSO 2). Thus, engagement in the project offered knowledge and skills women used to advance themselves.

What were the reasons for the SRC to engage with the community and pursue their "empowerment" approach? Community involvement in the CAH was paramount, since the goal of the project was to contribute to community capacity-building. Besides, the emphasis on community participation was consonant with the principles of the Swiss Agency for Development and Cooperation (SDC) and SRC, which financed and implemented the project, respectively. Still, instead of approaching community members as "free labor" (Earle et al., 2004) for project objectives, SRC approached them from an "empowerment" perspective. This was due to the project coordinator, Dr. Schüth. Having previously worked on a similar participatory community development project in Bangladesh (Schüth, 2011b), the project coordinator stressed the principle of "non-dominance" among SRC team members, the community, and state representatives.

The role of the project coordinator brings to light the significance of an individual, among other things, in understanding the outcomes of the organizational work. His background and perspective on community engagement were decisive to the "empowerment" approach pursued by the SRC in the CAH. Through his work, he established a close relationship with the communities. As one of the external evaluators noticed: "Indeed, it seems as if some villages will soon have little boys called Tobias" (Kickbusch, 2003, p. 13). The project coordinator spent thirteen years in the country and administered most of the CAH, leaving shortly before its completion. The VHC members were "upset" when the project coordinator was leaving the country, as they "considered him as their own son" (CSO 4). The community-based organizations interviewed for this research expressed their appreciation of the project coordinator's work and efforts (CSOs 2 and 7), emphasizing that "his work will never be forgotten" and the VHCs will not cease in their efforts (CSO 5).

## 7.2 Recipient State–CSOs: The "Utilitarian" Approach

The Ministry of Health, the HPUs, and local self-governments (LSGs) had a "utilitarian" approach toward community-based organizations, with collaboration primarily driven by promoting their own agendas rather than supporting VHCs and approaching them as equal partners.

Notably, stakeholders' roles during the project and structural factors did not point to a "utilitarian" approach. Both actors participated equally throughout the project realization process (Chap. 5), which could have been the basis for equal aid relationships. Similarly, the impact of structural factors on aid relationships was rather mixed. State organizations were not providers of aid during the CAH, but the continuous training and facilities they provided may suggest their roles as providers *after* the end of the project. In this sense, the recipient state offered limited flexibility in its assistance, which was largely limited to the areas the state itself prioritized (e.g., training), or the areas it could offer within the confines of its limited budget (e.g., office spaces or some funding). Political and economic instability in the country has also hindered the predictability of state support, though the areas prioritized in the national health care programs, such as "*Den Sooluk*" (2012–2018), were somewhat "secure" for the duration of the program. In terms of capacity and dependency, community-based organizations demonstrated exceptional leadership, endurance, and independence, contrasting with the frequent staff rotation and aid dependency on the side of the recipient state. These mixed outcomes from structural factors, in combination with the stakeholders' roles throughout the project, are open to interpretation.

The "utilitarian" approach owes to the power dynamics formed between stakeholders. The recipient state exercised two forms of power in relation to CBOs, namely the "power to" and "power over." The former occurred due to social consciousness, whereas the latter was contingent on the (non) transformation of tacit knowledge into its discursive form.

The recipient state provided the "power to" to CBOs through systems of thought. As noted in the project cycle, the government emphasized prevention over treatment and citizens' responsibilities for their own health. This idea, in a way, constrained the role of the state in health by providing a window of opportunity for community participation. Indeed, the idea of CAH was broader than state activities driven by retrenchment, but still the project complied with the state agenda. The systemic bias toward community participation in the project was based on the interpretative horizon advocated by the government, which provided power to population involvement in health (Haugaard, 2003, pp. 107–108). The VHCs, supported by the SRC and HPUs, used this opportunity to define and implement community initiatives in health. Similarly, the community-based organizations received acknowledgment and support through governmental decree, which solicited LSGs to collaborate with community-based organizations. The VHCs used this opportunity to expand and legitimate their activities through cooperation with local authorities.

The relationship between the tacit and discursive knowledge was decisive in creating the recipient state's "power over" community-based organizations, at the expense of supporting CBOs' "power to." Following Haugaard (2003, p. 108), the use of knowledge driven by practical consciousness results in the "power over," whereas its transformation and internalization into discursive knowledge create the "power to." The nondominance principle, which contributed to the SRC's empowerment approach toward CBOs (see the previous section), equally resulted in the relationship of CBOs with the recipient state being characterized by the "power to" and not "power over" VHCs.

However, the outcome was contingent on state actors' internalization of the nondominance principle beyond their mere compliance with this idea due to donor recommendations. This process scales down to an individual perception of this principle, emphasizing the relevance of and the difference between the analysis of actors at individual and organizational levels. At an individual level, support from the key employees from the Ministry was critical to the countrywide roll-out of the CAH, as these individuals advocated for increased community participation in health (see Schüth, 2011a). Similarly, VHC training, particularly at the end of the donor funding in the CAH, largely depended on the HPUs working with them, or rather their individual commitments to community empowerment. This significance of individual perspectives is also traceable to LSGs. As one interviewee noted, the community members manage to achieve more in locations where the LSGs support the VHCs' work (CSO 5). Emphasizing this significance of internalization of the norm at the individual level, this book makes no generalizations, but it highlights that in cases when this internalization occurs, the state actors pursue the "empowerment" approach toward community members, and where it does not, the "utilitarian."

Without internalization of the norm and transformation in a discursive form at an individual level, tacit knowledge will result in state organizations exercising "power over" the CBOs.

The Ministry of Health's agenda, not that of the communities, prevailed in the relationships between these actors. One vivid example of ministerial agenda guidance can be found in the cutting of the number of training areas after the end of CAH. During the project, the HPUs provided a broad range of training to the VHCs in the areas relevant to their work. After the end of the CAH, training shrunk to four areas prioritized by the national health care program: TB, HIV/AIDS, cardiovascular diseases, and mother and child health. The VHCs received no training outside these four areas, even though other issues might have been equally important to their communities or to their organizational capacity. In this regard, an interviewee closely working with the VHCs claimed that the Ministry "used" the CBOs to achieve its indicators on preventive activities without "acknowledging" the VHCs or their work. This way, although reporting on the engagement of VHCs, the Ministry does not provide institutional support for the VHCs' organizational capacity (CSO 1).

This cooperation, following the approach/agenda of the recipient state rather than that of communities, qualifies the relationships between the Ministry and the VHCs as a "utilitarian" approach toward the CBOs. The CBOs were "passive" recipients (Rasschaert et al., 2014, p. 7) of training courses and a "means" of implementation (Morgan, 2001, p. 221) for the Ministry, which was guided by its own agenda rather than the agenda of the VHCs. This style of relationships is drastically different from the "empowerment" approach of the SRC toward the VHCs, where the community-based organizations expressed their concerns and set priorities, acting as the key decision-makers.

Without the transformation of tacit knowledge into discursive, HPUs pursued a "utilitarian" approach toward community-based organizations. As part of primary health care, the HPUs have an increased workload, which, combined with low salaries, may contribute to pro forma rather than actual work with communities unless an individual HPU member decides otherwise (see Chap. 6). The "empowerment" approach toward CBOs also depends on the extent to which individual medical professionals internalize the principle of nondominance or maintain the hierarchical doctor-patient relations consonant with the *Semashko* health care system inherited from the Soviet Union. The project cycle offered a glimpse into issues, including the protective attitude of medical professionals questioning the exercise of VHCs and their activities in health. Thus, at an organizational level, HPUs may have limited incentives for pursuing the community empowerment approach.

Overall, the recipient state's "utilitarian" approach toward CBOs is primarily an outcome of power dynamics between these stakeholders. Interestingly, all other things being equal, internalization of the empowerment and nondominance norm at the individual level seems to be decisive in transforming the tacit knowledge into a discursive form, and thus creating the "power to" in place of "power over."

Correspondingly, the relationships between the LSGs and community-based organizations depended on the extent to which the former considered CBOs equal to them or merely instrumentalized VHCs for the sake of their own objectives. Engaging with the VHCs is essential for the work of the local authorities since the VHCs not only have the capacity for dissemination activities, but also have a certain status in their communities. One interviewee noted that not a single activity organized by the LSGs takes place without the VHC, which also helps the local authorities mobilize the local population (CSO 2). The community-based organizations express their concerns and participate in the LSGs' decision-making processes. However, their ability to set priorities on the agendas of the local authorities remains unclear. The VHC involvement in meetings seems to be limited to supporting the activities of the local self-government. In these circumstances, the VHCs remain the "means" of implementation (Morgan, 2001, p. 221), which qualifies the relationships between the two actors as a "utilitarian" approach on the part of the local self-government toward the community-based organizations.

What are the actors' interests in the "utilitarian" approach? Through cooperation with the VHCs, the Ministry of Health improves the performance of the national health care program by increasing the awareness of the population about the diseases relevant to the four areas prioritized in the program. Through the VHCs, the Ministry has the possibility to outsource disease prevention measures and health promotion activities from overloaded and understaffed primary health care personnel to the population itself. For the VHCs, the HPUs have remained the main source of training since the end of the CAH. Although they are not receiving training in other areas, the VHCs continue to improve their knowledge of the prevalence and prevention of the four diseases prioritized in the national health care program, which contributes to their expertise in disease prevention and health promotion. These reasons explain both the VHCs' and the Ministry of Health's interest in pursuing a "utilitarian" approach to the CSOs. HPUs, in their turn, engaged with the VHCs as part of their responsibilities.

Surely, the VHCs could have also benefited HPUs by providing outreach to local communities. Equally, the local authorities have a limited capacity for outreach among the community members in their villages (CSO 1). Therefore, the VHCs served as mediators between the recipient state and the local population. For the VHCs, collaboration with local self-governments offers limited financial incentives due to the budget deficit, but does provide administrative support for community activities. Notably, the VHCs were not financially dependent on any institution representing the recipient state.

## 7.3   Donor–Donor: Unequal Cooperation

In terms of structural factors, there is no explicit hierarchy in the relationship among donors, in contrast to donor–recipient relations. The unequal cooperation between donors formed primarily as a result of uneven involvement in the project and power dynamics.

Donor participation in the CAH was uneven. Absent during the initiation and design period, USAID and Sida joined the project during the implementation phase to support the countrywide roll-out of this initiative. Yet, the two donors had different forms of engagement: while USAID implemented the program activities in collaboration with the SRC trainers, Sida financed the SRC activities in the agreed areas without direct participation in the project. Despite this difference, both donors complied with the SRC's approach to communities, including the principle of "non-dominance" and evaluation of the VHCs and their work according to the criteria developed by the SRC. Thus, both implementation and evaluation phases were primarily guided by the SRC, with two actors following its framework.

This inequality also found its reflection in the power dynamics between the three donors, which combined attributes of both "power over" and "power to." In this context, the former is related to the preeminent position of some organizations, whereas the latter concerns the ability of organizations to work with each other.

First, the SRC exercised "power over" two other organizations through what Haugaard (2003, p. 108) called "reification." Reification occurs if stakeholders reinforce power relations because these relations are based on more than "simply arbitrary convention" (ibid.). In the case of the CAH, the reification concerns the "evidence-based" nature of arguments in favor of the SRC's approach. Though not explicitly focused on community engagement in health, the SRC has nevertheless demonstrated the effectiveness of its approach. These kinds of achievements supported the *evidence* for the "Jumgal model" (see Chap. 5) and contributed to USAID and Sida's compliance with the SRC's approach, including the nondominance principle, during the design and implementation phases as the project expanded beyond the selected regions.

Second, the "power to" was a result of the social order related to the ownership of the recipient country and harmonization among donors. It facilitated the collaboration between development partners guided by the global agenda on aid effectiveness. The principles of "ownership" and "harmonization" that would become almost synonymous with effective aid were accentuated in the Paris Declaration on Aid Effectiveness (2005) and the following Accra Agenda for Action (2008) (see S. Brown, 2020). The significance of these two principles is vividly demonstrated by the support USAID and Sida offered following the Ministry of Health's call for the expansion of the Jumgal model. The project life cycle vividly demonstrates that the commitment of the recipient state to provide Health Promotion Units encouraged donors to support the CAH. This response is consonant with the principle of "ownership" recalled in the Paris Declaration (OECD, n.d.). Similarly, a rapprochement between development organizations during the evaluation phase helped avoid duplications. Donors continued monitored project achievements, also by involving external consultants. Yet they seem to have agreed to retain the community capacity-building criteria developed by the SRC as the key approach to evaluating the VHCs and their activities in health.

At the same time, the social order has also contributed to the SRC's "power over" other organizations due to its awareness of the areas the recipient state was willing to expand. The idea of community participation in health has been discussed since the 1950s and culminated in the 1970s with the adoption of the Alma-Ata declaration (1978) (Morgan, 1993). This social order on community participation in health contributed to the emphasis on empowering local community members and community-based organizations. This bias has also allowed the SRC to implement the project and encouraged two other donors to join its expansion process in their efforts to contribute to the reform of the Kyrgyz health care system. Implemented in combination with reification, the "power over" created through social order was still different. Thus, in contrast to the evidence-based rationale of reification, the power here was an outcome of development organizations following the global agenda on community participation.

In both cases of social order creating the "power over" and the "power to," development organizations confirmed the meaning of community participation in health and the recipient state's ownership over the assistance. Both endorsed the Jumgal model, resulting in development organizations supporting the CAH as the initiative pursued by the Ministry of Health and giving the leading role to the SRC based on its experience and expertise on the desired topics.

Why did these three donor organizations engage in unequal cooperation? USAID and Sida agreed to unequal cooperation because of the SRC's expertise in community involvement. Overall, donors vary in their capacities, in their awareness of the context in recipient countries, and in other characteristics; however, the power dynamics between the donors are relatively equal (Chap. 2). A "dominant" donor only emerges if the other donors are "less motivated" or "financially less able" to compete (Bueno de Mesquita & Smith, 2016, p. 2). The leading role of the SRC in the CAH was related not to funding, but rather to its expertise in community involvement, which was also acknowledged by the Ministry of Health. Developing community capacity was the main area of activity for neither USAID nor Sida. Implementing the "ZdravPlus" (2000–2005) and "ZdravPlus II" (2005–2009) projects in five Central Asian countries (Abt Associates, 2023), USAID capitalized on primary health care development. As the core financier of the Sector-Wide Approach, Sida aimed to support health care reforms in Kyrgyzstan. In this way, the development of community capacity was only part of USAID and Sida's activities, which may explain their interest in going along with the SRC's approach instead of developing a new one. For the SRC, USAID and Sida involvement provided the necessary finance for the countrywide expansion of the pilot program.

## 7.4 Donor–Recipient State: (Contingent) Equal Cooperation

Both stakeholders participated throughout the project realization process. However, the formation of the type of aid relationships between them largely depended on the structural factors and power dynamics.

The relations between the SRC and the Ministry of Health, including the HPUs,[1] combined both the "power to" and the "power over." The SRC supported the ministerial social order, creating the "power to," but reification contributed to the SRC's "power over" the Ministry. Yet, the relationships between stakeholders also largely depended on the transformation of knowledge state representatives received from the SRC.

The Ministry of Health advocated for a new social order in which the population, and not only the state, was responsible for health. Following Haugaard (2003, p. 91), the social order creates power through predictability emerging from actors' "structuring" and "confirm-structuring" specific meanings. In other words, the social order is built on predictability assured through actors acting in compliance with ideas that support this social order (Haugaard, 2003, p. 90). The project life cycle and, more specifically, the initiation phase of the CAH vividly demonstrated the shift in the governmental agenda toward delegating part of the state's

---

[1] I do not include the local self-governments here as they did not directly work with the SRC. Affected by the decrees from the government, they have worked with CBOs but had only limited interaction with the SDC and SRC.

responsibilities for health care to its citizens. The Ministry conducted reforms according to the social order of increased population responsibility for its health (see Government of KR, 2006; WHO/Europe and UNDP, 1997). The government commenced optimization reforms in the hospital sector to increase primary health care funding. This emphasis on prevention over treatment and individual responsibility over health was also consistent with the Alma-Ata Declaration (1978) and the Ottawa Charter for Health Promotion (1986). The Charter highlighted that health was the responsibility of an individual and community, and not just one of the health sector alone (WHO/Europe, 1986). The Declaration stressed primary health care and called for the participation of individuals and communities in health care to overcome health inequalities within and between countries (WHO/Europe, n.d.).

The CAH was consonant with the social order the Ministry advocated for. The CAH aimed to "enable rural communities to act on their own for the improvement of their health" and support the state health care system "to work in partnership with communities for improving their health" (Kessler & Renggli, 2011, p. 24). These objectives were in harmony with the state agenda and international documents mentioned above. Through its objectives and activities, the CAH confirmed the social order promoted by the Ministry. The VHCs' activities allowed a nationwide expansion of awareness-raising campaigns and subsequent prevention of diseases significantly affecting the population, also according to the national health care strategies. In addition to cost-saving (see Chap. 6), these activities supported overloaded medical professionals in primary health care and expanded the coverage of prevention activities to rural areas, in which access to health care is particularly pressing.

However, in contrast to the state initiative, the project stipulated broader community participation in, and not only responsibility over, health. Having its roots in these international declarations, the state reforms were still driven by efficiency concerns. As one state official engaged in reforms at that time noted, several hospitals were demolished to reduce the excessive spending on maintenance, because they took up most of the public health financing (State Partner 1). Though highlighting responsibility, the state documents did not indicate the participation of communities or populations in health (e.g., Government of KR, 2006).

The CAH, in contrast, foresaw community participation in health. This difference in framing was crucial to community engagement and subsequent agenda-setting in health. This way, the community members and organizations were not mere implementers of state or donor-defined objectives, but also had the opportunity to define their own agenda.

In addition to supporting the Ministry's social order (the "power to" mentioned above), the SRC also had the "power over" the Ministry through reification. As discussed in donor–donor relationships in this chapter, the "Jumgal model" was a demonstration of the effectiveness of community participation in health. It demonstrated the effectiveness of the SRC's nondominance approach and became evidence in favor of it. This "evidence-based" rationale was the basis for reification, a situation in which stakeholders reinforce power relations, even if these relations are unequal, based on the belief that these relations rest on more than "simply arbitrary convention," in this case referring to science as the source of reification (Haugaard, 2003, pp. 104–105). It is important to remember that though consistent with the state agenda, the CAH went beyond it. In addition to targeting priority areas in the national health care program, the VHCs also pursued their own objectives (e.g., capacity-building) and problems their communities defined as significant. Thus, though supportive of state medicine, these activities also increased requests community-based organizations sent to state institutions, including the local self-governance organizations (see Chap. 5), which may not be ideal in the circumstance of a budget deficit.

Combined with reification, the tacit knowledge created the SRCs' "power over," though its transition to discursive also contributed to the "power to." The Ministry followed the SRC's approach in the project implementation and evaluation phases. Trained according to the principle of nondominance, the HPUs trained the VHCs accordingly and evaluated the VHCs using the community capacity-building criteria developed by the SRC. In so doing, the Ministry and HPUs "confirm-structured" the systems of thought on community participation advocated by the SRC. On the level of "practical consciousness," the knowledge the HPUs received from the SRC contributed to unequal power relations in which the recipient state followed the approach suggested by a donor. However, a transition of this knowledge into a "discursive form" empowered the HPUs and the

Ministry (Haugaard, 2003, p. 108). Accordingly, the HPUs did not apply this knowledge following donor recommendations or regulations from the Ministry. Instead, the HPU representative applied this knowledge due to personal vision or motivation. This perspective is likely behind the statement of an HPU representative, who noted that medical professionals "should not be teachers" but rather "equal to" the population they treat (CSO 5). Resulting from a personal vision or motivation and not inculcated from outside, the knowledge becomes discursive and enables the HPUs to build relations with communities. This significance of a personal perspective was also vivid at the end of the project, as the HPUs continued training community-based organizations without the SRC's support.

Overall, the relations between the SRC and the Ministry of Health, including the HPUs, combined both the "power to" and the "power over." However, the transition of tacit knowledge into discursive was critical to equal cooperation between the donor and recipient state. Thus, if the state representatives applied the knowledge based on personal vision and commitment and not following the regulations from "outside," this knowledge became discursive and empowered the recipient state instead of the donor.

Notably, the structural conditions, including predictability and flexibility of aid, capacity, and aid dependence of state institutions, did not have definite implications on the type of relationships formed between the actors. On the one hand, the Swiss aid was predictable and flexible, reflected in the duration of the CAH (thirteen years) and adjustment to the "wishes" of the Ministry of Health at the beginning of the project. On the other hand, the capacity of the recipient state, on the part of both the Ministry and the HPUs, remained somewhat limited (see Chap. 4). The state institutions also relied on the SRC's expertise in working with communities due to the lack of prior experience in this area. Thus, the structural factors could have contributed to both equal and unequal forms of aid relationships.

What was the Ministry of Health's interest in pursuing equal cooperation, which was contingent upon knowledge transformation? The Ministry intended to strengthen primary health care in the country and get citizens to take more responsibility for their own health, which is also reflected in the national health care reform programs. However, the Ministry had limited interaction with the population and no previous experience in working with communities. Acknowledging the success of the "Jumgal model," the Ministry requested the nationwide roll-out of this program and agreed to provide the HPUs for this purpose, following the SRC's request. The ministerial decision to establish the HPUs is remarkable, given the budget deficit in the country (and particularly in health care). This may, however, also have been a response to the SRC's flexibility and responsiveness to ministerial requests, particularly at the beginning of the project. Notably, the renovation of hospitals exclusively had not been part of the SRC or the SDC's vision (Schüth, 2011b, p. 23), and yet, despite this, the donor had supported the maintenance works in the Naryn region as demanded by the Ministry.

To understand the Ministry of Health's interest in committing itself to the CAH, it is also important to consider the relationships between the Ministry and the SDC, which financed the project. The SDC is one of the three donors, along with the German Development Bank and the World Bank, providing financial assistance for health care in Kyrgyzstan. The relationship of the Ministry with the donors, particularly with those providing the financial assistance, is unequal (see Isabekova & Pleines, 2021). According to a development partner, budgetary assistance allows development partners to promote their "parallel" projects (IO Partner 9) or the projects implemented along with the budget assistance. This way, the position of the SDC may have been a foreground for the Ministry to include the CAH in the Sector-Wide Approach. This inclusion contributed to coordination among development organizations and "more efficient use of available resources in support of the CAH model" (Gotsadze & Murzalieva, 2017, pp. 13–14).

Why did the SRC pursue equal cooperation? The main reason behind the SRC's interaction with the Ministry of Health and the HPUs was the sustainability of the CAH beyond the period of donor funding, which required the acknowledgment and commitment of the Ministry to the project activities. Another reason was the idea of "ownership," which intended to show the engagement of the recipient state and its ownership over the project, following the social order described in the "donor–donor" section of this chapter. Similar reasons guided stakeholders in the grants provided by the Global Fund to Fight AIDS, Tuberculosis and Malaria (see Chap. 10) to be introduced in the following chapter.

# References

Abt Associates. (2023). Improving the quality of health services in central Asia. Retrieved February 2, 2023, from https://www.abtassociates.com/projects/improving-the-quality-of-health-services-in-central-asia

Brown, S. (2020). The rise and fall of the aid effectiveness norm. *The European Journal of Development Research, 32*(4), 1230–1248. https://doi.org/10.1057/s41287-020-00272-1

Bueno de Mesquita, B., & Smith, A. (2016). Competition and collaboration in aid-for-policy deals. *International Studies Quarterly, 60*(3), 413–426. https://doi.org/10.1093/isq/sqw011

Dahl, R. A. (2005). *Who governs?: Democracy and power in an American City* (2nd ed.). Yale University Press.

Earle, L., Fozilhujaev, B., Tashbaeva, C., & Djamankulova, K. (2004). Community development in Kazakhstan, Kyrgyzstan and Uzbekistan: Lessons learnt from recent experience. *Occasional Papers Series, 40*, 1–63.

Fawcett, S. B., Paine-Andrews, A., Francisco, V. T., Schultz, J. A., Richter, K. P., Lewis, R. K., et al. (1995). Using empowerment theory in collaborative partnerships for community health and development. *American Journal of Community Psychology, 23*(5), 677–697. https://doi.org/10.1007/BF02506987

Gotsadze, T., & Murzalieva, G. (2017). *Impact evaluation of the community action for health (CAH) project in Kyrgyzstan: Phase I–VII (April 2002–March 2017) report* (pp. 1–44). n.p. Retrieved March 3, 2023, from https://www.newsd.admin.ch/newsd/NSBExterneStudien/880/attachment/en/3725.pdf

Government of KR. (2006). *Nacional'naâ programma reformy zdravoohraneniâ Kyrgyzskoj Respubliki "Manas taalimi" na 2006–2010 gody [National Health Care Reform Program "Manas Taalimi" for 2006–2010]: Utverždena postanovleniem Pravitel'stva Kyrgyzskoj Respubliki ot 16 fevralâ 2006 goda № 100 [Approved by the Decree of the Government of the Kyrgyz Republic dated February 16, 2006 No. 100].* Retrieved March 3, 2023, from http://cbd.minjust.gov.kg/act/view/ru-ru/57155

Haugaard, M. (2003). Reflections on seven ways of creating power. *European Journal of Social Theory, 6*(1), 87–113. https://doi.org/10.1177/1368431003006001562

Isabekova, G., & Pleines, H. (2021). Integrating development aid into social policy: Lessons on cooperation and its challenges learned from the example of health care in Kyrgyzstan. *Social Policy & Administration, 55*(6), 1082–1097. https://doi.org/10.1111/spol.12669

Jana, S., Basu, I., Rotheram-Borus, M. J., & Newman, P. A. (2004). The Sonagachi project: A sustainable community intervention program. *AIDS Education and Prevention, 16*(5), 405–414. https://doi.org/10.1521/aeap.16.5.405.48734

Kessler, C., & Renggli, V. (2011). *Health promotion: Concepts and practices. A key issue paper focusing on the relevance for international cooperation* (pp. 1–36). SDC.

Kickbusch, I. (2003). External review of the Kyrgyz-Swiss health reform support project (KSHRSP) phase II. In *Components strengthening primary health care and Public health* (pp. 1–37). n.p.

Morgan, L. M. (1993). *Community participation in health: Politics of primary Care in Costa Rica*. Cambridge University Press.

Morgan, L. M. (2001). Community participation in health: Perpetual allure, persistent challenge. *Health Policy and Planning, 16*(3), 221–230.

OECD. (n.d.). *The Paris declaration on aid effectiveness (2005) and Accra agenda for action (2008)*. Retrieved February 15, 2023, from http://www.oecd.org/dac/effectiveness/34428351.pdf

Rasschaert, F., Decroo, T., Remartinez, D., Telfer, B., Lessitala, F., Biot, M., et al. (2014). Sustainability of a community-based anti-retroviral care delivery model—a qualitative research study in Tete, Mozambique. *Journal of the International AIDS Society, 17*(18910), 1–10. https://doi.org/10.7448/IAS.17.1.18910

Schüth, T. (2011a). Glava 9 Dejstviâ soobŝestv po voprosam zdorov'â v Kyrgyzstane [Chapter nine community action for health in Kyrgyzstan]. In G. Laverack (Ed.), *Ukreplenie zdorov'â i rasširenie vozmožnostej [Health promotion and empowerment]: Translated into Russian with financial assistance from the SDC and technical support from the SRC* (pp. 132–175). Bishkek.

Schüth, T. (2011b). *Appreciative principles and appreciative inquiry in the Community Action for Health Programme in Kyrgyzstan*. Tilburg University, n.p. https://pure.uvt.nl/ws/portalfiles/portal/1359087/Schueth_appreciative_07-11-2011.pdf

Sending, O. J. (2015). *The politics of expertise: Competing for authority in global governance*. University of Michigan Press.

UNESCO Institute for Statistics. (2023). *Kyrgyzstan*. Retrieved March 3, 2023, from https://uis.unesco.org/en/country/kg

WHO. (2008). *Community involvement in tuberculosis care and prevention Towards partnerships for health: Guiding principles and recommendations based on a WHO review.* Retrieved February 28, 2023, from http://apps.who.int/iris/bitstream/10665/43842/1/9789241596404_eng.pdf

WHO/Europe. (1986). *Ottawa charter for health promotion.* Retrieved March 3, 2023, from https://www.euro.who.int/__data/assets/pdf_file/0004/129532/Ottawa_Charter.pdf

WHO/Europe. (n.d.). *Declaration of Alma-Ata.* Retrieved February 3, 2023, from https://www.unicef.org/media/85611/file/Alma-Ata-conference-1978-report.pdf

WHO/Europe, & UNDP. (1997). *Manas health care reform Programme of Kyrgyzstan.* Retrieved March 3, 2023, from https://apps.who.int/iris/bitstream/handle/10665/108088/EUR_KGZ_CARE_07_01_11.pdf?sequence=1&isAllowed=y

# 8

# The Global Fund Grants: Project Life Cycle

The Global Fund to Fight AIDS, Tuberculosis and Malaria (the Global Fund) delegates the realization of its project in Kyrgyzstan to the relevant national actors involved in tuberculosis (TB) and human immunodeficiency virus infection and acquired immune deficiency syndrome (HIV/AIDS) programs. However, external development actors are equally relevant since the Global Fund project is implemented in parallel with other health aid provided to the country. These actors are grouped into the following three analytical categories:

First, the *recipient state* refers to the Ministry of Health, represented by the National Center of Phthisiology and the Republican AIDS Center.
Second, *civil society organizations (CSOs)* are the local nongovernmental organizations (NGOs) receiving the Global Fund grants, but not the community-based organizations, as in the case of the "Community Action for Health" project in Kyrgyzstan.
Third, *donors* denotes the Global Fund and other international organizations working on TB and HIV/AIDS, such as the United States Agency for International Development (USAID); World Health Organization (WHO); German Development Bank (*die Kreditanstalt für*

© The Author(s) 2024
G. Isabekova, *Stakeholder Relationships And Sustainability*, Global Dynamics of Social Policy, https://doi.org/10.1007/978-3-031-31990-7_8

*Wiederaufbau*—KfW); Joint United Nations Programme on HIV/AIDS (UNAIDS); United Nations Population Fund (UNFPA); and United Nations Educational, Scientific and Cultural Organization (UNESCO).

In addition to these three analytical categories, the grants involve the Local Fund Agent (LFA) and the United Nations Development Programme (UNDP). LFA involvement is a standard part of all Global Fund financing, being responsible for validating the information provided by grant recipients. The UNDP is not on the list of donors, but rather became the first-line recipient of the Global Fund grants after the misappropriation of funds by the official(s) of the state organization. The following subsections expand on the roles of these national and international actors throughout the project cycle (i.e., the initiation, design, implementation, and monitoring phases).

# 8.1 Initiation

This phase is critical to understanding who stands behind the objectives targeted by health aid. According to Andrews (2013), ideally, assistance is driven by the pressing problems of the aid recipient and not by the objectives imposed by donors from outside. This section discusses whether TB and HIV/AIDS were perceived as significant issues in Kyrgyzstan before the assistance from the Global Fund.

Indeed, the problem of HIV/AIDS and TB was recognized as pressing in Kyrgyzstan long before the country received the Global Fund grant. To address the increasing HIV incidence in the country, the government initiated the National Program on Prevention of HIV and Sexually Transmitted Infections (STIs) (1997–2000). It restated the country's commitment to the Paris Declaration (1994) by recognizing the threat of the AIDS pandemic and the need to fight against HIV/AIDS (WHO, 1995). This program prioritized the prevention of Sexually Transmitted Infections (STIs), including HIV, through awareness-raising activities, improved blood donor screening, and distribution of condoms (Government of KR, 1997). However, it also acknowledged insufficient financing for health care and noted that even a small number of HIV cases represented a burden on the state budget (ibid.). Correspondingly, the national program highlighted the contributions of the UNDP, UNFPA, UNAIDS, UNESCO, and WHO to HIV/AIDS-related activities (ibid.). International development organizations were equally significant to the National Tuberculosis Program (1996–2000), which aimed to develop an affordable solution to the growing number of TB cases in the country. In addition to vaccinating newborns and children and identifying the sources of infection, this program intended to standardize chemotherapy treatment (Government of KR, 1995). The country adopted the WHO-recommended Directly Observed Treatment Short course (DOTS). A countrywide roll-out of the DOTS was possible due to the WHO's technical assistance and a continuous supply of medications from the German government. Still, problems with limited financing and a disparity between the activities and the TB epidemic in the country remained (Government of KR, 1995, 2008).

As seen above, TB and HIV/AIDS-related activities were present in the country before the Global Fund. The presence of these activities, despite their insufficient funding, hints at the political commitment of the government to combat the two diseases. Furthermore, its collaboration with multiple donor organizations points to the fact that the state initiative was in place long before the arrival of the first Global Fund grants to the country in 2004.

## 8.2   Design

The design phase expands on stakeholders' roles in defining the content of grant applications and their participation in the application process. This section discusses the compliance of grant applications with national programs; how recommendations and requirements of the Global Fund still shape the content of grants; and elaborates on the roles of national and international stakeholders in drafting the country's applications to the Global Fund.

First of all, in terms of the content, the grant applications are consonant with the health care objectives of recipient countries in the areas targeted by the Global Fund. A close overview of project activities and national TB and HIV/AIDS programs of the Kyrgyz Republic demonstrates adherence of activities to national goals. My interviewees support this observation (State Partners 4 and 9) and note that compliance with national strategies is the foremost evaluation criterion for the applications (IO Partner 4). Nevertheless, this adherence focuses on diseases targeted by the Global Fund (TB, HIV/AIDS, and malaria) and not necessarily on those causing the most deaths on the local level.

To illustrate, non-communicable diseases, such as ischemic heart disease and stroke, were the leading causes of death in the Kyrgyz Republic between 2009 and 2019 (Institute for Health Metrics and Evaluation, 2023). TB, by contrast, ranked 10th in 2009 and 16th in 2019, and HIV/AIDS was not among the most common causes of death (ibid.). One state official noted that, despite the clarity of this data, cardiovascular diseases received the least funding in the national health care program (2012–2018), whereas TB and HIV/AIDS received the most (State Partner 6). In fairness, it should be noted that the ranking mentioned above may be an outcome of the Global Fund grants improving the detection and treatment of the two diseases and, in doing so, extended the lives of persons affected by TB and HIV/AIDS—the relative fall in TB as a cause of death over the period 2009–2019 being one piece of evidence for this argument. Still, the extensive funding for two diseases alludes to the role of the organizational mandate in shaping grant applications. In this regard, a state representative notes that "technical specifications are developed by the donor, and all the rest is adjusted to its tune" (State Partner 4).

The Global Fund's requirements and recommendations also considerably shape the grant applications. Recommendations are not binding, but requirements are implicitly put forward via the explicit conditions the applicant is expected to meet to receive or continue receiving the grants.[1] Joint application and "dual-track financing" could be typical

---

[1] I will not focus on technical specifications, such as the provisions of the procurement plan (see Grant Performance. Report External Print Version. Kyrgyzstan KGZ-910-G07-T, 2016), submission of policies and procedures to evaluate them (Grant Performance Report External Print Version. Kyrgyzstan KGZ-607-G04-T, 2012), and appointment of an "independent auditor" to evaluate the program (Grant Performance Report External Print Version. Kyrgyzstan KGZ-202-G01-H-00, 2011), provision of updated plans on monitoring and evaluation (Grant Performance. Report External Print Version. Kyrgyzstan KGZ-910-G07-T, 2016), and others. For more information on these, see the documents related to the Global Fund grants to Kyrgyzstan.

instances of recommendations, whereas the incorporation of human rights, co-financing, and the Country Coordinating Mechanism (CCM) serves as illustrative examples of the Global Fund's requirements. However, in this regard, an interviewee notes that applicants, including Kyrgyzstan, equally follow both recommendations and requirements, though the former are not binding (IO Partner 20).

Kyrgyzstan submitted a joint application for two diseases. The Global Fund introduced the submission of joint applications as part of its New Funding Model (2012–2016) in order to facilitate dialogue and decision-making between the disease programs and ensure greater synergy and strategic use of funding (Global Fund, n.d.-c). The new scheme was designed exclusively for those countries with high TB-HIV coinfection rates, and encouraged those ineligible for further financing to submit joint applications for diseases (ibid.). Thus, in addition to applicants transitioning from the Global Fund grants (ibid.), joint proposals were relevant to many countries in Sub-Saharan Africa, where, for instance, nearly half of TB patients were HIV-positive in 2012 (Nelson, n.d., p. 4). Kyrgyzstan was neither transitioning from the grants nor did it have high TB-HIV coinfection rates. In 2012, for instance, only 2.2% of TB cases were found to be HIV-positive (Global Fund, 2014, p. 8). Still, the country submitted a joint TB/HIV application in 2015 for the first time since it received the Global Fund grants. This fact corresponds to the country's acquiescence to recommendations, suggested by the interviewee in the previous paragraph.

The country has also incorporated "dual-track financing" in the grant implementation process (see the section on "Implementation"). According to this approach, the Global Fund streams financing via "two tracks"—state and non-state actors—to strengthen the role of civil society and the private sector in grants (Global Fund, 2015, p. 3).[2] It encourages the CCM to use this approach to financing each disease and asks for an

---

[2] Involvement of the private sector in grant implementation in Kyrgyzstan is somewhat limited. According to UNDP (2015a, p. 38), eight private pharmacies, eight NGOs, and thirty-two state health care facilities offered HIV prevention, care, and support services to persons who inject drugs. Private sector involvement in the Global Fund grants included testing and treatment services by client-friendly clinics and a private family group practice in Issyk-Kul region, and a few other instances (Murzalieva et al., 2007, p. 41).

explanation if the CCM do not apply this approach (ibid.). Yet, unlike the CCM, dual-track financing is not a requirement, although my interviewees' perceptions of it varied: some saw it as a recommendation based on international practice (CSO 8), and others approached it as a condition for financing (State Partner 2).

Despite the country's compliance with both recommendations, this book differentiates between recommendations and requirements to distinguish their (non-)binding nature. Though possibly increasing the chances for funding, the recommendations are not preconditions for financing, unlike the Global Fund requirements discussed below.

First, the applicants are expected to incorporate human rights into their grant applications. The Global Fund denies supporting programs violating human rights ("Local Fund Agent manual. Section G—Global Fund essentials," 2014, p. 18). It asks applicants to target human rights and gender constraints on health care services (Global Fund, 2016a) and stipulates additional financing, also known as "catalytic investments," for this purpose. In the case of Kyrgyzstan, the catalytic investments in the amount of US $1 million focused on eliminating human rights constraints on HIV-related health care services (Global Fund, 2023b). The country took several steps to address human rights issues among groups vulnerable to HIV, including men who have sex with men (MSM), commercial sex workers (CSWs), persons who inject drugs (PWIDs), and others. The Government of the Kyrgyz Republic decriminalized sex between men and voluntary adult sex work and introduced changes to the law on possession and use of drugs (Ancker et al., 2017). Although these changes cannot be attributed to the Global Fund alone, they nevertheless constitute changes corresponding to the human rights perspective in the country's applications to the Global Fund. Thus, the country's joint HIV/TB proposal for 2017–2019 aimed to eliminate the "legal barriers to human rights-oriented services" (Zardiashvili & Garmaise, 2017, n.p.). Similarly, the country's previous applications stipulated training of law enforcement officers on stigma, discrimination, and HIV/AIDS prevention (see UNDP, 2015b).

Another example of Global Fund regulations followed by the applicants is co-financing or a domestic contribution by the grant recipient country in the form of government revenues, loans, health insurance, and others to the areas supported by the Global Fund (2016a, p. 12). The goal of this scheme is to demonstrate that the Global Fund grants are complementary to (Brown & Griekspoor, 2013) but do not replace state funding to relevant areas (Vujicic et al., 2011). Applicants are allowed to waive this requirement upon the provision of a detailed plan on how they intend to catch up with co-financing in the future (Global Fund, 2016a).

The Global Fund negotiates the share of cofinancing with each applicant individually, but sets general thresholds depending on income groups. Lower low-middle-income countries such as Kyrgyzstan are expected to cover at least 50% of financing for disease programs and progressively absorb the key program costs (Global Fund, 2016a, pp. 5, 16). Notably, 15% of grant disbursements are conditional on the fulfillment of this requirement (ibid.). In compliance with co-financing (State Partners 2 and 9), the country planned to increase the share of state funding for TB and HIV. More specifically, national stakeholders developed detailed plans, also known as roadmaps, to gradually transfer donor-funded services in these two areas to the state budget. My interviewees note that the cofinancing increased the state funding for TB and HIV/AIDS (CSO 3; IO Partner 3) and that without this condition, the government "would not even move a centimeter to look [for money] in its budget" (CSO 8). In this way, the country followed this explicit regulation from the Global Fund, similar to the accounting for human rights in grant applications.

Third, the most salient yet implicit requirement grant applicants, including Kyrgyzstan, comply with is the CCM. Applicants are expected to have a unit to supervise the planning, implementation, and use of grant resources (Global Fund, 2008, p. 9). Although not explicitly asking for the establishment of the CCM, the Global Fund accepts applications without CCM only in "exceptional circumstances," that is, from countries in conflict or without a legitimate government and those facing natural disasters and other emergencies (Global Fund, 2018, p. 21). Established during the country's first application to the Global Fund, the CCM in Kyrgyzstan has 23 members (nine CSOs, nine state, and five donor representatives) (Committee on TB and HIV under the Government of KR, 2023) and 23 alternates (State Partner 10 and Academic Partner 2). Alternates are primarily recipients of Global Fund grants. In contrast to members, they have observer status, which precludes them from voting or participating in grant monitoring due to conflicts of interest. CCM meetings are case-dependent, but their frequency increases during the development of grant applications (IO Partner 3).

The CCM in Kyrgyzstan underwent multiple changes because of a narrow focus on three diseases and duplication of existing institutions. The CCM added to at least three coordinating platforms the country had before the application to the Global Fund (see Table 8.1). In 2005, these platforms merged into the Country Multisectoral Coordinating Committee to fight HIV/AIDS, Tuberculosis, and Malaria (Murzalieva et al., 2007, p. 22). It engaged representatives of state agencies and ministries, international organizations, CSOs, and persons living with HIV (Government of KR, 2006) and dealt with the Global Fund grants exclusively (State Partner 2).

However, in 2007, the responsibilities of the Committee expanded to over 40 infectious human and animal diseases (Ancker et al., 2013,

**Table 8.1** Evolution of the Country Coordinating Mechanism in Kyrgyzstan

| | |
|---|---|
| The National Multisectoral Coordinating Committee for HIV/AIDS Prevention (1997) | The Sector-Wide Approach (SWAp) to Health care (preparations commenced 1996; SWAp formalized 2005) |
| The UN Thematic Group on HIV/AIDS (1996–2001) | |
| The Coordinating Committee for Prevention of Drug and Alcohol Abuse (2001) | |
| The Country Coordinating Committee on Prevention of STIs, HIV/AIDS and Tuberculosis (2002) | |
| Country Multisectoral Coordinating Committee to Fight HIV/AIDS, Tuberculosis and Malaria (2005) | |
| The Multisectoral Country Coordination Committee on Socially Significant and Especially Dangerous Communicable Diseases (2007) | |
| The Country Coordination Committee (2011) | The Coordination Council of Public Health (2014) |
| Integration of the Country Coordination Committee into the Coordination Council of Public Health under the Government of the Kyrgyz Republic (2017) | |

This table was compiled by the author based on Murzalieva et al. (2007, 2009), and other sources pp. 75–76). Though aimed at enlarging the Committee's competence beyond malaria, TB, and HIV/AIDS, this change jeopardized its ability to supervise the grants (Manukyan & Burrows, 2010) and resulted in the delegation of responsibility over three diseases to the Country Coordination Committee under the Ministry of Health. Yet, the Committee had limited impact beyond the supervision of grants. It did not participate in developing and evaluating national policies relevant to the three diseases (Ancker & Rechel, 2015b). Furthermore, grant-related issues required broader engagement of stakeholders. For instance, irrespective of the decision taken at the level of the Ministry of Health, discrimination against sex workers and prosecution of drug users by police forces continued due to their accountability to the Ministry of Internal Affairs and not the Ministry of Health (State Partner 4). The following reform aimed to address these issues.

Integration into the Country Coordination Council for Public Health under the Government of the Kyrgyz Republic elevated the authority of the Committee and partially addressed the issues with duplication of existing institutions. This integration guaranteed a high level of authority for the Committee's decisions (CSO 8) and compliance of stakeholders beyond the Ministry of Health (State Partner 4). The Council involves all relevant ministries and stakeholders, including the Parliament, under the chairmanship of the vice prime minister responsible for social affairs (WHO/Europe, 2019). This merger also aimed to reduce the duplication of organizational mandates (Government of KR, 2017a).

Similar to the Committee, the Council sought to coordinate the actors, though not only in the areas of three diseases but rather for Kyrgyz public health in general. Its functions include developing and implementing public health policy, monitoring and evaluating public health programs, and coordinating actors working in this area (Government of KR, 2014). It aimed to address the shortcomings of state stewardship in health care observed in the Sector-Wide Approach (SWAp) to health care implemented in Kyrgyzstan. Covering health care systems as a whole, the SWAp still faces multiple issues, including the limited capacity of the Ministry of Health and oversight of development organizations to provide the relevant support (see Isabekova & Pleines, 2021). Before incorporating the Committee into the Council, the Government of the Kyrgyz

Republic, with financial assistance from the Global Fund and SDC (Health Focus, 2020), initiated a study that ascertained the feasibility of integrating the Committee into the SWAp (Global Fund Office of the Inspector General, 2016). However, the decision was taken in favor of the Council due to, among other reasons, the limited representation (IO Partner 4) and participation of civil society organizations in SWAp (see Isabekova & Pleines, 2021).

During the process of applications, the CCM is intended to serve as an inclusive platform for stakeholders working in TB and HIV/AIDS.[3] It aims to facilitate the collaboration between stakeholders (Spicer et al., 2011a) and provide a broader representation of all relevant actors, including people affected by the diseases and representatives of the private sector, academia (IO Partner 20), international development organizations, state institutions, and civil society (State Partner 9). Civil society representation is one of the critical aspects of the CCM: grant applicants are expected to provide evidence for CCM membership of persons affected by diseases or their representation by NGOs and individuals advocating for their interests (Global Fund, 2018). Notably, delegate representation may be waived by the Global Fund's Secretariat to protect key populations (Global Fund, n.d.-c, p. 7), for instance, if direct participation of persons affected by diseases, and subsequent disclosure of their status or sexual orientation, may subject them to discrimination and criminalization. In any case, national civil society should compose at least 40% of the CCM, and CCM leadership (e.g., chair and vice-chairs) should be elected from state and non-state actors on a rotational basis (Global Fund, n.d.-c, p. 9).

However, the design of grant proposals in Kyrgyzstan shows that the ideal scenario does not always play out. One state representative notes that, in comparison to neighboring countries where CSOs have "no voice," the Kyrgyz state institutions take their opinions into account (State Partner 2). However, the literature on civil society organizations and development programs in Kyrgyzstan still points to the "tokenistic" participation of NGOs (Spicer et al., 2011b, p. 1752) and persons living with HIV (Ancker et al., 2013). Outnumbered by state representatives

---

[3] Kyrgyzstan was declared malaria-free in 2006.

(Harmer et al., 2013), civil society members have limited resources of their own to take part in meetings, and there are, in fact, no "effective mechanisms" in the CCM to support their participation (Spicer et al., 2011a, p. 10; Spicer et al., 2011b, p. 1752). This unequal distribution of powers between the state and civil society organizations urged CCM reforms in Kyrgyzstan (Manukyan & Burrows, 2010), particularly after the Global Fund's rejection of the country's application.

CCM reforms addressed insufficient civil society representation and participation in designing the grant applications. By 2014, the CCM included two persons living with HIV, one affected by TB, one by coinfection of TB and HIV, two persons who inject drugs, one commercial sex worker, and one MSM (UNAIDS, 2015a, p. 20). Despite this increase in numbers, civil society participation in the decision-making process remained limited. For instance, most civil society organization representatives had difficulties understanding the proposals written in English. Due to the lack of documents in Russian and Kyrgyz, CSOs had a limited understanding of the country's application to the Global Fund (Global Fund, 2016b). Following the Global Fund's rejection of the HIV proposal due to its noncompliance with the CCM eligibility criteria in 2014, the CCM members applied for the Community, Rights, and Gender Special Initiative of the Global Fund (ibid). This initiative covered extensive consultations with and capacity-building activities for CSOs, based on findings from situation analysis and review of the country's HIV proposal (ibid, pp. 12–15). Interviews with multiple stakeholders, including the state, NGOs, and international organizations, identified issues that were targeted during the capacity-building activities (ibid.), contributing to the improved participation of civil society in the following grant applications as well (Zardiashvili & Garmaise, 2017).

In addition to the Global Fund, multiple donor organizations support national stakeholders in designing the applications. The organizations providing technical assistance are the Stop TB Partnership, USAID, UNAIDS, WHO, BACKUP Health,[4] and others (Global Fund, 2023b).

---

[4] The initiative implemented by the German Corporation for International Cooperation (*die Deutsche Gesellschaft für Internationale Zusammenarbeit*—GIZ) and funded by the German Federal Ministry for Economic Cooperation and Development (*das Bundesministerium für wirtschaftliche Zusammenarbeit und Entwicklung*—BMZ).

A few individual examples of this assistance include the UNAIDS, USAID, and the United Kingdom's Department for International Development (DFID)[5] support to the state agencies and NGOs in preparing the country's HIV proposal and making the relevant budget calculations (Manukyan & Burrows, 2010). The WHO, in its turn, provided an evaluation of the HIV situation in the country to support the government in defining the priority areas for the HIV proposal (Mansfeld et al., 2015, p. 8).

Similar assistance was provided in the areas of TB. The German Corporation for International Cooperation (*die Deutsche Gesellschaft für Internationale Zusammenarbeit*—GIZ), for instance, offered training to national stakeholders on the development of a joint TB/HIV proposal (2018–2020) to the Global Fund (AFEW Kyrgyzstan, n.d.). In addition to technical assistance, donors support national actors in coordinating each other's activities. They use the CCM to inform about their plans, available budget (IO Partner 3), "preferences," and prospective projects (State Partner 10 and Academic Partner 2). This coordination's purpose is twofold. To avoid duplication of activities, the country's proposals to the Global Fund focus on the areas which are not covered by donors and are consistent with the mandate of the Global Fund. Donor organizations and the national government aim to cover the remaining areas (i.e., those excluded from the country's applications), though within the limits of the financial possibilities and interests of each donor (State Partners 4 and 9). These depend on organizational structure, earmarking, and geopolitical interests that vary across donors and considerably limit the flexibility of their assistance.

Overall, designing Kyrgyzstan's applications to the Global Fund project involves a large number of national and international actors working on TB and HIV/AIDS in Kyrgyzstan. Donor organizations participate in the country's proposal to the Global Fund by providing their technical assistance and taking over those areas not included in the proposal or state budget. Overall, the country's applications to the Global Fund comply with national health care programs.

---

[5] DFID was replaced by Foreign, Commonwealth and Development Office in 2020.

The grant applications are intended to cover the needs of all stakeholders in targeted areas. By providing a platform for civil society organizations and the persons affected by the relevant diseases, the Global Fund supports the representation of groups often excluded from decision-making. This support is demonstrated by the Global Fund's requirement to establish the relevant platform, rejection of proposals not complying with the civil society representation requirement, and provision of additional financing to strengthen the capacity of local CSOs. All these possibilities elevated the participation of often underrepresented and vulnerable stakeholders in Kyrgyzstan. Their engagement in drafting the grant applications in Kyrgyzstan was also intended to ensure that the applications were consonant with the needs and interests of target groups and not only with the aims of the recipient government and donor organizations.

However, the design phase also shows that civil society representation and its actual participation are still in their infancy. Hence, limited capacity and awareness of grant regulations hinder CSOs from fully participating and discussing the country's proposals. The Global Fund's assistance provided considerable support in this regard, but this was nevertheless limited to a one-time event, and does not represent the regular activity available to CSOs. The mature engagement of civil society is further hindered by state organizations that reckon with this requirement, mainly *pro forma*, to receive donor financing.

Along with promoting civil society representation and participation in designing country proposals, the CCM, along with other recommendations and requirements, demonstrated the pertinence of the Global Fund and its mandate in defining the content of the applications. The evolution of the CCM vividly showed that the country complied with requirements and recommendations, even if it meant duplicating existing institutions. Multiple changes in the CCM structure allude to the dilemma between ensuring the supervision of grants and integrating the platform into the broader context of infectious diseases and health care in general. Evidence suggests that Kyrgyzstan is not alone in these struggles: an audit of 50 sample CCMs in recipient countries showed that they all "partially or entirely" duplicated existing structures (Global Fund Office of the Inspector General, 2016, p. 13). This supports the assumption about the significance of the Global Fund recommendations and regulations to grant applicants.

## 8.3   Implementation

The Global Fund delegates implementation of its projects to the Principal Recipients (PRs) and Sub-Recipients (SRs) of its grants. Both are nominated by the CCM and approved by the Global Fund. Grant recipients could equally be state or nongovernmental organizations, as long as they have programmatic, financial, and management capacities (see Global Fund, 2015). Great emphasis is placed on PRs, responsible for assessing SRs, concluding contracts with them, and achieving the indicators stated in the grant agreement with the Global Fund. The PR also provides a procurement plan, reports on prices and quality of health products, coordinates with partners, and fulfills other functions (see Grant Performance Report External Print Version. Kyrgyzstan KGZ-202-G01-H-00, 2011; Grant Performance Report External Print Version. Kyrgyzstan KGZ-H-UNDP, 2016).

Not all actors are capable of accomplishing these responsibilities in a timely manner. Nine to sixteen months may pass from the commencement of a project until the arrival of the procured products (Global Fund n.d.-c, p. 31). Delays in tasks may cause disruptions in treatment or other services stipulated by grants. Ideally, the grant recipients are local public, private sector, or civil society organizations, although in "exceptional circumstances" (e.g., conflict, currency risks), the Global Fund may temporarily approve the nomination of a multilateral organization or an international NGO (Global Fund n.d.-c, p. 2). These organizations are then required to provide a capacity-building plan and a timeline for transferring their PR functions to national actors (ibid.).

In Kyrgyzstan, the Principal Recipients of the grants changed from government institutions to international nongovernmental and multilateral organizations in 2011 (Table 8.2). The following subsections discuss the reasons behind this transfer of PR functions that are also relevant to understanding the relations between the Global Fund, the state institutions, and the NGOs involved in the grant implementation process.

**Table 8.2** Principal Recipients of the Global Fund grants to Kyrgyzstan

| Period | Principal Recipient | Area |
|---|---|---|
| 2004–2011 | National Center of Phthisiology | TB |
| 2004–2009 | Republican AIDS Center | HIV |
| 2011–2015 | Project HOPE | TB |
| 2011-present | UNDP | TB and HIV |

Sources: The documents related to the Global Fund grants to Kyrgyzstan see Global Fund (n.d.-c)

Initially, the Principal Recipients of the first Global Fund grants to Kyrgyzstan were the National Center of Phthisiology (for TB) and the Republican AIDS Center (for HIV grants). Implementation of both grants was initially rated "strong" (Global Fund 2006a, p. 2, 2006b, p. 2). My interviewees note that both agencies procured health products according to the World Bank procedures (IO Partner 21), but the loopholes in the National Procurement Law still provided room for corruption schemes (State Partner 7). During the TB grant period (2007–2012), the Global Fund hinted at management issues and agreed to continue its funding primarily due to the engagement of the UNDP in building the capacity of the two state agencies (Grant Performance Report External Print Version. Kyrgyzstan KGZ-607-G04-T, 2012). Yet an anonymous call to Global Fund headquarters about the financial violations taking place in the country (IO Partners 11 and 21) resulted in a visit of its Audit Unit, which took place between November and December 2009 (Global Fund Office of the Inspector General, 2012). It found "significant financial irregularities" in the implementation process and urged an investigation into this matter (Global Fund Office of the Inspector General, 2013, p. 3).

This investigation, conducted by the Global Fund between February 2010 and August 2012, found multiple violations in the grant implementation process. There were violations in medical supply procurement (IO Partner 21), unauthorized cash advances, and transfers to unauthorized entities (Global Fund Office of the Inspector General, 2013). Preposterous justifications for the misuse of funds included the construction of a fish pond fencing to serve fish to TB patients (IO Partner 21). The head of the National Center of Phthisiology also used grant finances to buy a vehicle for his wife (Global Fund Office of the Inspector General,

2013, p. 3), with "maintenance costs" exceeding the value of the vehicle itself (IO Partner 21). Three out of four Sub-Recipient NGOs had family ties to the head of the National Center of Phthisiology, and one of these NGOs was used to misuse finances (Global Fund Office of the Inspector General, 2013, p. 3). Similar issues were found in the grant implemented by the Republican AIDS Center. My interviewees suggest that the initial amount of misused finances identified during the audit reached several million USD, but the state agencies provided supportive documentation in their own defense (State Partner 4; IO Partner 21). However, US $120,974 remained accounted for (Friends of the Global Fight Against AIDS, Tuberculosis and Malaria, 2018). Despite the National Center of Phthisiology's disagreement with the investigation results (Global Fund Office of the Inspector General, 2013, p. 68), the General Prosecutor's Office of the Kyrgyz Republic opened a criminal case on suspected misuse of position (Office of the Attorney General of KR, 2012). The head of the National Center of Phthisiology passed away before the investigations were concluded.

The Global Fund repeatedly asked the Ministry of Health to return finances that were unaccounted for. The Minister of Health neither replied to the Global Fund requests (Kasmalieva, 2015) nor returned the finances, referring to the budget deficit (Bengard, 2017). Notably, Kyrgyzstan was not the only case of grant mismanagement. "Misuse" of the grants was identified in Cameroon, Djibouti, Haiti, Mali, Mauritania, and Zambia (Benjamin, 2011, p. 3). In response, the High-Level Independent Review Panel on Fiduciary Controls and Oversight Mechanisms of the Global Fund to Fight AIDS, Tuberculosis and Malaria developed a report. Its recommendations included strengthening the capacity of the CSOs to ensure their supervisory roles as well as a closer evaluation of training activities in the grants. In Kyrgyzstan, the Global Fund neither discontinued the grants nor contacted the supranational authorities, as it usually does in corruption cases (see Global Fund, 2018). Instead, it took a disciplinary measure by deducting US $241,948 or "two dollars for every dollar that the Global Fund sought to recover" from the following grant to the country in 2017 (Friends of the Global Fight Against AIDS, Tuberculosis and Malaria, 2018).

After the mismanagement of finances by the state agency, the UNDP became the main recipient of Global Fund grants in Kyrgyzstan. It contracted with 33 local NGOs to work with persons living HIV, persons who inject drugs, commercial sex workers, men who have sex, and others (see UNDP, 2015a, pp. 34–47; 60–61). The organization also cooperates with state institutions, such as the National Center of Phthisiology, the Republican AIDS Center, the State Service for the Execution of Punishment, the Republican Center for Narcology, the Republican Center for Dermatovenereology, and others (UNDP, 2014). Notably, before assuming this new role, the UNDP implemented TB and HIV grants along with these state agencies and Project HOPE. The Global Fund rated the performance of these two organizations as "excellent," "exceeding expectations," "meeting expectations," and "adequate" (Grant Performance. Report External Print Version. Kyrgyzstan KGZ-910-G07-T, 2016, pp. 19–28; Grant Performance Report External Print Version. Kyrgyzstan KGZ-H-UNDP, 2016, p. 36). Yet, the nomination of UNDP by the CCM and its approval by the Global Fund was not random. Globally, the UNDPs are Primary Recipients of 31 Global Fund grants in 18 countries (UNDP, 2018). In Kyrgyzstan, the organization has worked on HIV issues since 1997 (Manukyan & Burrows, 2010). In other words, the UNDP received Primary Recipient status due to demonstrated country-based expertise and extensive experience with grants.

Still, both state and non-state actors were concerned with the transfer of PR functions to the UNDP. In 2015, the local NGOs appealed to the President, and the Parliament of the country, threatening to discontinue their activities if the national actors did not reconsider this transfer, which purportedly was not agreed with the CSOs (Ismanov, 2015). Similarly, the state actors criticized the transfer of PR functions to the UNDP, referring to the high administrative costs and loss of the country's ownership over the grants. According to state officials, about 20% of the grant funds were spent on administrative management due to the high salaries of foreign managers and project coordinators (State Partner 9), although state institutions could complete the same work (even with "good salaries") for one-ninth the cost, or about 2% of the grant value[6] (State Partner 2). Validating these estimates was not feasible within the

---

[6] Estimates are made by the author, based on approximate numbers provided by interviewees.

framework of this research: administrative expenditures are not visible in the Global Fund reports (e.g., Grant Performance Report External Print Version. Kyrgyzstan KGZ-H-UNDP, 2016), and the UNDP representatives in Kyrgyzstan (PR) did not answer research requests on multiple occasions. According to a news agency report, the total administrative costs were about US $3 million (Èrkebaeva, 2017).

In addition to increasing management expenses, the transfer of the PR functions to the UNDP allegedly jeopardized the country's ownership of the grants. Several interviewees emphasized the ownership of state institutions over the finances provided to the country (CSO 3; State Partner 9). Others noted that even though frequently argued by the Ministry of Health (State Partner 14), this notion of ownership does not prioritize the interests of the population affected by the diseases (IO Partner 4). This discussion raised the pertinent question of whether the recipient state's ownership over the grants represented the "country" and the interests of the population affected by the diseases.

In response to the allegations mentioned above, the UNDP pointed to grant savings and the small number of NGOs that signed the petition against it. The organization reported US $1.7 million in savings achieved through changing the suppliers and contractors previously involved through the state agencies (Eurasianet, 2012). Although more expensive, the UNDP represented a "safe" option for the Global Fund (IO Partner 20), notably due to the reliability of its procurement procedures. According to an anonymous "UN source" interviewed by an independent news organization, the costs of 13 essential items in the grants were 300% higher during the period of grant implementation by the state organizations (Eurasianet, 2012). Thus, despite the seemingly higher administrative costs, the UNDP assured the effective use of finances. In response to the CSO petition, the UNDP emphasized the small number of NGOs that signed the appeal, which merely attempted to "discredit" the organization's work (Ismanov, 2015). Yet, the small number of signatures could also relate to CSOs' aid dependency. Spicer et al. (2011b, p. 1752) note that the NGOs in Kyrgyzstan refrained from criticizing the PR (a state agency) due to the fear of not receiving further financing. This observation could, however, be equally relevant to the NGOs' relationship with PRs in general and not limited to the state PR.

Following the grant agreement, the UNDP committed itself to building the capacity of national actors. The replacement of NGOs previously involved in the grants implemented by the state agencies caused "serious protests," and in response, the new PR offered capacity-building activities to the excluded organizations to support their potential future return to grant activities (Grant Performance Report External Print Version. Kyrgyzstan KGZ-H-UNDP, 2016, p. 6). Twenty-one CSOs received training on quality of services, HIV prevention, adherence to treatment, and other areas (Grant Performance Report External Print Version. Kyrgyzstan KGZ-H-UNDP, 2016, p. 32). The UNDP has equally committed itself to building the capacity of state agencies and gradually transferring its PR functions to them (IO Partner 3).

Nevertheless, the Global Fund grants remained with a multilateral organization. Government organizations repeatedly emphasized their willingness to resume their roles as PRs (CSO 9; IO Partner 4), and in 2014, the CCM voted in favor of this resumption. To enable this, the Ministry of Health had to fulfill several conditions, namely, to develop the necessary mechanisms for contracting the local NGOs and to register the medications currently procured by the UNDP as humanitarian assistance (Minus Virus, 2017). The Ministry was also expected to provide timely reporting and financial management within the grants. The Global Fund and USAID provided US $600,000 to establish the Project Implementation Group under the Ministry of Health to support it in these tasks. However, there were multiple inefficiencies in its work. For instance, a supervisor of this group, appointed by the Minister of Health, ended up sending personal acquaintances for training abroad (ibid.). The Ministry also once delayed its report to the UNDP for two months, subsequently delaying for six months the payout of financial incentives for adherence to treatment for persons living with HIV for six months (ibid.). One and a half years after its establishment, the Project Implementation Group did not achieve all of the agreed goals, fulfilling eight of eleven indicators (Bengard, 2017). The Global Fund evaluated the Ministry as not yet ready to take over the PR functions (State Partners 4 and 9). The Ministry of Health continued negotiating the transfer of PR functions and reductions in the UNDP's administrative costs (Èrkebaeva, 2017).

There are in theory no restrictions on the types of organizations receiving the grants, but the Global Fund's requirements for grant implementation in practice result in the selection of organizations with specific qualifications. Following the Global Fund requirements, the UNDP has also developed a transition plan to transfer PR responsibilities to the national stakeholders. However, a state official noted that the donor procedures do not specify the period within which the organization is expected to transfer PR functions to national actors (State Partner 4). During both field trips to Kyrgyzstan in 2016 and 2018, multiple interviewees expected the near-term transfer of PR functions to state organizations. Yet, to this day, the UNDP remains the PR of grants.

It should be noted that regardless of other actors taking over the PR functions, local NGOs remained Sub-Recipients of the Global Fund grants. NGOs' interaction with the donor is limited to meetings with the portfolio manager of the Global Fund. There are no statistics about the frequency of these meetings, but in 2014 alone, the portfolio manager visited Kyrgyzstan at least three times (UNDP, 2015b, 2015c). Encompassing multiple actors, including the Primary and Sub-Recipients of grants, members of the Parliament, and others, these meetings are used to discuss the issues and achievements in the grant implementation process, the administrative, financial, and management systems of the Global Fund, and other matters (ibid.). The portfolio manager also answers questions and explains the changes (if any) in the Global Fund policies and regulations (ibid.). Still, the interaction between the local NGOs and the financiers beyond these meetings remains limited. In contrast, the financier seems to have continuous communication with the PR of the grants (IO Partner 4), which is the main point of contact for the local NGOs.

Still, the Global Fund and PR have hierarchical relations with local NGOs. According to one NGO representative interviewed for this research, donors greatly vary in their approach toward NGOs. She pointed to hierarchical relations in the Global Fund grants and stated that during the interaction with donors and project managers, the SR was frequently reminded of grant objectives and indicators that prevailed over the changes and suggestions made by the NGO (CSO 6). The interviewee noted that as "implementers," they were well aware of their "functions" and target groups, and their inability to go beyond these (ibid.). The interviewee contrasted this experience with her work on another health project. There, project managers "listened to" and considered the NGO's suggestions because, working on the ground, they had first-hand knowledge on how to improve the situation (ibid.). The interviewee was "astonished" by the appreciation and respect she experienced in this project, which aimed to introduce, not reject, the NGO's suggested changes (ibid.). This interviewee's perspective is not generalizable, but it does echo certain issues raised in the literature on health aid to Kyrgyzstan.

Multiple studies point to the limited flexibility of donor organizations. According to Benjamin (2011), the Global Fund assessment criteria focus on input and output indicators but leave little space for qualitative information. Yet this openness to suggestions is essential to the responsiveness of health assistance to local needs. For instance, multiple studies note increased emphasis on prevention (Murzalieva et al., 2009) but not advocacy in health care programs (Harmer et al., 2013; Spicer et al., 2011b). However, this may not reflect the priorities of target groups, such as commercial sex workers, who consider police harassment as their most significant problem (Ancker & Rechel, 2015a). Some interviewees in the study by Burrows et al. (2018) go even further by partially relating the increased violence and hostility toward the groups vulnerable to HIV to the reductions in donor funding and its growing emphasis on testing and treatment instead of advocacy for human rights. Designed by local stakeholders, the Global Fund grants ideally target issues identified by them. Yet, further openness to suggestions by local implementers would ensure the responsiveness of the assistance to the changing realities on the ground.

Local NGOs implemented the grants in collaboration with state agencies—former PRs of the Global Fund grants. Joint project implementation by state organizations and NGOs was possible due to the "dual-track financing" of the Global Fund, which contributed to collaboration between these actors. According to Harmer et al. (2013), this cooperation laid down the basis for overcoming the stereotypes actors had of each other. Yet, the sections below show multiple issues encountered during the joint implementation, which may not have overcome these stereotypes but did become the basis for collaboration beyond the grants.

State and civil society organizations found common ground for collaboration. The actors jointly develop the clinical protocols, organize round tables (State Partner 4), and implement harm reduction programs (Murzalieva et al., 2009) and awareness-raising activities throughout the country (CSO 3). State organizations largely provide the treatment of TB and HIV/AIDS, and NGOs complement these activities by reaching out to groups vulnerable to HIV out of reach to the state health care system (e.g., PWIDs, CSWs, MSM, and others). NGOs primarily work on disease prevention, the distribution of information materials, outreach, and care for the abovementioned population groups (Ancker et al., 2013). The state officials interviewed for this research claimed a "quite good" relationship and close collaboration with NGOs (State Partners 2 and 4). A civil society representative emphasized the significance of working with state officials, but stressed the importance of "speaking the language of state officials" by highlighting the general benefits of the services to the city and population instead of talking about the patients' needs (CSO 6). This framing seems to have contributed to the changing attitudes of state officials toward groups vulnerable to HIV and to their readiness to make the relevant changes (ibid.).

Still, tensions, particularly regarding the role of NGOs and their expertise in health, remained. Spicer et al. (2011b, pp. 1751–1752) note that state officials merely tolerate the CSOs' advocacy work and essentially perceive them as "helpers" rather than (equal) partners. The authors conclude that state institutions are not ready to consider NGOs' opinions and are cautious of their growing influence on social policy (ibid., p. 1754). Indeed, often overloaded with a large number of patients, health care workers have limited capacity to work with groups vulnerable to TB and HIV that tend to avoid state health care systems due to the fear of stigma, discrimination, and anonymity concerns. NGO social workers commonly come from the groups they are working with, which contributes to the trust between the social workers and these groups (CSO 6). By filling in the gaps in the state health care system (Semerik et al., 2014), NGOs, in a way, take over some state responsibilities (Ancker & Rechel, 2015a). However, their expertise in working with vulnerable groups is not necessarily acknowledged by state officials. One interviewee pointed to the discussions in the Ministry of Health regarding the abilities and qualifications of NGO employees to deal with health care issues without having relevant medical education (State Partner 4). This finding corresponds to the statements of the former Minister of Health (2014–2018), who portrayed the Ministry of Health as the primary actor in health care and advocated for ministerial control over NGO financing and activities in this field (Majdan.kg, 2018).

Nevertheless, the collaboration between the NGO and state organizations continued, particularly in preparing for the country's transition from Global Fund grants. The government adopted a "roadmap," in which it committed to increasing its share of HIV-related financing to 80% during the 2017 to 2021 period (State Partner 2). The Ministry of Health "worked closely" with the NGOs on the development of a roadmap, demonstrating the gradual transition of the activities currently financed by donors to the state budget (ibid.). My interviewees emphasized civil society organizations' role, including active lobbying efforts, in increasing state financing for HIV (State Partner 4). In addition to justifying the relevance of the roadmap before the Ministry of Finance (State Partner 2), CSOs advocated for increased funding and their role in monitoring the use of HIV-related resources. These activities found their reflection in the national program (see Government of KR, 2017a, 2017b), hinting at future collaboration between state and civil society actors.

Similar to the relationship between the recipient state and CSOs, limited financing seems to have intensified the coordination among donors. The Global Fund pays particular attention to coordination with American institutions, such as USAID, the Centers for Disease Control and Prevention (CDC), and the President's Emergency Plan for AIDS Relief (PEPFAR) (IO Partner 20). Still, interviewees noted that coordination among donors intensified mainly due to decreased financing (IO Partner 3; State Partner 2). According to state officials, previously, a project beneficiary may have received the same service from three organizations (State Partner 2), but de-duplication was finally achieved in the recent National HIV Program (2017–2021) (State Partner 4). Yet a civil society representative notes that donor coordination intensified only due to a "catastrophic shortage of finances":

*The money was so little that if you take it here, [a gap] opens there, [if] you take it there [a gap] opens here. For this reason, they are now endlessly meeting to review [the spending] and to try to cover these holes.* (CSO 8)

In addition to complementarity concerns, donor coordination during implementation is driven by attempts to de-duplicate efforts. Though expected to prevent the duplication of donor activities (IO Partner 20), the CCM may always not be able to coordinate the donors or have a complete picture of the programs implemented in the country (IO Partner 4). A single health care worker may simultaneously have contractual agreements with multiple donor organizations (Semerik et al., 2014). Data gathering in these circumstances is exceptionally challenging (see the following section on monitoring). Therefore, the Global Fund additionally meets with the relevant donor organizations, also during the visits of the portfolio manager to a grant-recipient country (see UNDP, 2015b). Through coordination with major partners, the Global Fund avoids the duplication of efforts and substantial gaps in aid-recipient countries (IO Partner 20) to ensure the continuity of services.

Overall, the roles of actors and their relations to each other during implementation demonstrated multiple differences to those of the design stage, except for the relations between the Global Fund and other donors remaining equal and driven by coordination of efforts to avoid duplication and gaps in services. However, there were considerable changes in state/civil society organization, donor/CSO, and donor/recipient state relations.

First, the relationship between the CSOs and state agencies implementing the Global Fund grants remained strained but equal. The vision of individual ministers on the mandate of the Ministry of Health and its prerogative to supervise and control all organizations working in health care complemented the general discourse about the inefficient use of finances by NGOs. The purely medical perception of health care by individual state authorities has led to additional questioning of the expertise of NGOs and their ability to work with target groups. Common to the post-Soviet region, this perspective is not unique to Kyrgyzstan. Despite these concerns, actors still continued jointly implementing the grants and lobbying for future financing. In contrast to the design phase, the local NGOs were not outnumbered by state organizations and seemed contested but equal partners here.

Second, relations of the Global Fund with the local NGOs were hierarchical. Despite its contribution to civil society participation in the implementation of grants, the financier seems to provide little space for SRs' suggestions. With their roles defined and little space for change, local NGOs are seen merely as implementers of grant activities. This approach is different from the promotion of active participation of NGOs in drafting the country's applications we observed in the design phase.

Third, though complying with the Global Fund's decisions, state organizations demonstrated some resistance during the implementation phase, in contrast to the acceptance without reservations we observed during the design stage. State organizations complied with the Global Fund's decision to keep the UNDP as the PR, as the Ministry of Health could not demonstrate its ability to do so. Still, organizations repeatedly requested the transfer of functions to state institutions and discussed the potential cost-saving in administration by returning the administration of grants to state organizations. Moreover, in response to the Global Fund's repeated request to return the unaccounted-for finances, the Ministry of Health neither acknowledged the inquiries nor returned the missing finances. Unable to obtain finances from the recipient state, the donor cut this amount from its follow-up grant. Though the theoretical assumption about changing power assumes high provider leverage at the beginning of the grant process, this particular finding suggests an increased role of the recipient state in the project implementation phase, as well.

## 8.4   Monitoring

The Global Fund outsources project monitoring to the Local Fund Agent, the Principal Recipient, and the Country Coordinating Mechanism[7]:

---

[7] In addition to these actors, the Global Fund (2003) involves an external auditor that conducts an independent audit of the grants and reports back to the Principal Recipient, Local Fund Agent, and the CCM. This section, however, focuses on the role of the national and international actors working on TB and HIV/AIDS in Kyrgyzstan. For more information about the auditor, see Global Fund (2019).

The Country Coordinating Mechanism is expected to have "strategic oversight" over the grants (IO Partner 4), but this ability depends on the CCM's capacity to do so. The Oversight Committee of the CCM conducts field trips to observe the implementation of the Global Fund project (UNAIDS, 2015b) and discusses the Primary Recipients' progress with programmatic, procurement, and financial indicators (UNDP, 2015a). For this, CCM members are expected to be aware of Global Fund policies and procedures, as well as the financial, procurement, and implementation details of the grant operation process (Sands, 2019). Yet, a study of 50 CCMs (including the one in Kyrgyzstan) found their oversight function "weak," with a need for further improvements (Global Fund Office of the Inspector General, 2016, p. 11). The Kyrgyz CCM received technical and financial assistance from multiple donors, including the European Union, DFID, PEPFAR, USAID, and others (Manukyan & Burrows, 2010). Nevertheless, the CCM's ability to monitor the grants remained relatively weak. Studies on health care aid to Kyrgyzstan point to lack of work plans, problems with analytical work (ibid. p. 14), and CCM members' unawareness of their functions (Spicer et al., 2010, pp. 11–12). These issues culminated in the CCM's inability to oversee the Global Fund grants, resulting in the mismanagement of finances discussed in the "Implementation" section.

The Local Fund Agent (LFA) monitors the grant implementation process by the PR and SRs and reports directly to the Global Fund.[8] Known as the "eyes and ears" of the Global Fund (IO Partner 21), the LFA participates in the CCM meetings, but its interaction with grant implementers remains somewhat limited to ensure the neutrality of its assessment reports. More specifically, the LFA verifies the prices, quantities, and salaries indicated in the programmatic and financial reports of the Primary Recipient (IO Partners 4, 20 and 21). In addition to desk research, it also conducts field trips to evaluate the service coverage and the end receipt of procured goods by grant beneficiaries. In Kyrgyzstan, for instance, there were instances in which commercial sex workers had to pay for the condoms they were entitled to receive for free, and cases where condoms procured within the grants and marked "the Global Fund, not for sale"

---

[8] For more information on LFA selection, see Global Fund (2007).

were sold in local kiosks (IO Partner 21). Based on these accounts, the LFA reports to the Global Fund with suggestions for further grant-related disbursements (Global Fund, 2007). The LFA monitoring results are critical to the continuity of the grants.

In contrast to the LFA, the Principal Recipient participates in designing and implementing the grants, but also monitors the achievement of indicators and takes corrective actions to address the relevant issues. The PR visits the Sub-Recipients of grants to meet grant beneficiaries and identify and solve issues, including those related to the quality of reported data, patient adherence to treatment, and other aspects relevant to the grant indicators (e.g., UNDP, 2015b, 2015d, 2015e). During these meetings, the PR also validates the programmatic and financial data reported by the SRs. There are concerns that the local NGOs misrepresent and manipulate data in their reports (Ancker & Rechel, 2015a). There are no statistics about the frequency of PR visits to Sub-Recipient NGOs, but in 2014 alone, the UNDP conducted 63 field trips to the SRs (UNDP, 2014, p. 21). Based on the monitoring and SRs' reports, the PR submits programmatic and financial reports to the Global Fund, the LFA, and the CCM (Global Fund, 2003) on a quarterly to biannual basis (Grant Performance. Report External Print Version. Kyrgyzstan KGZ-910-G07-T, 2016; Grant Performance Report External Print Version. Kyrgyzstan KGZ-H-UNDP, 2016). These reports aim to demonstrate the progress against the indicators stated in the grant agreement, which is essential to continuous financing from the Global Fund.

Overall, the Sub-Recipients, including state and civil society organizations, provide data for monitoring activities but do not participate otherwise to avoid conflicts of interest. Still, state and civil society organizations monitor each other's activities.[9]

State and civil society organizations share information about each other's activities, except for data on NGO financing. CSOs participate in SWAp meetings in which the Ministry of Health reports about achievements and issues in the national health care program (see Isabekova & Pleines, 2021). A state official interviewed for this research emphasized

---

[9] In addition, the organizations have their own monitoring to assess the achievement of stated indicators, which is not discussed here.

NGOs' reciprocal responsiveness, openness, and readiness to provide the requested material (State Partner 2). However, actors' access to financial information on each other varied. If necessary, the CSOs could request the information, also in terms of public financing, from the relevant ministries (CSO 8). In contrast, government organizations had no right to scrutinize NGO funding until 2021. The former Minister of Health (2014–2018) accused NGOs of receiving almost half of the Global Fund grants but not reporting on their use of funding (Malyševa, 2018). The state official interviewed for this research similarly resented having no right to access the funding information, noting that NGOs are "only accountable to those who finance them" (State Partner 2). Similar concerns were raised in the literature on health aid to Kyrgyzstan, suggesting that local NGOs are accountable to donor organizations that finance and monitor their activities (Spicer et al., 2010) but not project beneficiaries or the government (Ancker et al., 2013; Ancker & Rechel, 2015a).

State organizations are mistrustful of the use of finances by local NGOs. Government organizations perceive CSOs as "foreign agents" and "grant eaters" (CSO 8) rather than equal partners (Murzalieva et al., 2009, p. 55). Exacerbated by the limited access to the data on NGO financing, these accusations are based on two main reasons:

First, there is alleged disproportionality of payments for services provided by NGOs. Several interviewees noted that the salary rates of government officials were not consistent with their workload (State Partner 10 and Academic Partner 2), and that NGO staff received higher salaries compared to health care workers (State Partner 2). To be fair, in addition to their routine workload, state officials may indeed have additional tasks related to health aid provided by donor organizations. The intensive workload, in combination with low salaries, contributes to understaffing and high staff turnover rates, also in the Ministry of Health (see Isabekova & Pleines, 2021). Health care workers in public facilities face similar issues (see the subsection below).

NGOs justified the proportionality of payment to services by referring to the "difficult cases" they take over from the state health care system and the irregular working hours these require. In contrast to general practitioners providing health care services to the general population, NGOs have a small number of patients. Yet these are the "most difficult" cases, including patients with addiction problems (drugs or alcohol) (CSO 8), as well as the homeless (State Partner 4). As a rule, these patients avoid state health care facilities and require more time for care. Therefore, the costs of finding, persuading, and supporting these patients are not comparable to the costs of patients willingly coming to health care facilities (ibid.). The latter will, as a rule, adhere to treatment, but the former require the continuous engagement of health care professionals to do so. A civil society representative in this regard notes that, in contrast to state employees, NGO staff have irregular working hours depending on the project needs and the groups they are targeting (CSO 8).

Second, the mistrust toward NGOs is also driven by the perception that the state institutions "should control" health aid. One former Minister of Health repeatedly restated the role of the state in all matters of citizens' health (Mályševa, 2018) and emphasized that the Ministry of Health had the authority to "control any organization working in health care independently of its form of ownership" (Majdan.kg, 2018). Another state official noted that the Minister's concern over NGO accountability mainly refers to finances because all NGO activities and indicators fully comply with the national health care program (State Partner 4). In any case, the discourse about governmental control over health assistance contributed to continuous discussions about the role of the government in scrutinizing NGOs, resulting in the amendments to the Law of the Kyrgyz Republic "On non-profit organizations." Since June 26, 2021, NGOs are required to report on sources of their financing and the use of these funds (Government of KR, 2021).

Regarding content, the Global Fund aims to coordinate its monitoring activities and indicators with the grant-recipient government and other donor organizations.

First, the Global Fund integrates the monitoring of its grants into national systems by aligning its monitoring requirements with the monitoring and evaluation (M&E) system of a grant-recipient country. The organization asks project implementers to provide national rather than grant-specific M&E to demonstrate the project impact, coverage, and outcome indicators ("Local Fund Agent manual. Section G—Global Fund essentials," 2014). In doing so, it encourages the use of data already gathered by government institutions. State agencies and ministries routinely collect and report the information related to the realization of national health care programs to the Ministry of Health (see Majtieva et al., 2015). The PR is free to use this data as long as it clearly demonstrates the indicators and objectives stated in the project. Further exceptions to the use of the national M&E are the cases with no national system or in which the system is not relevant to the Global Fund grants ("Local Fund Agent manual. Section G—Global Fund essentials," 2014).

The use of national indicators is also intended to strengthen the national M&E systems, though this support is in practice limited to the areas relevant to the grants. The applicants are also encouraged to include support for the national M&E systems in project proposals. The Global Fund may provide assistance in the amount of 5–10% of the total grant financing for data systems, registration, analytical skill development, and other purposes ("Local Fund Agent manual. Section G—Global Fund essentials," 2014, p. 8). In the context of low- and middle-income countries, this assistance may be pivotal to strengthening the national systems. In Kyrgyzstan, for instance, this support resulted in a unified database with common indicators, data collection, and analytical mechanisms (Ancker & Rechel, 2015a). The Government planned to further increase the funding for the national M&E and provide continuous training to specialists. However, in the face of budget deficits, the national system remained "weak" and largely dependent on donor funding (see Government of KR, 2017b, n.p. Majtieva et al., 2015, p. 29).

Second, the Global Fund aims to coordinate its monitoring requirements and activities with other donor organizations to avoid duplications. Still, its emphasis on the visibility of its contribution jeopardizes these attempts. To decrease the burden on grant recipients having to report to multiple donors using different indicators, the organization negotiates the list of common indicators with the WHO, USAID, PEPFAR, and other actors ("Local Fund Agent manual. Section G—Global Fund essentials," 2014). Furthermore, suppose the Global Fund contributes to the national program by pooling its finances together with other donors: in that case, the Primary Recipient of grants may provide a single audit report with all other donors, as long as this audit explicitly indicates the Global Fund's contribution (Global Fund, 2019, p. 12). However, as a rule, the organization does not merge its finances with other donors due to the difficulties with tracking and validating the use of its resources (IO Partner 20). This notion of transparency hinders the Global Fund's attempts to coordinate its monitoring activities with other donors.

Donor visibility and tracking requirements contribute to counting irregularities in the NGO sector. In Kyrgyzstan, there have been cases of double-counting of the target groups due to the multiplicity of donor approaches to the registration of project beneficiaries (Murzalieva et al., 2009). These irregularities in counting may artificially inflate the number of people covered by the services and contribute to inaccurate estimation of the size of the groups targeted by projects (e.g., commercial sex workers, persons living with HIV, and others). Local NGOs register their clients (e.g., project beneficiaries) by using a universal identification code, but the organizations do not share these data with each other and mainly concentrate on collecting the data requested by donors (ibid.). In other words, a person may have received analogous services from multiple NGOs that registered him/her in parallel to each other. As neither NGOs nor donor organizations comprehensively share the reporting data with each other, this double-counting may remain hidden in reports submitted to, and later by, development organizations.

Limited coordination among donors in terms of their monitoring requirements overwhelms civil society organizations, having to deal with various, at times contradictory criteria. After the misappropriation of grant disbursements in multiple countries, including Kyrgyzstan, the Global Fund introduced several changes in its financial reporting requirements. The increased control over finances resulted in the grant recipients spending extensive time and effort on reporting, which affected their grant implementation functions (Benjamin, 2011). Ancker and Rechel (2015a) went even further, suggesting that the NGOs spent more time reporting on projects than actually implementing them. This was true particularly for those that received financing from multiple organizations, and therefore had to comply with various project cycles, reporting forms, indicators, and other requirements of each donor (ibid.). The authors noted that the NGOs felt "torn" between the multiplicity of donor requirements that at times contradicted each other. For instance, the Global Fund stipulated 100% coverage of the groups vulnerable to

HIV, while the United Nations General Assembly Special Session on HIV/AIDS defined a 60% target (ibid.). Still, the organizations were expected to fulfill the indicators to continue receiving finances.

In the government sector, donor visibility and tracking requirements caused problems with quantifying and forecasting demand for medications. The vivid examples hereof were documented in relation to TB medications. There were problems with forecasting and quantifying the drugs in Kyrgyzstan (Manukyan & Burrows, 2010) because the medications are stored, recorded, and reported in separate registers according to their sources of supply (van den Boom et al., 2015). The WHO study suggests the presence of nine registers in one health facility, which made the accurate review of the total quantity of the relevant medications impossible, and due to the lack of a unified electronic database, the personnel in this health institution recorded and reported the quantities by hand (ibid.). Unfortunately, this example is not limited to a single facility. According to a development partner interviewed for this research project, it took almost a year to monitor the overall stock of medications in the country due to the "parallel reporting systems" used by health facilities (IO Partner 4). Overall, donor visibility and tracking requirements increase the burden on health care workers already overwhelmed with routine tasks and responsibilities. It also complicates the quantification and forecasting of medications, in doing so jeopardizing the continuity of treatment.

Overall, donor coordination of monitoring activities remains limited. One state representative notes that organizations do not duplicate each other in terms of their objectives and geographic coverage, but their monitoring visits often repeat each other's efforts. The interviewee reported receiving multiple invitations from various donors to joint monitoring visits to the same area. The state representative agreed to participate in some cases but not in others (State Partner 3).

Similar to implementation, the Global Fund delegates monitoring of its project to local stakeholders in Kyrgyzstan. The Country Coordinating Mechanism, Local Fund Agent, and Primary Recipient complement each other and provide comprehensive coverage of stakeholders (see Diagram 8.1). As Sub-Recipients of grants, the state agencies and local NGOs do not directly participate in the monitoring process to avoid conflicts of

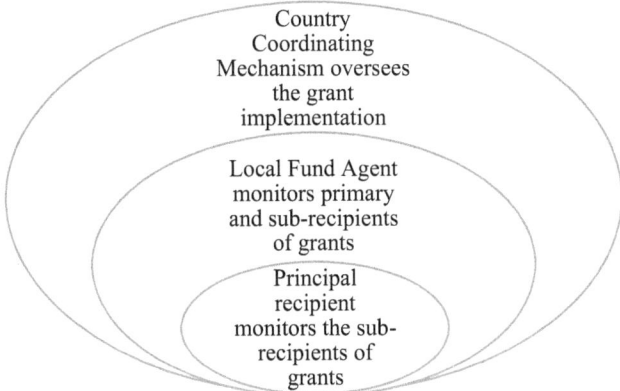

**Diagram 8.1**   The realm of actors monitoring the Global Fund grants

interest. Instead, they end up monitoring each other. As in the implementation phase, NGOs' use of financing remains an issue between the state and civil society organizations.

The Global Fund attempts to coordinate its monitoring with the national M&E system and evaluation activities of other donors. It aligns its monitoring plans with the national programs on TB and HIV/AIDS and contributes to the development of the national M&E system. Though considerable in some areas, this support obviously failed to solve Kyrgyzstan's systemic issues related to staff capacity and budget deficit. Moreover, this support also seems to primarily facilitate the alignment of national systems with grant indicators.

The Global Fund attempts to coordinate its monitoring activities with other donors, but its requirement for the visibility of its contribution hinders these efforts. Other donors have similar issues, demonstrated by the presence of nine registers in one health facility. The actors vest different interests and standards in their M&E, but until donor commitments to harmonize this area materialize, aid recipients continue bearing most of the related costs (Holzscheiter et al., 2012). In Kyrgyzstan, the multiplicity of donor requirements not only increased the burden on the state and civil society organizations, but also affected their abilities to forecast need for medications and record the project beneficiaries.

# References

AFEW Kyrgyzstan. (n.d.). *Garmonizatsiya i konsolidatsiya usiliy dlya bor'by s VICH-infektsiyey i tuberkulezom [Harmonization and consolidation of efforts in fighting HIV infection and tuberculosis]*. Retrieved March 3, 203AD, from http://www.afew.kg/project_GIZ_end_ru.html

Ancker, S., McKee, M., & Rechel, B. (2017). HIV/AIDS discourses in Kyrgyzstan's policy arena. *Global Public Health, 12*(10), 1242–1253. https://doi.org/10.1080/17441692.2017.1344285

Ancker, S., & Rechel, B. (2015a). 'Donors are not interested in reality': The interplay between international donors and local NGOs in Kyrgyzstan's HIV/AIDS sector. *Central Asian Survey, 34*(4), 516–530. https://doi.org/10.1080/02634937.2015.1091682

Ancker, S., & Rechel, B. (2015b). HIV/AIDS policy-making in Kyrgyzstan: A stakeholder analysis. *Health Policy and Planning, 30*, 8–18. https://doi.org/10.1093/heapol/czt092

Ancker, S., Rechel, B., McKee, M., & Spicer, N. (2013). Kyrgyzstan: Still a regional 'pioneer' in HIV/AIDS or living on its reputation? *Central Asian Survey, 32*(1), 66–84. https://doi.org/10.1080/02634937.2013.771965

Andrews, M. (2013). *The limits of institutional reform in development: Changing rules for realistic solutions* (Illustrated Ed.). Cambridge University Press.

Bengard, A. (2017). *Bor'ba za grant dlâ pacientov s VIČ i tuberkulezo [Fight for a grant for patients with HIV and tuberculosis]*. Retrieved February 3, 2023, from https://24.kg/obschestvo/61042_borba_zagrant_dlya_patsientov_svich_ituberkulezom/

Benjamin, H. (2011). *Examining the impact of global fund reforms on implementation: Results of the global fund implementers survey*. Open Society Foundations. Retrieved February 3, 2023, from https://www.opensocietyfoundations.org/sites/default/files/global-fund-implementers-20120305_0.pdf

Brown, J. C., & Griekspoor, W. (2013). Fraud at the Global Fund? A viewpoint. *The International Journal of Health Planning and Management, 28*(1), 138–143. https://doi.org/10.1002/hpm.2152

Burrows, D., Bolotbaeva, A., Sydykanov, B., Iriskulbekov, E., & Dastan uulu Ulan. (2018). *Baseline assessment—Kyrgyzstan: Scaling up programs to reduce human rights-related barriers to HIV and TB services* (pp. 1–94). Global Fund. Retrieved February 4, 2023, from https://www.theglobalfund.org/

media/8145/crg_humanrightsbaselineassessmentkyrgyzstan_report_en.pd
f?u=636809011150000000

Committee on TB and HIV under the Government of KR. (2023). *Sostav
Komiteta—SKK [Composition of the CCM Committee]*. Retrieved March 2,
2023, from http://hivtbcc.kg/pages/members.html

Èrkebaeva, A. (2017). *Global'nyj fond ne doveril Minzdravu grant v $23,5 mln
iz-za «korrupcionnyh riskov» [The Global Fund has not entrusted a $ 23.5 mil
lion grant to the Ministry of Health due to "corruption risks"]*. Retrieved February
3, 2023, from https://kloop.kg/blog/2017/12/07/minzdrav-ne-budet-
upravlyat-grantom-globalnogo-fonda-v-23-5-mln-na-borbu-s-vich-i-
tuberkulezom/

Eurasianet. (2012). *Kyrgyzstan: Donor-funded AIDS project shines light on corruption issue*. Retrieved March 3, 2023, from https://eurasianet.org/
kyrgyzstan-donor-funded-aids-project-shines-light-on-corruption-issue

Friends of the Global Fight Against AIDS, Tuberculosis and Malaria. (2018).
*Global fund accountability mechanisms*. Retrieved February 3, 2023, from
https://www.theglobalfight.org/global-fund-accountability-mechanisms/

Global Fund. (2003). *Fiduciary arrangements for grant recipients*. Retrieve
October 15, 2019, from https://www.theglobalfund.org/media/6025/core_
fiduciary_arrangements_en.pdf?u=636917016190000000

Global Fund. (2006a). *Grant scorecard KGZ-202-G01-H-00* (pp. 1–19).
Retrieved February 3, 2023, from http://docs.theglobalfund.org/program-
documents/GF_PD_004_4eb67380-6953-4743-a19b-63afa9b63159.pdf

Global Fund. (2006b). *Grant scorecard KGZ-202-G02-T-00* (pp. 1–18).
Retrieved March 3, 2023, from http://docs.theglobalfund.org/program-
documents/GF_PD_004_838aba70-e225-427f-8bb6-653793958160.pdf

Global Fund. (2007, December). *The role of the local fund agent (LFA)*. Workshop
on grant negotiation and implementation of TB grants. Retrieved March 30,
2020, from https://www.who.int/tb/events/archive/gf_presentations/14_
lfa_role.pdf

Global Fund. (2008). *Country coordinating mechanisms. Partnership and leadership* (pp. 1–14). Retrieved March 28, 2020, from https://www.theglobal-
fund.org/media/5476/ccm_2008thematicpartnershipleadership_report_en.
pdf?u=637066568340000000

Global Fund. (2014). *Effective TB and HIV control project in Kyrgyzstan. Investing
for Impact against tuberculosis and HIV* (pp. 1–57). Retrieved March 4, 2023,
from https://ecuo.org/mvdev/wp-content/uploads/sites/4/2016/09/
KGZ-C_ConceptNote_0_en.pdf

Global Fund. (2015). *Guidelines on implementers of Global Fund grants.* Retrieved March 28, 2020, from https://www.theglobalfund.org/media/5663/core_guidelinesonimplementers_guideline_en.pdf

Global Fund. (2016a). *35th board meeting. The Global Fund sustainability, transition and co-financing policy.* GF/B35/04—Revision 1 Board Decision (pp. 1–16). Retrieved March 2, 2023, from https://www.theglobalfund.org/media/4221/bm35_04-sustainabilitytransitionandcofinancing_policy_en.pdf

Global Fund. (2016b). *How we engage. Stories of effective community engagement on AIDS, tuberculosis and malaria* (pp. 1–32). Retrieved March 3, 2023, from https://www.theglobalfund.org/media/1547/publication_howweengage_report_en.pdf?u=637066568370000000

Global Fund. (2018). *Country coordinating mechanism policy including principles and requirements.* As approved by the global fund board on 10 May 2018. Retrieved February 3, 2023, from https://www.theglobalfund.org/media/7421/ccm_countrycoordinatingmechanism_policy_en.pdf?u=637066568320000000

Global Fund. (2019). *Guidelines for annual audit of global fund grants* (pp. 1–30). Retrieved February 6, 2023, from https://www.theglobalfund.org/media/6041/core_annualauditsoffinancialstatements_guideline_en.pdf?u=636917015500000000

Global Fund. (2023b). *Technical cooperation.* Retrieved March 3, 2023, from https://www.theglobalfund.org/en/funding-model/throughout-the-cycle/technical-cooperation/

Global Fund. (n.d.-c). *Introduction to the 2017–2019 funding cycle and the differentiated funding application process.* Retrieved February 3, 2023, from http://www.stoptb.org/assets/documents/global/fund/Differentiated%20Approaches%20for%20Countries%20to%20Access%20Funding_Panel.pdf

Global Fund Office of the Inspector General. (2012). *Audit of global fund grants to the Kyrgyz Republic. Executive summary* (pp. ES-1-ES-2). Retrieved February 3, 2023, from https://www.theglobalfund.org/media/2668/oig_gfoig09012auditkyrgyzstan_executivesummary_en.pdf?u=637166002670000000

Global Fund Office of the Inspector General. (2013). *Investigation of global fund grants to Kyrgyzstan. Global fund tuberculosis grants managed by the national center of phthisiology under the ministry of health of the Kyrgyz republic.* GF-OIG-13-023 28 August 2013 (pp. 1–75). Retrieved Accessed March 3,

2023, from https://www.theglobalfund.org/media/2761/oig_gfoig13023in-vestigationkyrgyzstan_report_en.pdf?u=636852021560000000

Global Fund Office of the Inspector General. (2016). *Audit report. The global fund country coordinating mechanism* (pp. 1–24). Retrieve March 3, 2023, from https://www.theglobalfund.org/media/2645/oig_gf-oig-16-004_report_en.pdf

Government of KR. (1995). *Nacional''naâ Programma Kyrgyzskoj Respubliki "Tuberkulez" na 1996–2000 gody [National Program Kyrgyz Republic "Tuberculosis" for 1996–2000].* Retrieved February 3, 2023, from http://cbd.minjust.gov.kg/act/view/ru-ru/36659

Government of KR. (1997). *Nacional''naâ Programma Kyrgyzskoj Respubliki po profilaktike SPIDa i boleznej, peredaûŝihsâ polovym putem, na 1997–2000 gody [National Program of the Kyrgyz Republic on the prevention of AIDS and sexually transmitted diseases for 1997–2000]: Postanovlenie Pravitel'stva Kyrgyzskoj Respubliki ot 1 sentâbrâ 1997 goda № 507 [Decree of the Government of the Kyrgyz Republic dated September 1, 1997 No. 507].* Retrieved February 17, 2023, from http://cbd.minjust.gov.kg/act/view/ru-ru/34692

Government of KR. (2006). *Gosudarstvennaâ programma po preduprеždeniû épidemii VIČ/SPIDa i ee social'no-èkonomičeskih posledstvij v Kyrgyzskoj Respublike na 2006–2010 gody [State Programme on prevention of HIV/AIDS epidemic and social and economic consequences in the Kyrgyz Republic for 2006–2010]: Postanovlenie Pravitel'stva Kyrgyzskoj Respubliki ot 6 iûlâ 2006 goda N 498 [Decree of the Government of the Kyrgyz Republic dated July 6, 2006 N 498].* Retrieved February 17, 2023, from http://cbd.minjust.gov.kg/act/view/ru-ru/57612

Government of KR. (2008). *O hode realizacii postanovleniâ Pravitel'stva Kyrgyzskoj Respubliki ot 6 maâ 2006 goda N 331 "O Nacional'noj programme 'Tuberkulez-III' na 2006–2010 gody" [On the implementation of the Decree of the Government of the Kyrgyz Republic dated May 6, 2006 N 331 "On the National Program 'Tuberculosis-III' for 2006–2010"]: Postanovlenie Pravitel''stva Kyrgyzskoj Respubliki ot 27 avgusta 2008 goda № 476 [The Decree of the Government of the Kyrgyz Republic dated August 27, 2008 N 476].* Retrieved March 5.2 023, from http://cbd.minjust.gov.kg/act/view/ru-ru/6838?cl=kg-kg

Government of KR. (2014). *Postanovlenie Pravitel'stva Kyrgyzskoj Respubliki ot 26 iûnâ 2014 goda № 352 O Koordinacionnom sovete po obŝestvennomu zdravoohraneniû pri Pravitel'stve Kyrgyzskoj Respubliki [Decree of the Government of the Kyrgyz Republic dated June 26, 2014 No. 352 On the Coordinating Council for Public Health under the Government of the Kyrgyz Republic].* Retrieved February 3, 2023, from http://cbd.minjust.gov.kg/act/view/ru-ru/96604? cl=ru-ru

Government of KR. (2017a). *Programma Pravitel'stva Kyrgyzskoj Respubliki po preodoleniû VIČ-infekcii v Kyrgyzskoj Respublike na 2017–2021 gody [The Government of the Kyrgyz Republic Program on Overcoming HIV Infection in the Kyrgyz Republic for 2017–2021]: Priloženie 1 Utverždeno postanovleniem Pravitel'stva Kyrgyzskoj Respubliki ot 30 dekabrâ 2017 goda № 852 [Annex 1 approved by the Decree of the Government of the Kyrgyz Republic dated December 30, 2017 No. 852].* Retrieved February 17, 2023, from http://cbd.minjust. gov.kg/act/view/ru-ru/11590

Government of KR. (2017b). *O Programme Pravitel'stva Kyrgyzskoj Respubliki po preodoleniû VIČ-infekcii v Kyrgyzskoj Respublike na 2017–2021 gody [The program of the Government of the Kyrgyz Republic sight to overcome HIV infection in the Kyrgyz Republic for 2017–2021]: Postanovlenie Pravitel'stva Kyrgyzskoj Respubliki ot 30 dekabrâ 2017 goda № 852 [Decree of the Government of the Kyrgyz Republic dated December 30, 2017 No. 852].* Retrieved February 3, 2023, from http://cbd.minjust.gov.kg/act/view/ru-ru/11589

Government of KR. (2021). *Zakon Kyrgyzskoj Respubliki ot 15 oktâbrâ 1999 goda № 111 "O nekommerčeskih organizaciâh" [Law of the Kyrgyz Republic dated October 15, 1999 No. 111 "On non-profit organizations"].* Retrieved February 4, 2023, from http://cbd.minjust.gov.kg/act/view/ru-ru/274

Grant Performance Report External Print Version. Kyrgyzstan KGZ-202-G01-H-00. (2011) (pp. 1–34). Retrieved March 3, 2023, from http://docs. theglobalfund.org/program-documents/GF_PD_003_d916e133-ccae-4f6d-b57a-29e5ac579c65.pdf

Grant Performance Report External Print Version. Kyrgyzstan KGZ-607-G04-T. (2012) (pp. 1–28). Retrieved March 3, 2023, from http://docs.theglobalfund.org/program-documents/GF_PD_003_e64411eb-4f8d-4d71-a61a-8c627880dcfd.pdf

Grant Performance Report External Print Version. Kyrgyzstan KGZ-H-UNDP. (2016) (pp. 1–44). Retrieved March 3. 2023, from http://docs. theglobalfund.org/program-documents/GF_PD_003_51112a72-8240-4690-985b-ebc1c63e5618.pdf

Grant Performance. Report External Print Version. Kyrgyzstan KGZ-910-G07-T. (2016) (pp. 1–30). Retrieved March 3, 2023, from http://docs.theglobalfund.org/program-documents/GF_PD_003_e99065eb-b1c1-409a-a5f5-e0db338541f2.pdf

Harmer, A., Spicer, N., Aleshkina, J., Bogdan, D., Chkhatarashvili, K., Murzalieva, G., Rukhadze, N., Samiev, A., & Walt, G. (2013). Has global fund support for civil society advocacy in the former Soviet Union established meaningful engagement or "a lot of jabber about nothing"? *Health Policy and Planning, 28*, 299–308. https://doi.org/10.1093/heapol/czs060

Health Focus. (2020). *Joint research project: Improving governance and strategic coordination of HIV and TB programmes in Kyrgyzstan.* Retrieved February 3, 2023, from https://www.health-focus.de/completed/asia-2/joint-research-project-improving-governance-and-strategic-coordination-of-hiv-and-tb-programmes-in-kyrgyzstan

Holzscheiter, A., Walt, G., & Brugha, R. (2012). Monitoring and evaluation in global HIV/AIDS control—Weighing incentives and disincentives for coordination among global and local actors. *Journal of International Development, 24*(1), 61–76. https://doi.org/10.1002/jid.1705

Institute for Health Metrics and Evaluation. (2023). Retrieved March 5, 2023, from https://www.healthdata.org/kyrgyzstan

Isabekova, G., & Pleines, H. (2021). Integrating development aid into social policy: Lessons on cooperation and its challenges learned from the example of health care in Kyrgyzstan. *Social Policy & Administration, 55*(6), 1082–1097. https://doi.org/10.1111/spol.12669

Ismanov, A. (2015). *Global'nyj fond bor'by so SPIDom trebuet, čtoby Kyrgyzstana vozmestil 120 tysâč dollarov [The Global Fund requests Kyrgyzstan to reimburse 120 thousand dollars].* Retrieved March 3, 2023, from http://www.nlkg.kg/ru/society/health/globalnyj-fond-borby-so-spidom-trebuet_-chtoby-kyrgyzstana-vozmestil-120-tysyach-dollarov

Kasmalieva, A. (2015). *Global'nyj fond trebuet ot KR pogašeniâ dolga [The Global Fund demands the Kyrgyz Republic to repay the debt].* Radio Azattyk, p. n.p.

Local Fund Agent manual. Section G—Global Fund essentials. (2014). Retrieved March 2, 2023, from https://www.theglobalfund.org/media/3238/lfa_manual09sectiong_manual_en.pdf?u=636709997320000000

Majdan.kg. (2018). *«Interv'û s ministrom goda—2017». Talantbek Batyraliev, ministr zdravoohraneniâ KR: «Dolžen že byt' v sfere zdravoohraneniâ čelovek, kotoryj obrušit gniûŝuû sistemu i postroit novuû!» [Interview with the Minister of the Year 2017". Talantbek Batyraliev, Minister of Health of the Kyrgyz Republic: "There must be a person in the health sector who will bring down the decaying system and build a new one!"].* gezitter.org. Retrieved March 5, 2023, from https://www.gezitter.org/interviews/66841_intervyu_s_ministrom_goda_-_2017_talantbek_batyiraliev_ministr_zdravoohraneniya_kr_doljen_ je_byit_v_sfere_zdravoohraneniya_chelovek_kotoryiy_obrushit_gniy-uschuyu_sistemu_i_postroit_novuyu/

Majtieva, V. S., Čokmorova, U. Zh., Ismailova, A. D., Asybalieva, N. A., Ânbuhtina, L. F., Sarybayeva, M. E., et al. (2015). *Stranovoj otčet o dostignu-tom progresse v osuŝestvlenii global'nyh mer v otvet na vič-infekciû za 2014 god [Kyrgyzskaâ Respublika] [2014 Country Progress Report on the Global Response to HIV [Kyrgyz Republic]]* (pp. 1–29). Ministry of Health, Republican AIDS Center, UNAIDS, WHO, UNICEF. Retrieved February 3, 2023, from http://www.unaids.org/sites/default/files/country/documents/KGZ_narra-tive_report_2015.pdf

Malyševa, V. (2018). *Polovina grantov, postupaûŝih na zdravoohranenie, «s" edaetsâ» [Half of the grants to healthcare are 'eaten']*. VESTI.KG. Retrieved February 4, 2023, from https://vesti.kg/politika/item/49867-polovina-grantov-postupayushchikh-na-zdravookhranenie-s-edaetsya.html

Mansfeld, M., Ristola, M., & Likatavicius, G. (2015). *HIV/AIDS programme in Kyrgyzstan. Evaluation report* (pp. 1–84). WHO/Europe; Centre for Health and Infectious Disease Research. Retrieved February 17, 2023, from http://www.euro.who.int/__data/assets/pdf_file/0005/273308/HIV-Programme-Review-in-Kyrgyzstan.pdf?ua=1

Manukyan, A., & Burrows, D. (2010). *Country-level partnership case study—Kyrgyzstan. For The Global Fund to Fight AIDS, TB and Malaria* (pp. 1–27). AIDS Projects Management Group. Retrieved November 10, 2019, from http://apmglobalhealth.com/project/country-case-study-partnerships-kyrgyzstan

Minus Virus. (2017). *Ajbar Sultangaziev: U Minzdrava Kyrgyzskoj Respubliki byli vse šansy načat' upravlât' sredstvami Global'nogo fonda [Aibar Sultangaziyev: The Ministry of Health of the Kyrgyz Republic had all chances to start managing the Global Fund's funds]*. Retrieved February 2, 2023, from http://mv.ecuo.org/ajbar-sultangaziev-u-minzdrava-kyrgyzskoj-respubliki-byli-vse-shansy-nachat-upravlyat-sredstvami-globalnogo-fonda/

Murzalieva, G., Aleshkina, J., Temirov, A., Samiev, A., Kartanbaeva, N., Jakab, M., Spicer, N., & Network, G. H. (2009). *Tracking global HIV/AIDS initia-tives and their impact on the health system: The experience of the Kyrgyz Republic: Final report* (pp. 1–89). Royal College of Surgeons in Ireland. Retrieved March 4, 2023, from https://repository.rcsi.com/articles/report/Tracking_Global_HIV_AIDS_Initiatives_and_their_Impact_on_the_Health_System_the_experience_of_the_Kyrgyz_Republic/10776524/1

Murzalieva, G., Kojokeev, K., Manjieva, E., Akkazieva, B., Samiev, A., Botoeva, G., Ablezova, M., & Jakab, M. (2007). *Tracking global HIV/AIDS initiatives and their impact on the health system: The experience of the Kyrgyz Republic: Context report* (pp. 1–48). Center for Health System Development; American University of Central Asia. Retrieved March 3, 2023, from http://elibrary.auca.kg/bitstream/123456789/220/1/Tracking%20Global%20HIV-AIDS%20Initiatives_AUCA.pdf

Nelson, L. J. (n.d.). *Preparing for a single TB and HIV concept note: What is new in the global fund and opportunities?* Retrieved March 27, 2020, from https://www.who.int/tb/challenges/hiv/joint_planning_and_single_tb_and_hiv_concept_note.pdf

Office of the Attorney General of KR. (2012). *General'noj prokuraturoj Kyrgyzskoj Respubliki vozbuždeno ugolovnoe delo v otnošenii dolžnostnyh lic Ministerstva zdravoohraneniâ KR [The General Prosecutor's Office of the Kyrgyz Republic has initiated criminal proceedings against the officials of the Ministry of Health].* Retrieved March 28, 2020, from https://www.prokuror.kg/news/422-generalnoj-prokuraturoj-kyrgyzskoj-respubliki-vozbuzhdeno-ugolovnoe-delo-v-otnoshenii-dolzhnostnykh-lits-ministerstva-zdravookhra neniya-kr.html

Sands, P. (2019). Putting country ownership into practice: The Global Fund and country coordinating mechanisms. *Health Systems & Reform, 5*(2), 100–103. https://doi.org/10.1080/23288604.2019.1589831

Semerik, O., Berdsli, K., Datar, A., & Dad'yan, M. (2014). *Analiticheskiy obzor rekomendatsiy v sfere VICH-infektsii dlya Kazakhstana, Kyrgyzskoy Respubliki i Tadzhikistana (2007–2012).* Health Policy Project Futures Group. Retrieved February 5, 2023, from https://www.healthpolicyproject.com/pubs/205_RusHPPFinaldraftFORMATTED.pdf

Spicer, N., Aleshkina, J., Biesma, R., Brugha, R., Caceres, C., Chilundo, B., Chkhatarashvili, K., Harmer, A., Miege, P., Murzalieva, G., & Ndubani, P. (2010). National and subnational HIV/AIDS coordination: Are global health initiatives closing the gap between intent and practice? *Globalization and Health, 6*, 3. https://doi.org/10.1186/1744-8603-6-3

Spicer, N., Bogdan, D., Brugha, R., Harmer, A., Murzalieva, G., & Semigina, T. (2011a). "It's risky to walk in the city with syringes": Understanding access to HIV/AIDS services for injecting drug users in the former Soviet Union countries of Ukraine and Kyrgyzstan. *Globalization and Health, 7*, 22. https://doi.org/10.1186/1744-8603-7-22

Spicer, N., Harmer, A., Aleshkina, J., Bogdan, D., Chkhatarashvili, K., Murzalieva, G., Rukhadze, N., Samiev, A., & Walt, G. (2011b). Circus monkeys or change agents? Civil society advocacy for HIV/AIDS in adverse policy environments. *Social Science & Medicine, 73*(12), 1748–1755. https://doi.org/10.1016/j.socscimed.2011.08.024

UNAIDS. (2015a). *How AIDS changed everything. MDG 6: 15 years, 15 lessons of hope from the AIDS response* (pp. 1–543). Retrieved March 2, 2023, from https://www.unaids.org/sites/default/files/media_asset/MDG6 Report_en.pdf

UNAIDS. (2015b). *Stranovoy otchet o dostignutom progresse v osushchestvlenii global'nykh mer v otvet na vich infektsiyu za 2014 god. Kyrgyzskaya respublika, otčetnyj period: ânvar'—dekabr' 2014g. [Country report on the achieved progress in implementing global measures in response to HIV infection for the year 2014. The Kyrgyz Republic, reporting period: January–December 2014]* (pp. 1–29). Retrieved February 4, 2023, from https://www.unaids.org/sites/default/files/country/documents/KGZ_narrative_report_2015.pdf

UNDP. (2014). *Annual report on the implementation of grants provided by the Global Fund to fight AIDS, Tuberculosis and Malaria in Kyrgyzstan—2013* (pp. 1–70). UNDP. Retrieved February 3, 2023, from https://www.kg.undp.org/content/kyrgyzstan/en/home/library/hiv_aids/annual-report-on-the-implementation-of-grants-provided-by-the-gl.html

UNDP. (2015a). *Annual report on the implementation of UNDP project in support of the Government of the Kyrgyz Republic, funded by The Global Fund to Fight AIDS, Tuberculosis and Malaria—2014* (pp. 1–108). UNDP. Retrieved February 3, 2023, from https://www.kg.undp.org/content/kyrgyzstan/en/home/library/hiv_aids/gfatmannualreport_eng.html

UNDP. (2015b). *Newsletter: Grants on HIV, TB and malaria | UNDP in Kyrgyz Republic.* November 2014 (pp. 1–11). Retrieved February 4, 2023, from https://www.kg.undp.org/content/kyrgyzstan/en/home/library/hiv_aids/april-2014-newsletter%2D%2Dgrants-on-hiv%2D%2Dtb-and-malaria1.html

UNDP. (2015c). *Newsletter: Grants for HIV, tuberculosis and Malaria.* January 2014 (pp. 1–10). Retrieved March 26, 2020, from https://www.undp.org/content/dam/kyrgyzstan/Publications/hiv-tb-malaria/2014/kgz_Newsletter%20UNDP%20GF_January%202014_ENG.pdf

UNDP. (2015d). *Newsletter grants for HIV, tuberculosis and Malaria.* July 2014 (pp. 1–11). Retrieved March 26, 2020, from https://www.undp.org/content/dam/kyrgyzstan/Publications/hiv-tb-malaria/2014/kgz_Newsletter%20UNDP%20GF_July%202014_ENG.pdf

UNDP. (2015e). *Newsletter: Grants on HIV, TB and malaria.* June 2014 (pp. 1–10). Retrieved February 7, 2023, from https://www.kg.undp.org/content/kyrgyzstan/en/home/library/hiv_aids/june-2014-newsletter%2D%2Dgrants-on-hiv%2D%2Dtb-and-malaria.html

UNDP. (2018). *Partnership with the global fund.* Retrieved October 11, 2019, from https://www.undp-capacitydevelopment-health.org/en/about-us/partners/global-fund-partnership/

van den Boom, M., Mkrtchyan, Z., & Nasidze, N. (2015). *Review of tuberculosis prevention and care services in Kyrgyzstan 30 June–5 July 2014 Mission report* (pp. 1–95). Retrieved February 17, 2023, from http://www.euro.who.int/__data/assets/pdf_file/0010/287803/Review-of-tuberculosis-prevention-and-care-services-in-Kyrgyzstan.pdf?ua=1

Vujicic, M., Weber, S. E., Nikolic, I. A., Atun, R., & Kumar, R. (2011). GAVI, the global fund and the world bank support for human resources for health in developing countries. *HNP Discussion Paper,* 1–16.

WHO. (1995). *Paris AIDS summit (1 December 1994): Report by the director-general.* Retrieved March 4, 2023, from https://apps.who.int/iris/bitstream/handle/10665/172199/EB95_60_eng.pdf?sequence=1

WHO/Europe. (2019). *Governance snapshot: Whole-of-government approach.* Coordinating Council on Public Health (pp. 1–4). Retrieved February 3, 2023, from https://www.euro.who.int/__data/assets/pdf_file/0017/412820/Kyrgyzstan-snapshot-Coordinating-Council-for-Public-Health-CCPH.pdf

Zardiashvili, T., & Garmaise, D. (2017). *Kyrgyzstan's program continuation funding request to the Global Fund provides little information on the proposed program.* Retrieved April 19, 2019, from http://www.aidspan.org/gfo_article/kyrgyzstan%E2%80%99s-program-continuation-funding-request-global-fund-provides-little

# 9

# Sustainability of Global Fund Grants

The grants of the Global Fund to Fight AIDS, Tuberculosis and Malaria (the Global Fund) provided to the Kyrgyz Republic are still ongoing, but the country is preparing to transition from its assistance. This chapter discusses the sustainability of Global Fund grants provided to the country by expanding on the continuity of project activities, maintenance of benefits, and community capacity building within the grants. As discussed in Chap. 3, within ongoing projects, the continuity of project activities and maintenance of benefits received by the targeted population refer to the services taken over by a donor, recipient state, or civil society organization. Community capacity building, in its turn, implies the leadership of civil society organizations, their ability to continue their work and mobilize the necessary resources for it. In addition to discussing the three components of sustainability, this chapter also examines the significance of the factors relevant to these components, such as the commitment of the recipient state, quality of services, and financing. This chapter commences with a description of the grants and major activities stipulated by them.

© The Author(s) 2024

G. Isabekova, *Stakeholder Relationships And Sustainability*, Global Dynamics of Social Policy, https://doi.org/10.1007/978-3-031-31990-7_9

## 9.1   Description of Grants

The Global Fund is among the largest financiers of activities targeting tuberculosis (TB), human immunodeficiency virus infection, and acquired immunodeficiency syndrome (HIV/AIDS) problems worldwide. In the Kyrgyz Republic, it covered the costs of antiretroviral therapy (ART), treatment of TB and HIV/AIDS coinfection, prevention of mother-to-child transmission of HIV (Ancker et al., 2013), TB medications, laboratory reagents, and more (State Partner 9). The organization also provided equipment and staff training to support health systems strengthening in the country (Murzalieva et al., 2009).

The Global Fund provided multiple grants to facilitate TB control in the country.[1] The first grant (2004–2009) aimed to prevent the disease by training medical specialists, providing treatment and detection of TB, increasing the awareness of TB in the civilian and penitentiary sectors, and other activities (Grant Performance Report External Print Version. Kyrgyzstan KGZ-202-G01-H-00, 2011). The following grant (2007–2012) provided training to health care workers, social support[2] to TB patients, and quality control in the labs to integrate TB services into primary health care (PHC) (Grant Performance Report External Print Version. Kyrgyzstan KGZ-607-G04-T, 2012). It also offered directly observed treatment short-course (DOTS) for both drug-susceptible and drug-resistant forms of TB for the patients in prisons (ibid.). The third grant (2011–2015) intended to increase TB detection and treatment in the country by implementing drug-resistance surveillance and improving the regulatory basis for service delivery (Grant Performance. Report External Print Version. Kyrgyzstan KGZ-910-G07-T, 2016). It stipulated social support for TB patients, training of medical workers and the Village Health Committees, and other activities (ibid.). The following grant (2011–2016) intended to consolidate the DOTS framework by improving the detection and treatment of all forms of TB, providing quality control in the labs, training TB specialists, and so on (Grant

---

[1] For more information on the list of grants included in the analysis, see Appendix 1.

[2] This includes psychological support as well as financial and in-kind benefits (reimbursement of travel costs, food and hygiene parcels, etc.).

Performance Report External Print Version. Kyrgyzstan KGZ-S10-G08-T, 2016). Kyrgyzstan was also among the eleven countries benefiting from the TB Regional Eastern Europe and Central Asia Project on Strengthening Health Systems for Effective TB and drug-resistant TB Control, financed by the Global Fund. This grant aimed to reduce TB-related stigma and discrimination, promote people-centered outpatient care, and facilitate the participation of persons affected by this disease in the decision-making process (Amanzholov et al., 2018). The Kyrgyz NGO "Association of AIDS Service NGOs of the Kyrgyz Republic Anti-Aids" promoted outpatient TB care among the decision-makers, the general public, and health care professionals (ibid.).

In parallel with TB grants, the Global Fund also provided grants to control HIV/AIDS in the country. The first grant (2004–2009) focused on HIV prevention among the general population and the groups most vulnerable to this disease, including commercial sex workers (CSWs), persons who inject drugs (PWID), men who have sex with men (MSM), and others (Global Fund, 2006a). The grant stipulated a wide range of activities, such as establishing needle-exchange points, providing methadone treatment and antiretroviral therapy (ART), HIV prevention and counseling, training for journalists and health care workers, and other activities (Grant Performance Report External Print Version. Kyrgyzstan KGZ-202-G01-H-00, 2011). The following grant (2011–2016) aimed to achieve universal access to prevention, diagnosis, and treatment, particularly among vulnerable groups (Grant Performance Report External Print Version. Kyrgyzstan KGZ-H-UNDP, 2016). It stipulated HIV prevention, ART, training of health care workers, and capacity building of communities affected by the disease (ibid.).[3] In 2016, the country received a joint grant for TB and HIV/AIDS (2016–2023) aimed at universal access to TB and HIV diagnostics, treatment, and care (UNDP, 2023). In addition to TB and HIV/AIDS prevention, testing, and treatment, this grant emphasized treating coinfection of these two diseases (Grant Performance Report External Print Version: Kyrgyzstan

---

[3] Communities/community organizations/community-based organizations in the Global Fund grants refer to persons affected by TB and HIV/AIDS and the organizations working with them. This operationalization used by the Global Fund is different from the one presented in relation to the "Community Action for Health" project, in which communities are persons living in the same area.

KGZ-C-UNDP, 2016) and achieving sustainability of the national programs targeting them (UNDP, 2023).

## 9.2   Continuity of Project Activities

This section discusses the continuity of grant activities by elaborating on the types (what) and the extent of activities related to the treatment and prevention of TB and HIV/AIDS. It also discusses the factors relevant to the continuity of activities, namely, the formal character of state support, financing, the epidemiological situation in the country, and the quality of services.

Regarding HIV/AIDS, I focus on grant activities related to prevention (testing, condom distribution, needle- and syringe-exchange program, opioid substitution therapy) and treatment services (ART and treatment of sexually transmitted infections).

First, the Global Fund grants increased the breadth of prevention activities. The grants contributed to the establishment of testing services at NGOs and state health care facilities. This included the provision of consent-based testing for pregnant women and children under five at health care facilities (Murzalieva et al., 2009) and the establishment of rapid saliva-based HIV testing at nongovernmental organizations (NGOs) for the groups unwilling to receive the services at state health care facilities. As of 2015, 20 NGOs and 63 state health care institutions provided saliva and capillary blood-based HIV testing free of charge (Mansfeld et al., 2015, pp. 9, 16). As a result, the amount of HIV testing in the country increased (European Centre for Disease Prevention and Control and WHO/Europe, 2019). The number of HIV tests in 2020 alone reached 32,299 (Global Fund, n.d.-b). There are concerns that this increase is primarily attributed to extensive testing among pregnant women, not of the groups most vulnerable to HIV (Mansfeld et al., 2015; Semerik et al., 2014). Nevertheless, in 2020, HIV tests taken among MSM, CWS, and PWIDs cumulatively represented 82% of the total number of HIV tests (Global Fund, n.d.-b). Therefore, the plausibility of concerns about insufficient testing among vulnerable groups requires further research.

Second, the grants contributed to condom distribution among the groups most vulnerable to HIV, but the availability and use of condoms in the country remained limited. In 2014 alone, about 1.5 million condoms were distributed at the expense of the Global Fund grants. Yet, these condoms had a "supplemental" character, and they did not meet the needs of grant beneficiaries (UNDP, 2015d, p. 4). The availability of condoms in prison settings also remained limited (Burrows et al., 2018). Unmet needs contributed to the irregular use of condoms. For example, a survey conducted within the framework of the Global Fund grants showed that the CSWs did not use condoms with their regular sexual partners or if a client paid extra (UNDP, 2015d).

Third, the Global Fund expanded the needle and syringe exchange program (NSP) in the country, but the coverage of this program remained limited due to the criminalization of drug use. Kyrgyzstan introduced the NSP in 1999 with the support of the Soros Foundation Kyrgyzstan, the United Nations Development Programme (UNDP), and Joint United Nations Programme on HIV/AIDS (UNAIDS) (Wolfe, 2005). By the end of 2004, the program covered twelve prisons and two large cities— Bishkek and Osh (ibid.). The Global Fund grants expanded the NSP further by including all pre-trial detention centers, open prisons, and ten large and ten small cities into the program (Murzalieva et al., 2009). As of 2015, there were 31 state and 15 nongovernmental organizations and eight pharmacies offering NSP services to the population (Foundation for AIDS Research, 2015, p. 18). In 2014 alone, seven million syringes and needles were distributed at the expense of the grants (UNAIDS, 2015).

Nevertheless, the NSP services covered only 36% of PWID (Grant Performance Report External Print Version. Kyrgyzstan KGZ-202-G01-H-00, 2011, p. 24). Despite the high demand (Murzalieva et al., 2009), the actual use of services was limited due to the criminalization of PWID for possession of used syringes (Spicer et al., 2011a). There were cases in which police officers confiscated NSP supplies of outreach workers (Wolfe, 2005),[4] arrested them for carrying needles (Mansfeld et al., 2015), and asked for bribes (Spicer et al., 2011a) and/or information

---

[4] Outreach workers are the NGO employees providing HIV testing, NSP, and other services to the groups most vulnerable to HIV.

about the grant beneficiaries using the NSP services (Murzalieva et al., 2009). Police officers' activities contributed to the high attrition of outreach workers (ibid.) and the distribution of new syringes without exchanging the old ones (Spicer et al., 2011a). Despite the countrywide expansion, the actual use of NSP services by persons who inject drugs remained limited.

Similarly, the use of opioid substitution therapy (OST) services remained limited due to the harassment of and discrimination against persons who inject drugs. Kyrgyzstan is among the few post-Soviet countries offering methadone maintenance treatment to opiate addicts (Wolfe, 2005). The Global Fund expanded the maintenance therapy, which was initially financed by the Soros Foundation and the UNDP (ibid.), by ensuring continuous financing and supply of methadone (Murzalieva et al., 2009). Between 2008 and 2015, the number of health care facilities providing OST in civil and penitentiary sectors more than doubled, increasing from 13 to 31 (Subata et al., 2016, pp. 1–4).

Still, methadone remains "an extremely controlled substance" (Wolfe, 2005, pp. 46–47), since possession of drugs is illegal and may result in a fine or imprisonment of up to four years (Foundation for AIDS Research, 2015). PWID willing to receive the OST are required to undergo registration at narcological centers (ibid.) and come to health care facilities on a daily basis for the therapy (Mansfeld et al., 2015). The coverage of OST services remains low (18% of all PWID) due to the negative attitude of medical staff, policy-makers, and some PWID toward these services (ibid., pp. 33–34) as well as social disapproval (Semerik et al., 2014). There are also cases of harassment (Subata et al., 2016), detention (Spicer et al., 2011a), and police officers' use of withdrawal syndrome to torture the PWID receiving the OST services (Foundation for AIDS Research, 2015).

Overall, the Global Fund grants contributed to the expansion of HIV prevention activities, such as HIV testing, condom distribution, NSP, and OST, in the Kyrgyz Republic. But the outreach of these activities, particularly among the groups most vulnerable to HIV, seems unclear.

In addition to preventive services, the Global Fund grants contributed to the treatment of sexually transmitted infections (STIs) and the introduction of antiretroviral therapy in Kyrgyzstan, though the ART coverage and patients' adherence to it are still low. The Global Fund also contributed to the introduction of ART in 2005, which was not previously available in the country (Murzalieva et al., 2009). The National AIDS centers offered limited immune monitoring to persons living with HIV (Wolfe, 2005). The Global Fund supported the revision of clinical protocols on HIV treatment and the provision of relevant training to medical workers (Murzalieva et al., 2009). ART is provided by all AIDS centers and 76 PHC facilities throughout the country (UNDP, 2015a, p. 29). In 2020, 4435 persons received ART (Global Fund, n.d.-b). 100% of pregnant women with HIV and 72% of children born to them receive ART (Grant Performance Report External Print Version: Kyrgyzstan KGZ-C-UNDP, 2016, p. 25), yet only half of registered HIV cases are covered by the therapy (Government of KR, 2017a). Persons living with HIV (PLHIV) often reject the treatment (UNDP, 2015a) due to its side effects, potential interruption of drug supplies, and misperception of ART as a "new drug trial" (Murzalieva et al., 2009, p. 82). In addition to limited coverage, there are issues with poor knowledge of PHC workers about the therapy (Mansfeld et al., 2015), stigma around and discrimination against PLHIV (Murzalieva et al., 2009), patients' non-adherence to treatment (Semerik et al., 2014) and development of acquired antiretroviral drug resistance (Masikini & Mpondo, 2015). Similar to prevention, the outreach of treatment activities remains an issue, often due to factors lying beyond grant activities.

Along with targeting HIV/AIDS, the Global Fund grants contributed to TB prevention in Kyrgyzstan by improving lab services, training health care workers, and increasing the awareness of the population about this disease. The grants stipulated equipment (Grant Performance Report External Print Version. Kyrgyzstan KGZ-202-G02-T-00, 2011) and quality assurance measures in the labs, including improved lab safety, appropriate collection and analysis of specimens (Grant Performance Report External Print Version. Kyrgyzstan KGZ-607-G04-T, 2012; Grant Performance Report External Print Version. Kyrgyzstan KGZ-S10-G08-T, 2016), and training for lab technicians (Grant Performance Report External Print Version. Kyrgyzstan KGZ-202-G02-T-00, 2011). The grants aimed to improve TB detection at PHC facilities by providing the relevant training to general practitioners (Grant Performance Report

External Print Version. Kyrgyzstan KGZ-607-G04-T, 2012; Grant Performance Report External Print Version. Kyrgyzstan KGZ-S10-G08-T, 2016). Similar activities were initiated in the health care facilities in prisons to improve the identification of TB patients among detainees. These activities contributed to the detection of about 1700 new smear-positive TB cases annually (Grant Performance Report External Print Version. Kyrgyzstan KGZ-202-G02-T-00, 2011; Grant Performance Report External Print Version. Kyrgyzstan KGZ-607-G04-T, 2012). The grants also covered information and educational campaigns on TB among the population through media outlets, schools, and detention centers (Grant Performance Report External Print Version. Kyrgyzstan KGZ-202-G02-T-00, 2011). It should, however, be noted that awareness-raising activities fighting against the stigmatization of and discrimination against TB patients were not explicitly stated in the grants.

In addition to prevention, the Global Fund grants contributed to the consolidation of the DOTS throughout the country. The grants stipulated the expansion of DOTS (against drug-susceptible TB) and DOTS-plus (against drug-resistant forms) in the civilian and penitentiary sectors (Grant Performance Report External Print Version. Kyrgyzstan KGZ-607-G04-T, 2012; Grant Performance Report External Print Version. Kyrgyzstan KGZ-S10-G08-T, 2016). The Global Fund guaranteed a continuous supply of TB medications, restructured storage facilities (Government of KR, 2013), and provided training to TB specialists on storage, quantification, and forecasting of drugs. It also financed the establishment and refurbishment of PHC service delivery points (Grant Performance Report External Print Version. Kyrgyzstan KGZ-607-G04-T, 2012). This integration of TB services into primary health care facilities contributed to the development and availability of outpatient care throughout the country. In addition to achieving timely detection and quality treatment (ibid.), the grants aimed to increase patients' adherence to TB treatment through counseling and follow-up of patients by NGO volunteers and medical workers (Grant Performance Report External Print Version. Kyrgyzstan KGZ-202-G02-T-00, 2011; Grant Performance Report External Print Version. Kyrgyzstan KGZ-S10-G08-T, 2016). In 2020, 4435 individuals with TB received treatment (Global Fund, n.d.-b).

Regarding the extent of activities ("to what extent"), TB and HIV/ AIDS programs vary in their readiness to transition to purely state-budget funding. The following sub-sections take a closer look at the factors affecting the government's compliance with its commitment to continue TB and HIV/AIDS-related services beyond the duration of the grants.

HIV prevention activities largely depend on the Global Fund, but the government took multiple steps to take over the financing. The initial state contribution to HIV prevention was insignificant. It included some parts of lab services (Gulgun Murzalieva et al., 2007), operation and maintenance costs of health care facilities, and medical workers' salaries (Maytiyeva et al., 2015). State financing did not extend to HIV prevention among vulnerable groups (International Charitable Organization "East Europe and Central Asia Union of People Living with HIV," n.d., p. 13). Condom distribution, NSP, and OST relied entirely upon the Global Fund (see Mansfeld et al., 2015), also illustrated by the interruption of services and supplies during the delays of grant disbursements (Murzalieva et al., 2009; Semerik et al., 2014). However, following the Global Fund's request to gradually transfer the grant activities to domestic or "alternative" sources of financing (Global Fund, n.d.-a, pp. 13–14), the government started increasing its contribution to HIV.

The government's commitments to HIV/AIDS services are outlined in related state programs. However, as of the beginning of December 2022, the Draft Programme of the Kyrgyz Republic on Combating HIV Infection for 2022–2026 was still not available. Therefore, in addition to interviews with stakeholders conducted in 2016 and 2018 (see Chap. 1), the analysis is based on the previous national program (2017–2021) and recent sustainability assessments provided by organizations such as the Eurasian Harm Reduction Association.

It should be noted that the government took extensive responsibility to increase its contribution to HIV prevention and treatment. In terms of preventive activities, the National HIV Program for 2017–2021 stipulated increased state financing for methadone (from 50 to 100%) and distribution of condoms among the groups vulnerable to HIV to cover at least half of their needs (Government of KR, 2017b). In terms of treatment, the Kyrgyz government has committed itself to providing ART and STI treatment to groups vulnerable to HIV and ART to HIV-positive pregnant women and children born to them (ibid.). Overall, the government aimed to increase the number of individuals on ART fourfold (from 2109 to 8644) and achieve adherence to treatment for no less than 12 months for 90% of patients on ART (ibid.).

Among the sources of financing, the government defined the contributions of national and local authorities. Thus, the Mandatory Health Insurance Fund (MHIF) and the Ministry of Health were responsible for procuring methadone (Government of KR, 2017a). Similarly, the local self-governments in the Osh and Chui regions contributed to HIV prevention services in their areas by providing 20% of necessary funding (by agreement) (Government of KR, 2017b). In addition, the program stipulated an increase in financing for antiretroviral (ARV) drugs and test systems from 10% to 50% between 2018 and 2020 (Government of KR, 2017a).

However, despite multiple sources of financing, the program was accompanied by a considerable budget deficit (33%) (Eurasian Harm Reduction Association, 2021, pp. 21–22) due to a substantial decrease in Global Fund grants. Nevertheless, despite the reductions by almost half, Global Fund grants represented 48% of funding, followed by the state budget (23%), the President's Emergency Plan for AIDS Relief (PEPFAR) (15%), and other donors (13%) (ibid.). The state acknowledged that insufficient financing due to reductions in and possible termination of Global Fund grants and other donors' assistance might jeopardize HIV services in the country (Government of KR, 2017a).

The state fulfilled its commitments but with mixed results. Eurasian Harm Reduction Association (2021) assessment indicates considerable progress in HIV diagnosis and treatment, human rights, and related barriers. This included improvement in HIV-related incidence and morbidity, awareness of HIV status (also among vulnerable groups), and the share of PLHIV on ART who have suppressed viral loads at the end of the reporting period (Eurasian Harm Reduction Association, 2021). The assessment also demonstrated improvements in reducing the stigma and discrimination against groups vulnerable to HIV and improved coverage of HIV prevention services (ibid., pp. 33–48). However, the achievements in other areas were less impressive. For example, the awareness of HIV status among CSWs and the share of CSWs receiving ART remained low, and the use of opioid agonist therapy in vulnerable groups decreased (ibid.). In this regard, the assessment notes that despite the improvements, stigma and discrimination continue to jeopardize access to health, vividly demonstrated by low coverage of treatment and prevention services, particularly among some groups (e.g., CSWs) (ibid.).

It should be noted that the government lived up to its financial commitments. It increased the funding for TB and HIV by 169 million KGS (around €1,812,236) in 2017–2020 and committed itself to providing an additional 280 million KGS (€3,002,521) for the 2021–2023 period (Eurasian Harm Reduction Association, 2021, pp. 23–25).[5] HIV expenses represented 80 million in additional funding, which allowed for the procurement of some ARV drugs, payments to medical professionals, and social contracting (ibid.). However, despite the considerable increase, state funding is insufficient to purchase second-line ARV drugs, rapid tests, and CD4 tests used to assess viral load (ibid.). Moreover, prevention services in vulnerable groups are still largely financed by donors (Eurasian Harm Reduction Association, 2021, p. 61).

As in the case of HIV/AIDS, the government committed itself to taking over TB activities. It should be noted that the government lived up to its financial commitments. It increased the funding for TB and HIV by 169 million KGS (€1,812,235) in 2017–2020 and committed itself to providing an additional 280 million KGS (€3,002,521) for the 2021–2023 period (Eurasian Harm Reduction Association, 2021, pp. 23–25). HIV expenses represented 80 million in additional funding, which allowed for the procurement of some ARV drugs, payments to medical professionals, and social contracting (ibid.). However, despite the considerable increase, state funding is insufficient to purchase second-line ARV drugs, rapid tests, and CD4 tests used to assess viral load (ibid.). Moreover, prevention services in vulnerable groups are still largely financed by donors (Eurasian Harm Reduction Association, 2021, p. 61).

As in the case of HIV/AIDS, the government committed itself to taking over TB activities. In its strategy for Eastern Europe and Central Asia, the Global Fund explicitly asked countries to take over the provision of first-line medications for drug-susceptible TB by 2017 and develop a

---

[5] The exchange rate, as of March 17, 2023, was applied throughout this book.

plan for a similar transition of second-line drugs for MDR-TB (WHO/ Europe, 2014). Since 2015, the government of Kyrgyzstan has fully financed first-line medications (State Partner 9), and it plans to increase its contribution to second-line drugs to 20% in 2023 (Global Fund Office of the Inspector General, 2022, p. 4).

In addition to medications, the Global Fund (n.d.-a, p. 10) also expects grant-recipient countries to transfer the costs of laboratory reagents and consumables, maintenance of equipment, and services to domestic or "alternative" sources of funding. As of 2015, TB diagnostics were "almost exclusively" financed by the Global Fund (Mansfeld et al., 2015, p. 9), which also covered laboratory supplies, equipment (van den Boom et al., 2015), co-payments to specialists working with hazardous materials and other costs (State Partner 9). The government intends to increase its contribution to these areas as well, but fulfilling these commitments in the context of a budget deficit is challenging. Thus, the financial gap in the national health care reform program (2019–2030) is approximately 45% or approximately 2.3 billion KGS (approximately € 24,663,564), with the optimization and redirection of resources ensuring 57 million KGS (about €611,227); the rest is foreseen from other sources (Government of KR, 2018).

The country initiated optimization reforms to ensure additional financing. The TB Roadmap for 2016–2025 aims to decrease unnecessary hospitalizations by 5–8% annually, reduce bed capacities by 60% compared to 2015, and increase coverage with full ambulatory treatment by 60% by 2025 (Ministry of Health of KR, n.d.). These reforms mainly target problems related to excessive hospitalization (also among patients whose TB diagnosis is not confirmed) to reduce the length of hospital stay, which could last up to two to three months (ibid.). A savings of 137.7 million KGS (approximately €1,476,597) resulting from these reforms are to be spent on PHC strengthening, procurement of medications, laboratory supplies, and reagents, and improving the conditions of buildings (ibid.). Nevertheless, the Government of Kyrgyzstan (2017a) acknowledged that these savings were insufficient to meet the country's TB needs. Thus, procurement of second-line medications, laboratory maintenance, and supplies remain dependent on external support.

Despite explicit commitments and reforms initiated by the government, there is skepticism regarding the actual fulfillment of its obligations. My interviewees noted that the country was "unique" in the sense that there were many "good" laws and decrees that nevertheless fizzled out with time (CSO 3). They questioned the actual implementation of the "written promises" (State Partner 4) and suggested that many documents were not further realized (State Partner 6). Studies by Ancker and Rechel (2015a, 2015b) similarly suggest a "declaratory manner" of state policies, targeting donors rather than actually guaranteeing the continued implementation of the programs.

I suggest that the actual implementation of commitments depends on two factors, namely, national priorities and the choices of decision-makers. TB and HIV were explicitly prioritized and delineated in previous national health care reform programs (2005–2018), but the new program, "Healthy People, Prosperous Country" (2019–2030), incorporates these two into broader priority areas, such as public health and primary health care. Indeed, improvements in other areas highlighted in the program, including laboratory services, access to medications, human resources, information systems, eHealth, and an increase in state financing, equally benefit TB and HIV (Government of KR, 2018). The program pursues an interdisciplinary approach to health and intends to take this perspective to a new level by harmonizing legislation, engaging a broad spectrum of stakeholders, and emphasizing their responsibilities in health (ibid.). The program still targets TB and HIV but in an integrated manner.

This comprehensive and non-disease-specific focus nevertheless has implications for prioritization. For instance, the new program defines the following TB- and HIV-related indicators: the percentage of patients successfully completing TB treatment at the PHC level and the number of HIV notifications and TB prevalence per 100,000 people (Government of KR, 2018). "Den Sooluk" delineated TB and HIV and defined six indicators (three for each) (Government of KR, 2012). Indeed, this program aimed to strengthen the health care system by targeting key barriers, including public health, financing, and stewardship, but it still delineated cardiovascular disease, mother and child health, TB, and HIV as "core services" (ibid.). Surely, indicators in the new program hint at the prioritization of these diseases but in the context of broader health care reforms. This difference in the approaches of the two programs also relates to broader changes beyond the country (Chap. 12).

In addition to national priorities, the continuity of the project activities, among others, depends on the decision-makers' personal interests and beliefs. Increased state financing of health care programs requires an "active position" of the Ministry of Health and the relevant state agencies (IO Partner 3). However, the leaders, often political appointees, may not necessarily be committed to TB and HIV/AIDS or other services (IO Partner 4). On the contrary, some members of the parliament and government seem to have "a detrimental or disruptive effect" on the HIV/AIDS policies due to their "moral" beliefs and conventional positions toward CSWs, MSM, and PWID (Ancker & Rechel, 2015b, pp. 8–16).

Although legal commitments hint at the government's intention to continue HIV/AIDS and TB services, the extent and depth of these services depend on other factors. Among others, there are procurement costs and opportunities, the epidemiological situation, the political environment in the country, the availability of trained personnel, and other factors (e.g., the COVID-19 pandemic).

First, the continuity of services is conducive to procurement costs and opportunities. Accordingly, the low costs of methadone (approximately US $0.10 per day (Subata et al., 2016)) and condoms (approximately US $0.18 per unit (Stover et al., 2011)) may be advantageous to their continuity beyond the duration of the Global Fund grants. However, the limited state contribution to HIV prevention for vulnerable groups hints at the relevance of other factors, such as prioritization or stigma and discrimination against these groups (Chap. 3). Nevertheless, costs matter, particularly in the context of limited financial capacity, and vivid examples thereof are ARV and TB medications. ART is a lifelong therapy, and the annual cost of ARV drugs per patient ranges between US $490 for first- and US $1520 for second-line medications (Stover et al., 2011, p. 3). According to state partners interviewed for this research, the estimated costs of a single course of TB treatment in Kyrgyzstan vary from US $50–107 for drug-susceptible (State Partner 9) and US $4–15,000 for multi- and extensively drug-resistant TB per patient (State Partner 6). In this way, although taking over first-line treatment, the government may find it challenging to finance second-line medications against drug-resistant strains of TB/HIV.

Surely, treatment costs are changing following scientific progress. A state interviewee hoped that the emergence of generic drugs and license expiration of some items might contribute to the affordability of medications (State Partner 2). As of 2019, ARVs are included in the list of essential medicines intended to ensure their accessibility (UNAIDS, 2020) due to the state control of the prices of items on this list. Moreover, prices for ARVs are gradually falling, and there are several alternatives to ART, including preexposure prophylaxis products offered by pharmaceutical companies and charities, as well as the potential use of mRNA vaccines against HIV (Economist, 2022b). Following the COVID-19 pandemic, there are also plans to commence trials of TB mRNA vaccines (Economist, 2022a). Trail results may provide alternatives to existing treatment regimens for TB and HIV.

Procurement opportunities are inherent to medication costs and treatment outcomes by ensuring the continuous supply of medications necessary for quality treatment. Kyrgyzstan procures TB and HIV/AIDS-related products via the voluntary pooled procurement mechanisms of the Global Fund and the Global Drug Facility (GDF), which allow aggregation of orders and negotiation of better prices and delivery conditions (Gotsadze et al., 2019). Upon its transition from the Global Fund grants, the country is expected to procure health products on its own. However, individual country procurement will result in an increase in prices (ibid.). It may also result in situations in which manufacturers are not interested in supplying health care products to the country due to the small size of the order. One of my interviewees recalled the "bitter experience" of Kazakhstan, which encountered a drastic increase in prices after the country's transition from the Global Fund grants (State Partner 2). To avoid this situation and allow the continued procurement via international organizations, the government actors initiated relevant amendments to the national legislation (ibid.).

In addition to access to international procurement mechanisms, procurement of health products beyond the duration of Global Fund grants requires their registration in the country. Most of the grant-recipient countries (including Kyrgyzstan) used one-time waivers for the drugs procured via the Global Fund (Gotsadze et al., 2019) by qualifying them as "humanitarian assistance." By contrast, the medications purchased at the expense of the state budget must be registered in the country. As of 2018, most of the medications used for ART and treatment of multidrug-resistant (MDR-TB) and extensively drug-resistant tuberculosis (XDR-TB) were not registered in the country (Mandel, 2018). In other words, their procurement by the government was not possible. Accelerated registration of medications has improved since then (see Eurasian Harm Reduction Association, 2021). One interviewee, however, expressed concerns that the country may switch to drugs with treatment outcomes different from those provided in the grants (IO Partner 4). As the situation evolves, procurement requires closer consideration and research on its own.

Second, in addition to the use of certified medications, the quality of TB and HIV/AIDS services depends on the availability of qualified medical workers. To improve the quality of services, the Global Fund financed multiple training seminars on infection control, quality of lab services, management of medical waste (UNDP, 2015e), and HIV prevention. The seminars also targeted health workforce management by providing training on electronic spreadsheets, management of payments to medical workers, and other areas (UNDP, 2015c). Yet, the long-term impact of these training activities is jeopardized by high staff rotation. A health care professional interviewed for this research suggests that training of one lab specialist takes around six months and costs KGS 36,000 (about €386 Euro). However, after a year, this specialist leaves the state hospital to work in a private lab due to the better salary rates offered there (Health worker 1). Therefore, the long-term impact of training activities on the qualifications/competencies of health care personnel involved in the TB and HIV/AIDS programs remains unclear due to the structural problems in the health care system (Chap. 4).

Third, the continuity of services also relates to the epidemiological situation in the country. Despite coverage issues, there is a growing demand for HIV and TB treatment. After a nosocomial outbreak of HIV in the south of the country, the procurement of antiretroviral medications in Global Fund grants changed from an annual to biannual basis (Murzalieva et al., 2009). Correspondingly, the number of people on

ART has doubled annually since 2011 (Mansfeld et al., 2015). Similarly, there is a growing demand for TB treatment, particularly in the context of the drug-resistant forms of this disease (Chap. 1). There is a considerable financial gap in the treatment of drug-resistant TB (WHO/Europe, 2011), and even donor financing cannot meet the increasing demand for treatment. The Global Fund grants and Doctors Without Borders/*Médecins Sans Frontières* (MSF) covered 609 out of 1136 cases of multidrug and 36 out of 60 cases of extensively drug-resistant TB (van den Boom et al., 2015, p. 88). All savings in the grants have been used to provide treatment, but the Global Fund could still not cover the existing needs of the National TB Control Program, much less meet the growing demand for treatment of drug-resistant forms of tuberculosis (see UNDP, 2013). Predicting changes in the epidemiological situation in the country goes beyond the scope of this research. However, financial struggles in meeting the growing demand for treatment suggest grim perspectives for the continuity of services.

Fourth, the epidemiological situation is closely related to other factors. For instance, the COVID-19 pandemic was an unexpected challenge that strained the health care system and led to a diversion of resources (e.g., facilities, health personnel, and finances) from other diseases (Davis et al., 2021). Health care providers and civil society volunteers demonstrated unprecedented solidarity, dedication, and commitment in tackling the pandemic. The Global Fund, along with the World Bank, the German Federal Ministry for Economic Cooperation and Development (*das Bundesministerium für wirtschaftliche Zusammenarbeit und Entwicklung*—BMZ), and others, supported the country in its immediate response to COVID-19.

Global Fund grants adapted to the unexpected situation by providing a range of services to mitigate the impact of the pandemic. These included mobile brigades, online services, video supervision, and opening centers and shelters for individuals to continue their TB and HIV-related treatments (see UNDP, 2020a). The grants also supported medical workers who were at the forefront of the epidemic and worked for long hours and often without access to protective equipment.

Although it reduced deaths, the lockdown imposed by the Kyrgyz government from March 24 to May 10, 2020, caused a decrease in testing and prevention and limited access to care for both TB and HIV services (Alliance for Public Health et al., 2021). The medication supply was uninterrupted, and the share of outpatient services increased, but mobility restrictions affected health care-seeking habits for both diseases (ibid.). The medium- and long-term consequences of disruptions caused by the pandemic remain to be seen. The country has no catch-up plan against the impact of the pandemic on health (Global Fund Office of the Inspector General, 2022). However, at the global level, COVID-19 set back the global achievements made in TB and HIV/AIDS in recent decades (Economist, 2022a, 2022b). Reductions in global TB and AIDS-related funding combined with the ongoing pandemic clearly jeopardize the continuity of TB and HIV/AIDS services. How much the Global Fund manages to raise in its seventh round and how it distributes these finances remain to be seen.

To summarize, this section analyzed the continuity of the Global Fund project activities after the country's transition by elaborating on the types (what) and the extent of the services currently provided within the grants. As demonstrated above, the Global Fund increased the geographic coverage and the type of HIV/AIDS services in Kyrgyzstan by contributing to HIV testing, distribution of condoms, opioid substitution therapy, needle and syringe exchange program, and antiretroviral therapy. The Global Fund has similarly contributed to the prevention and treatment of tuberculosis by ensuring the countrywide availability of MDR-TB treatment. Nevertheless, reaching out to the groups affected by TB and HIV/AIDS and patient adherence to treatment remained problematic.

The government, in turn, demonstrated its commitment to continuing TB and HIV services by increasing its financial contribution and indicating its responsibilities in relevant legislation. Following skepticism that some interviewees and related studies expressed, this section listed factors critical to fulfilling commitments. These included changes in national health care priorities, the choices of decision-makers, medication prices and procurement, health personnel availability, the country's epidemiological situation, and the COVID-19 pandemic. The continuity of services in the long-term also depends on how these factors evolve.

## 9.3    Maintaining Benefits

In addition to diagnosis and treatment, Global Fund grants provided incentives to patients and health care workers to increase patient adherence to treatment. This section discusses the types of incentives and the maintenance of patient and health care workers' benefits beyond the duration of the Global Fund grants to Kyrgyzstan.

First, the benefits for patients with TB included reimbursement of travel expenses (UNDP, 2014) and provision of hygiene and food parcels (Government of KR, 2013). Because of stigma and discrimination related to TB and confidentiality concerns, patients often prefer to receive their treatment in health care facilities outside their area of residence. The Global Fund reimburses all travel expenses a patient incurs on the way to examination and treatment (UNDP, 2014). Furthermore, patients received food and hygiene parcels to incentivize treatment adherence. For those in inpatient care, these parcels included butter, condensed milk, black tea, sugar, biscuits, laundry and toilet soaps, shampoo, toothpaste, and toilet paper (UNDP, 2013). The patients in outpatient care received vegetable oil, rice, pasta, grain sugar, black tea, washing powder, toilet, and laundry soaps (ibid.). These parcels could be received by TB patients themselves or their family members. A CSO representative, who distributed these parcels, noted that there were only seven rejections during the five years. Generally, people accepted these parcels as a contribution to the family budget since TB patients could not work for the duration of their treatment. An interviewee notes that the low number of rejections vividly demonstrates the economic hardships encountered by TB patients and the population in general (CSO 3).

Later, the hygiene and food parcels were replaced with vouchers and money transfers. The national actors implementing the Global Fund grants concluded an agreement with supermarkets and provided training to their staff members on how to work with the vouchers issued within the framework of the grants. The vouchers intended to provide a more "client-oriented approach," which allowed the patients to choose the necessary food and hygiene items. The patients received the change if the amount of purchase was less than or paid extra if it exceeded the value of the voucher (CSO 9). The vouchers were later replaced with money transfers, which patients received for their adherence to treatment after being confirmed by a health care worker providing the DOTS (CSOs 3 and 9). Monthly money transfers ranged from 1300 KGS (about €14) for patients with drug-susceptible to 1800 KGS (€19) for patients with drug-resistant TB (Health Worker 1). The decrease in the amount of the Global Fund grants to Kyrgyzstan also affected the patient benefits. Since 2018, only patients with drug-resistant TB receive money transfers (ibid.), though reimbursement of travel expenses has remained available to all TB patients.

Second, patients with HIV/AIDS were entitled to reimbursement of their travel expenses related to treatment and examination. They also received psychological and peer-to-peer support from the NGOs implementing HIV-prevention activities. These NGOs supported persons living with HIV in administrative and legal issues related to obtaining identity documentation, applying for social benefits, and others (CSO 6). Adults are entitled to disability pensions, depending on their clinical stage, although the amounts remain low (not just for HIV) and insufficient to cover actual needs (ibid.). Children with HIV are entitled to monthly motivational payments of 1000 KGS (approximately €11) (see UNDP, 2021a). In 2019, for instance, 80% of registered HIV-positive children received monthly support (UNAIDS, 2020). However, there is still a problem with the coverage of support activities due to rejections from PLHIV. The main reason is the fear of their status becoming known during the preparation and request of documents necessary for receiving these benefits (ibid.). A few activists, such as Baktygul Shukurova, disclose their HIV status to draw attention to this problem (Akipress.org, 2017).

Third, in addition to patient benefits, the Global Fund grants stipulated additional incentives for health care workers working in TB and HIV/AIDS. In TB, medical workers receive bonuses for the achievement of the "favorable treatment outcome," sputum conversion at the six-month point after the initiation of treatment (UNDP, 2015b). According to a health care worker interviewed for this research, 12,000 KGS (about €129) is awarded for the successful treatment of a patient with drug-susceptible and 24,000 KGS (around €257) for a patient with drug-resistant TB (Health Worker 1). The interviewee noted that the bonuses were divided between the health care workers participating in the treatment. Seventy-five percent is provided to a nurse supervising the patient, 15% to a head doctor, and the rest is awarded to the director and deputy director of the family group practice, the coordinating TB specialist, and others (ibid.). Similar co-payments are stipulated for health care workers in HIV/AIDS. For instance, a narcologist, nurse, and social worker are entitled to base salary and additional payments for every patient enrolled in methadone substitution therapy. A narcologist/nurse receives US $50 base monthly salary and an additional US $3 for each patient, while a social worker receives US $80 monthly and an additional US $2 per patient, respectively (UNDP, 2015d).

It should be noted that support for key groups was further reemphasized during the COVID-19 pandemic. For instance, 480 children with HIV received tablets to continue their school education during the pandemic (UNDP, 2021b). A total of 2577 individuals from vulnerable groups, including PLHIV, patients with TB, and those in precarious life situations, received food parcels (ibid.). In addition, the grants continued providing shelter opportunities, stipulating the provision of meals for vulnerable groups (ibid.). Training activities intended to inform the LGBTQ community and NGOs about COVID-19 and their rights, including access to health care and other issues, were conducted (UNDP, 2020b).

My interviewees emphasize that the discontinuity of some benefits was clear from the beginning, but the grants continued providing them to facilitate the fight against the two diseases (IO Partner 20). Nevertheless, some activities, such as the outpatient treatment of drug-resistant TB, may evolve into a "time-bomb" if the state or another donor will not take over the patient benefits to ensure their adherence to treatment (State Partner 10 and Academic Partner 2).

The Global Fund (n.d.-a) stipulates a gradual transition of expenses for human resources and social support from grants to the state budget. Some interviewees were skeptical in this regard. One noted that the discontinuity of some benefits was clear from the beginning, but the grants continued providing them to facilitate the fight against the two diseases (IO Partner 20). Another warned that some activities, such as outpatient treatment of drug-resistant TB, might evolve into a "time bomb" if the state or another donor would not take over patient benefits to ensure their adherence to treatment (State Partner 10 and Academic Partner 2).

Overall, the state committed itself to continuing the reimbursement of travel expenses, provision of social support to children with HIV, and financial incentives for health care workers. However, budget deficits, stigma, and discrimination against individuals affected by TB and HIV jeopardize the actual implementation of these commitments. Thus, local self-governments are expected to cover the travel costs of TB and HIV patients residing in their area (State Partner 6). However, as the majority of regions are subsidized by the national government (State Partner 9), local self-governments' ability to fulfill this function is unclear. In addition to travel expenses, the national government committed itself to providing social support to 90% of children with HIV (Government of KR, 2017b). Since 2020, the government has stipulated a lump-sum cash compensation in cases of nosocomial HIV infection. Individuals who are 18 or parents of children under this age are entitled to compensation in an amount not less than 1000 calculation indices (Government of KR, 2005). Yet, stigmatization and discrimination of persons living with HIV, bureaucracy, and unawareness about the entitlements hinder access to these benefits (Murzalieva et al., 2009). The national government has also stipulated co-payments to primary health care workers, particularly nurses providing DOTS, to ensure the patients' adherence to TB treatment (Health Worker 1). Currently, the nurses' monthly salary of 12,000 KGS (around €129) is below the average national wage rate of 16,427 KGS (about €176), and nurses have no incentives to follow-up on patients defaulting from treatment (IO Partner 17).

Overall, the Global Fund provided multiple benefits to patients and health care workers to facilitate the prevention and treatment of TB and HIV/AIDS. These benefits included the reimbursement of travel expenses, provision of hygiene and food parcels, vouchers and money transfers, and co-payments to medical workers involved in TB and HIV/AIDS services. However, the majority of these benefits are unlikely to be maintained beyond the duration of the grants. The government has committed itself to reimbursing the travel expenses incurred by patients on their way to TB and HIV/AIDS-related services. It has also promised to provide social support to 90% of children with HIV and pensions to adults with HIV (as discussed above). Nevertheless, the budget deficit jeopardizes the actual implementation of these commitments. In addition, due to stigma and discrimination, individuals with TB and HIV reject state support due to the fear of exposure.

## 9.3.1 Community Capacity Building

This section examines the survival of the NGOs involved in the Global Fund grants, their leadership, and resource mobilization beyond the duration of the grants.

First of all, in terms of survival, a decrease in the Global Fund grants affects the NGOs working in TB and HIV/AIDS, although to different extents. NGOs compete for the "scarce resources" (Spicer et al., 2011b, p. 1753) and some organizations currently working in the Global Fund grants will have problems with finding alternative sources of funding (Zardiashvili & Garmaise, 2017). Yet, the decrease in donor financing will have a differentiated impact on NGOs.

The organizations (including those involved in the Global Fund grants) vary greatly in their human resources and work experience. For instance, "AIDS Foundation East-West in the Kyrgyz Republic" ("AFEW-Kyrgyzstan") (n.d.) registered itself as a local Kyrgyz NGO in 2015, but it commenced its work in the country already in 2004 as part of the projects financed by AFEW-International. The organization inherited the standard operating procedures of the international organization, which ensured its strong capacity in comparison to other local NGOs (CSO 3). Another organization, "Socium," commenced its activities in 1996 as a public association working on social development and adaptation of individuals with drug and alcohol addiction (CSO 8). Similar to "AFEW-Kyrgyzstan" (2023), "Socium" collaborated with multiple donors, such as the Soros Foundation Kyrgyzstan, the Global Fund, USAID, UNAIDS, and others. This cooperation and long-term experience ensured the relative independence of these NGOs from the Global Fund grants.

However, smaller grant recipients largely depend on the Global Fund (Nasakt, 2015). These organizations concentrate on specific groups particularly vulnerable to HIV/AIDS, such as MSM, CSWs, and PWID. This specialization contributed to the selection of NGOs receiving Global Fund grant funding, as the financier typically differentiates between the vulnerable groups and assigns organizations to work with each of them. However, in the long-term perspective, this narrow specialization negatively affects the organizations' abilities to adjust to the changing environment of development assistance. For this reason, larger NGOs with multiple sources of financing are likely to survive, in contrast to smaller organizations working with specific groups and dependent on a single donor (IO Partner 4).

Furthermore, the leadership of the NGOs working in the Global Fund grants remains unclear due to their dependence on donor financing and the limited training provided within the grants. Leadership can be defined as the organizations' ability to define the problems, suggest solutions, and critically reflect on the general issues relevant to their work. In the NGOs' case, it closely relates to their ability to advocate for the issues pressing to them and the groups of the population they aim to represent. The studies on NGOs working in TB and HIV/AIDS in Kyrgyzstan note that the organizations generally refrain from criticizing the donors (Murzalieva et al., 2009) and acting independently from them (Ancker & Rechel, 2015b) due to fear of losing financing (Spicer et al., 2011b). Smaller NGOs solely dependent on the Global Fund seem to be particularly vulnerable in this regard (Harmer et al., 2013) and less likely to criticize the donor.

It should also be noted that the Global Fund stipulates limited support of the NGOs' advocacy work. Relevant in the early 2000s, the advocacy for treatment and human rights does not seem to be relevant to the donor anymore (IO Partner 20). An NGO representative also notes that most of the training activities targeted service provision (CSO 8). Mainly focusing on prevention (Murzalieva et al., 2009) and treatment, the Global Fund grants devote limited attention (Harmer et al., 2013) and resources to advocacy for the rights of the persons affected by the diseases (Spicer et al., 2011b). Burrows et al. (2018) go further by linking the increase in violence and hostility toward vulnerable groups to the reductions in external funding for advocacy. The general dependence of the organizations on donor financing and the limited support for advocacy within the Global Fund grants do not stimulate leadership among the

NGOs involved in the grants. However, a more specific estimation requires a closer look at individual NGOs, since the quality of their leadership greatly varies depending on the size and experience of organizations and their access (or lack thereof) to multiple sources of financing.

Equally, mobilization of resources via donor and state financing or fundraising is an essential component of sustainability, as it closely relates to the continuity of civil society organizations and their activities beyond the individual donor-funded projects. As mentioned above, the NGOs involved in the Global Fund grants greatly vary in terms of their access to alternative sources of financing. Therefore, instead of discussing the mobilization of resources by individual NGOs, this subsection focuses on social contracting, a source of financing for all NGOs developed as a result of the Global Fund grants to the country.

The Global Fund (n.d.-a) asks grant-recipient countries to develop social contracting to secure financing for NGOs and their activities in TB and HIV/AIDS after the end of the grant period. Social contracting presumes NGO contracting by government agencies. The Government of the Kyrgyz Republic (2017b) committed itself to developing the normative legal basis necessary for social contracting by the end of 2021. Seminars for representatives of state and nongovernmental organizations offered within the grants intended to support these aspirations (see UNDP, 2021c). Overall, by integrating social contracting into its legislation, the government aimed to continue the work of the NGOs with vulnerable groups through its integration with primary health care (Government of KR, 2017a). UNAIDS (2020, p. 6) notes that three million KGS (€32,170) were used to pilot social contracting projects on support and care to PLHIV. However, the issues and achievements in this process require further research, particularly in the face of challenges presented by COVID-19.

It is important to note that the introduction of social contracting will have further implications for the services provided by NGO social workers and the accountability of NGOs. First, as one interviewee noted, unlike the psychological and peer-to-peer support in the Global Fund grants, which targeted individuals affected by TB and HIV, an NGO social worker contracted by the government is expected to cover all patients with "socially significant diseases," including but not limited to TB and HIV/AIDS (State Partner 2). This will affect the quality and quantity of consultations provided to each patient. Furthermore, according to the same interviewee, the breadth of services offered by a social worker (e.g., peer-to-peer support, follow-up, and outreach to target groups) will also decrease, as the government cannot maintain the breadth of services offered through Global Fund grants (ibid.). In addition, an interviewee added that social contracting would result in NGOs' accountability to the government for the services financed by it (ibid.). This may have further implications for NGOs' ability to criticize the government, as in the case of NGOs and donor organizations.

At the same time, the feasibility of social contracting in practice remains to be seen. Indeed, social contracting may be the only opportunity (other than donor financing) to continue the services provided by NGOs (State Partner 2). "Unavoidable" and "possibly good," social contracting is probably something that the civil society organizations strived for (CSO 6). For the government, social contracting offers the possibility to involve knowledgeable and experienced social workers, and for NGOs, it promises some security in the context of decreasing donor funding (ibid.). Nevertheless, not all NGO employees may be willing to participate in social contracting due to salary differences. A state representative interviewed for this research suggests that project coordinators in NGOs may earn around US \$500–600 per month, while a family doctor might earn about US \$150 (State Partner 6). This difference in salary, along with accountability to the government, may affect the NGO employees' willingness to conclude social contracting with state institutions.

Overall, the Global Fund has contributed to community capacity building in multiple ways. It increased the number of NGOs and facilitated the development of social contracting to guarantee the NGOs' access to state financing after the country's transition from the Global Fund grants. These benefits notwithstanding, the increased number of NGOs and the scarcity of resources resulted in competition among the organizations and their dependence on donor financing. The NGOs working with the Global Fund grants seem to restrain themselves from criticizing the donor or the primary recipients of the grants due to the fear of losing access to financing. This dependency, along with the limited focus of the Global Fund on advocacy work, discouraged the leadership of civil society organizations from working with the grants.

## 9.4   Summary

The COVID-19 pandemic demonstrated an unprecedented challenge to the sustainability of grant activities. With its medium- and long-term implications still to be seen, the impact of the pandemic on each dimension of sustainability requires further research. Although it reflected on some initial implications, this chapter was nevertheless bound to provide a more general analysis over a longer time period. This chapter reviewed the sustainability of the Global Fund project in Kyrgyzstan by focusing on the continuity of activities, maintenance of benefits, and community capacity building beyond the duration of the project.

First, the Global Fund increased the type and geographic coverage of preventive and treatment services related to TB and HIV/AIDS. More specifically, it consolidated HIV testing and TB detection, and expanded access to opioid substitution therapy and needle-exchange programs. The Global Fund introduced antiretroviral therapy, previously inaccessible to persons living with HIV. It has also contributed to the provision and expansion of treatment of multi- and extensive drug-resistant forms of tuberculosis in Kyrgyzstan. Despite these improvements, the Global Fund grants neither reached out to all persons most vulnerable to HIV nor provided treatment to all MDR-TB patients. The author reckons with these issues as not to give a false impression about the extent of project activities financed by the Fund. Since the grants are still ongoing, the continuity of the Global Fund project remains an open question, which is also vividly demonstrated by the lack of consensus on this subject among the stakeholders involved in TB and HIV/AIDS. The government committed itself to continuing most of the project activities. However, the actual fulfillment of these commitments largely depends on policy-makers' interests and beliefs, further availability of state financing, the epidemiological situation in the country, access to certified medications, and trained health care personnel.

Second, the Global Fund provided extensive social support to patients and health care workers to increase patient adherence to treatment. The benefits included reimbursement of travel expenses, provision of hygiene and food parcels, vouchers, monetary incentives, and co-payments.

Although unequally distributed among the TB and HIV patients, these benefits supported preventive and treatment activities covered by the grants. The government has committed itself to taking over the reimbursement of travel expenses, co-payments to health care workers, and social support for children with HIV. However, the actual use of these benefits is unclear due to lack of awareness of the population about these entitlements, state bureaucracy, stigma and discrimination, and budget deficits at the level of local self-governments. The maintenance of benefits not taken over by the government seems implausible without donor assistance.

Third, the Global Fund contributed to community capacity building by ensuring the NGOs' access to state financing, as well as increasing the number of NGOs and number of NGO staff. Yet, the limited focus of the Global Fund on NGO advocacy, along with the civil society organizations' dependence on the grants, emasculated the leadership of the NGOs working with the grants. Notably, the transition of the country from the Global Fund grants will have a differentiated impact on the survival of civil society organizations, depending on their size, experience, and collaboration with other actors.

# References

AFEW Kyrgyzstan. (n.d.). *Strategicheskiy Plan Obshchestvennogo Fonda «Spid Vostok-Zapad V Kyrgyzskoy Respublike» (AFEW Kyrgyzstan) 2017–2020 [Strategic plan of the Public Foundation 'AIDS East-West in the Kyrgyz Republic' (AFEW Kyrgyzstan) 2017–2020]*. Retrieved February 15, 2023, from http://www.afew.kg/upload/userfiles/%D0%A1%D1%82%D1%80%D0%B0%D1%82%D0%BF%D0%BB%D0%B0%D0%BD.pdf

AFEW-Kyrgyzstan. (2023). *Naši donory [Our donors]*. Retrieved March 14, 2023, from http://www.afew.kg/donors.html

Akipress.org (2017). *First HIV-infected woman openly discloses her status in Kyrgyzstan*. Retrieved March 10, 2023, from https://akipress.com/news:599341:First_HIV-infected_woman_openly_discloses_her_status_in_Kyrgyzstan/

Alliance for Public Health, SoS_Project, & Matahari Women Workers' Center. (2021). *COVID-19 response and impact on HIV and TB services* (pp. 1–23).

Retrieved February 12, 2023, from https://aph.org.ua/wp-content/uploads/2021/04/kyrgyzstan-red.pdf

Amanzholov, N., Bakirova, C., Harutyunyan, A., Karliyeva, O., Kryshtafovich, N., Mukhtarli, E., et al. (2018). *Achievements in promoting people-centered TB care: Successes of civil society organizations within the TB-REP project (2016–2018)* (pp. 1–40). Alliance for Public Health. Retrieved February 6, 2023, from http://aph.org.ua/wp-content/uploads/2019/01/TB-REP__EN__SAIT.pdf

Ancker, S., & Rechel, B. (2015a). 'Donors are not interested in reality': The interplay between international donors and local NGOs in Kyrgyzstan's HIV/AIDS sector. *Central Asian Survey, 34*(4), 516–530. https://doi.org/10.1080/02634937.2015.1091682

Ancker, S., & Rechel, B. (2015b). HIV/AIDS policy-making in Kyrgyzstan: A stakeholder analysis. *Health Policy and Planning, 30*, 8–18. https://doi.org/10.1093/heapol/czt092

Ancker, S., Rechel, B., McKee, M., & Spicer, N. (2013). Kyrgyzstan: Still a regional 'pioneer' in HIV/AIDS or living on its reputation? *Central Asian Survey, 32*(1), 66–84. https://doi.org/10.1080/02634937.2013.771965

Burrows, D., Bolotbaeva, A., Sydykanov, B., Iriskulbekov, E., & Dastan uulu Ulan. (2018). *Baseline assessment—Kyrgyzstan: Scaling up programs to reduce human rights-related barriers to HIV and TB services* (pp. 1–94). Global Fund. Retrieved February 4, 2023, from https://www.theglobalfund.org/media/8145/crg_humanrightsbaselineassessmentkyrgyzstan_report_en.pdf?u=636809011150000000

Davis, N., Huffman, S., Rogers, D., Byrne, A., Oyediran, K., Ibraeva, G., et al. (2021). *Comparison of COVID-19-related tuberculosis resource reallocation in Afghanistan and Kyrgyzstan.* Retrieved February 21, 2023, from https://www.tbdiah.org/wp-content/uploads/2022/02/Union_E-Poster_2021_Comparison-of-COVID-19-related-TB-resrouce-reallocation-in-Afghanistan-and-Kyrgyzstan.pdf

Economist. (2022a). How one pandemic made another one worse. Covid-19 set back the battle against tuberculosis. But it also points the way forward. *The Economist.* Retrieved February 14, 2023, from https://www.economist.com/international/2022/10/27/how-one-pandemic-made-another-one-worse

Economist. (2022b). Despite setbacks, HIV can be beaten. But doing so will take patience and money. *The Economist.* Retrieved February 14, 2023, from https://www.economist.com/science-and-technology/2022/08/02/despite-setbacks-hiv-can-be-beaten

Eurasian Harm Reduction Association. (2021). *Kyrgyzskaâ Respublika: Ocenka ustojčivosti otveta na VIČ sredi klûčevyh grupp naseleniâ v kontekste perehoda ot podderžki Global'nogo fonda na gosudarstvennoe finansirovanie [Kyrgyz Republic: Assessing the sustainability of the HIV response among key populations in the context of the transition from Global Fund support to public funding]* (pp. 1–80). Retrieved February 23, 2023, from https://harmreductioneurasia.org/wp-content/uploads/2022/06/TMT-Assessment-Report-Kyrgyzstan-EHRA-2021-RUS.pdf

European Centre for Disease Prevention and Control, & WHO/Europe. (2019). *HIV/AIDS surveillance in Europe 2019. 2018 data* (pp. 1–95). Retrieved February 17, 2023, from https://www.ecdc.europa.eu/sites/default/files/documents/HIV-annual-surveillance-report-2019.pdf

Foundation for AIDS Research. (2015). *Harm reduction and the global HIV epidemic. Interventions to prevent and treat HIV among people who inject drugs* (pp. 1–22). Foundation for AIDS Research Public Policy Office. Retrieved March 27, 2019, from https://www.amfar.org/uploadedFiles/_amfarorg/Articles/On_The_Hill/2015/DC-PWID-Policy-Report_08-31-15v205.pdf

Global Fund. (n.d.-a). *Turning the tide against HIV and tuberculosis. Global fund investment guidance for Eastern Europe and Central Asia* (pp. 1–19). Retrieved February 17, 2023, from https://www.globalfundadvocatesnetwork.org/wp-content/uploads/2015/03/Global-Fund-Investment-Guidance-for-EECA_en.pdf

Global Fund. (n.d.-b). *Data explorer. Results. Year 2020.* Retrieved February 15, 2023, from https://data.theglobalfund.org/results?locations=KGZ

Global Fund Office of the Inspector General. (2022). *Global Fund supported TB/MDR-TB programs in Eastern Europe and Central Asia—Focus on Uzbekistan, Kyrgyzstan and Tajikistan.* Audit Report. Retrieved February 17, 2023, from https://www.theglobalfund.org/media/12251/oig_gf-oig-22-013_report_en.pdf

Gotsadze, G., Chikovani, I., Sulaberidze, L., Gotsadze, T., Goguadze, K., & Tavanxhi, N. (2019). The challenges of transition from donor-funded programs: Results from a theory-driven multi-country comparative case study of programs in Eastern Europe and Central Asia supported by the Global Fund. *Global Health, Science and Practice, 7*(2), 258–272. https://doi.org/10.9745/GHSP-D-18-00425

Government of KR. (2005). *Zakon Kyrgyzskoj Respubliki o VIČ/SPIDe v Kyrgyzskoj Respublike ot 13 avgusta 2005 goda № 149 [Law of the Kyrgyz Republic On HIV/AIDS in the Kyrgyz Republic from August 13, 2005 # 149].* Retrieved February 20, 2023, from http://cbd.minjust.gov.kg/act/preview/ru-ru/1747/50?mode=tekst

Government of KR. (2012). *Postanovlenie ot 24 maya 2012 goda № 309 O Natsional'noy programme reformirovaniya zdravookhraneniya Kyrgyzskoy Respubliki "Den sooluk" na 2012–2016 gody [Decree dated May 24, 2012 No. 309 On the National Healthcare Reform Program of the Kyrgyz Republic "Den Sooluk" for 2012–2016].* Retrieved February 2, 2023, from http://cbd.min-just.gov.kg/act/view/ru-ru/93628?cl=ru-ru

Government of KR. (2013). *Ob utverzhdenii Programmy "Tuberkulez-IV" na 2013-2016 gody: Postanovlenie Pravitel'stva Kyrgyzskoy Respubliki ot 10 iyunya 2013 goda № 325.* Retrieved February 2, 2023, from http://cbd.minjust.gov.kg/act/view/ru-ru/94467

Government of KR. (2017a). *Programma Pravitel'stva Kyrgyzskoj Respubliki po preodoleniû VIČ-infekcii v Kyrgyzskoj Respublike na 2017–2021 gody [The Government of the Kyrgyz Republic Program on Overcoming HIV Infection in the Kyrgyz Republic for 2017–2021]: Priloženie 1 Utverždeno postanovleniem Pravitel'stva Kyrgyzskoj Respubliki ot 30 dekabrâ 2017 goda № 852 [Annex 1 approved by the Decree of the Government of the Kyrgyz Republic dated December 30, 2017 No. 852].* Retrieved February 17, 2023, from http://cbd.minjust.gov.kg/act/view/ru-ru/11590

Government of KR. (2017b). *O Programme Pravitel'stva Kyrgyzskoj Respubliki po preodoleniû VIČ-infekcii v Kyrgyzskoj Respublike na 2017–2021 gody [The program of the Government of the Kyrgyz Republic sight to overcome HIV infection in the Kyrgyz Republic for 2017–2021]: Postanovlenie Pravitel'stva Kyrgyzskoj Respubliki ot 30 dekabrâ 2017 goda № 852 [Decree of the Government of the Kyrgyz Republic dated December 30, 2017 No. 852].* Retrieved February 3, 2023, from http://cbd.minjust.gov.kg/act/view/ru-ru/11589

Government of KR. (2018). *Programma Pravitel'stva Kyrgyzskoj Respubliki po ohrane zdorov'â naseleniâ i razvitiû sistemy zdravoohraneniâ na 2019–2030 gody "Zdorovyj čelovek—procvetaûŝaâ strana" [Program of the Government of the Kyrgyz Republic for the protection of public health and the development of the healthcare system for 2019–2030 "Healthy Person—Prosperous Country"]: Priloženie 1 (k postanovleniû Pravitel'stva Kyrgyzskoj Respubliki ot 20 dekabrâ 2018 goda № 600) [Annex 1 (to the Decree of the Government of the Kyrgyz Republic dated December 20, 2018 No. 600)].* Retrieved February 2, 2023, from http://cbd.minjust.gov.kg/act/preview/ru-ru/12976/10?mode=tekst

Grant Performance Report External Print Version. Kyrgyzstan KGZ-202-G01-H-00. (2011). (pp. 1–34). Retrieved March 3, 2023, from http://docs.theglobalfund.org/program-documents/GF_PD_003_d916e133-ccae-4f6d-b57a-29e5ac579c65.pdf

Grant Performance Report External Print Version. Kyrgyzstan KGZ-202-G02-T-00. (2011). (pp. 1–25). Retrieved February 17, 2023, from http://docs.theglobalfund.org/program-documents/GF_PD_003_a4eda052-5fe4-4048-86b7-58f8a7bc534f.pdf

Grant Performance Report External Print Version. Kyrgyzstan KGZ-607-G04-T. (2012). (pp. 1–28). Retrieved March 3, 2023, from http://docs.theglobalfund.org/program-documents/GF_PD_003_e64411eb-4f8d-4d71-a61a-8c627880dcfd.pdf

Grant Performance Report External Print Version. Kyrgyzstan KGZ-H-UNDP. (2016). (pp. 1–44). Retrieved March 3, 2023, from http://docs.theglobalfund.org/program-documents/GF_PD_003_51112a72-8240-4690-985b-ebc1c63e5618.pdf

Grant Performance Report External Print Version. Kyrgyzstan KGZ-S10-G08-T. (2016). (pp. 1–34). Retrieved February 5, 2023, from http://docs.theglobalfund.org/program-documents/GF_PD_003_f2a5f5ee-0eec-4d0c-8738-bd8b8b862e2c.pdf

Grant Performance Report External Print Version: Kyrgyzstan KGZ-C-UNDP. (2016). (pp. 1–31). Retrieved February 16, 2023, from http://docs.theglobalfund.org/program-documents/GF_PD_003_5a4ffe49-3db7-4fc6-8b67-a7a0fb69df71.pdf

Grant Performance. Report External Print Version. Kyrgyzstan KGZ-910-G07-T. (2016). (pp. 1–30). Retrieved March 3, 2023, from http://docs.theglobalfund.org/program-documents/GF_PD_003_e99065eb-b1c1-409a-a5f5-e0db338541f2.pdf

Harmer, A., Spicer, N., Aleshkina, J., Bogdan, D., Chkhatarashvili, K., Murzalieva, G., et al. (2013). Has global fund support for civil society advocacy in the former Soviet Union established meaningful engagement or "a lot of jabber about nothing"? *Health Policy and Planning, 28*, 299–308. https://doi.org/10.1093/heapol/czs060

International Charitable Organization "East Europe and Central Asia Union of People Living with HIV." (n.d.). *Otsenka sovremennoy situatsii: Kyrgyzstan: Analiz vtorichnykh dannykh [Assessment of the current situation: Kyrgyzstan. Analysis of secondary data].* n.p. Retrieved February 17, 2023, from http://ecuo.org/wp-content/uploads/2016/12/Otsenka-situatsii-analiz-vtorichnyh-dannyh-13-MB.pdf

Mandel, L. (2018). *Regional Green Light Committee for the WHO European Region face-to-face meeting and workshop with high drug-resistant tuberculosis burden countries: Programmatic aspects of the implementation of new tuberculo-*

*sis drugs and regimens* (pp. 1–39). WHO/Europe. Retrieved February 21, 2023, from https://www.euro.who.int/__data/assets/pdf_file/0008/379511/Report_GLC_18_20June_2018.pdf?ua=1

Mansfeld, M., Ristola, M., & Likatavicius, G. (2015). *HIV/AIDS Programme in Kyrgyzstan. Evaluation report* (pp. 1–84). WHO/Europe; Centre for Health and Infectious Disease Research. Retrieved February 17, 2023, from http://www.euro.who.int/__data/assets/pdf_file/0005/273308/HIV-Programme-Review-in-Kyrgyzstan.pdf?ua=1

Masikini, P., & Mpondo, B. C. T. (2015). HIV drug resistance mutations following poor adherence in HIV-infected patient: A case report. *Clinical case reports, 3*(6), 353–356. https://doi.org/10.1002/ccr3.254

Maytiyeva, V. S., Chokmorova, U. Zh., Ismailova, A. D., Asybaliyeva, N. A., Yanbukhtina, L. F., Sarybayeva, M. E., et al. (2015). *Stranovoy otchet o dostignutom progresse v osushchestvlenii global'nykh mer v otvet na vich-infektsiyu za 2014 god Kyrgyzskaya Respublika [Country report on progress in implementation of the 2014 global response to HIV the Kyrgyz Republic]* (pp. 1–29). Retrieved February 17, 2023, from https://www.unaids.org/sites/default/files/country/documents/KGZ_narrative_report_2015.pdf

Ministry of Health of KR. (n.d.). *Dorožnaâ karta Po optimizacii sistemy okazaniâ protivotuberkuleznoj pomoci v Kyrgyzskoj Respublike na 2016–2025 gody [Roadmap for optimizing the TB care system in the Kyrgyz Republic for 2016–2025]*.

Murzalieva, G., Aleshkina, J., Temirov, A., Samiev, A., Kartanbaeva, N., Jakab, M., et al. (2009). *Tracking global HIV/AIDS initiatives and their impact on the health system: The experience of the Kyrgyz Republic: Final report* (pp. 1–89). Royal College of Surgeons in Ireland. Retrieved March 4, 2023, from https://repository.rcsi.com/articles/report/Tracking_Global_HIV_AIDS_Initiatives_and_their_Impact_on_the_Health_System_the_experience_of_the_Kyrgyz_Republic/10776524/1

Murzalieva, G., Kojokeev, K., Manjieva, E., Akkazieva, B., Samiev, A., Botoeva, G., et al. (2007). *Tracking global HIV/AIDS initiatives and their impact on the health system: The experience of the Kyrgyz Republic: Context report* (pp. 1–48). Center for Health System Development; American University of Central Asia. Retrieved March 3, 2023, from http://elibrary.auca.kg/bitstream/123456789/220/1/Tracking%20Global%20HIV-AIDS%20Initiatives_AUCA.pdf

Nasakt. (2015). *Assessment of needle and syringe exchange programs in Kyrgyzstan* (3). Tangled Vines. Retrieved February 14, 2023, from https://liketangledvines.wordpress.com/2015/07/31/assessment-of-needle-and-syringe-exchange-programs-in-kyrgyzstan-3/

Semerik, O., Berdsli, K., Datar, A., & Dad'yan, M. (2014). *Analiticheskiy obzor rekomendatsiy v sfere VICH-infektsii dlya Kazakhstana, Kyrgyzskoy Respubliki i Tadzhikistana (2007-2012)*. Health Policy Project Futures Group. Retrieved February 5, 2023, from https://www.healthpolicyproject.com/pubs/205_RusHPPFinaldraftFORMATTED.pdf

Spicer, N., Bogdan, D., Brugha, R., Harmer, A., Murzalieva, G., & Semigina, T. (2011a). "It's risky to walk in the city with syringes": Understanding access to HIV/AIDS services for injecting drug users in the former Soviet Union countries of Ukraine and Kyrgyzstan. *Globalization and Health, 7*, 22. https://doi.org/10.1186/1744-8603-7-22

Spicer, N., Harmer, A., Aleshkina, J., Bogdan, D., Chkhatarashvili, K., Murzalieva, G., et al. (2011b). Circus monkeys or change agents? Civil society advocacy for HIV/AIDS in adverse policy environments. *Social Science & Medicine, 73*(12), 1748–1755. https://doi.org/10.1016/j.socscimed.2011.08.024

Stover, J., Korenromp, E. L., Blakley, M., Komatsu, R., Viisainen, K., Bollinger, L., & Atun, R. (2011). Long-term costs and health impact of continued global fund support for antiretroviral therapy. *PLoS One, 6*(6), 1–7. https://doi.org/10.1371/journal.pone.0021048

Subata, E., Moller, L., & Karymbaeva, S. (2016). *Evaluation of opioid substitution therapy in Kyrgyzstan* (pp. 1–10). WHO/Europe. Retrieved February 17, 2023, from https://www.researchgate.net/publication/307966135_Evaluation_of_Opioid_Substitution_Therapy_in_Kyrgyzstan

UNAIDS. (2015). *How AIDS changed everything. MDG 6: 15 years, 15 lessons of hope from the AIDS response* (pp. 1–543). Retrieved March 2, 2023, from https://www.unaids.org/sites/default/files/media_asset/MDG6Report_en.pdf

UNAIDS. (2020). *Stranovoj otčet o dostignutom progresse—Kyrgyzstan Global'nyj monitoring èpidemii SPIDa 2020 [Country progress report—Kyrgyzstan global AIDS monitoring 2020]* (p. n.p.). Retrieved February 17, 2023, from https://www.unaids.org/sites/default/files/country/documents/KGZ_2020_countryreport.pdf

UNDP. (2013). *Annual report on implementation of grants provided by the Global Fund to fight AIDS, Tuberculosis and Malaria in Kyrgyzstan—2012* (pp. 1–52). Retrieved February 17, 2023, from https://www.kg.undp.org/content/kyrgyzstan/en/home/library/hiv_aids/annual-report-on-the-implementation-of-grants-provided-by-the-gl1.html

UNDP. (2014). *Annual report on the implementation of grants provided by the Global Fund to fight AIDS, Tuberculosis and Malaria in Kyrgyzstan—2013* (pp. 1–70). UNDP. Retrieved February 3, 2023, from https://www.kg.undp.org/content/kyrgyzstan/en/home/library/hiv_aids/annual-report-on-the-implementation-of-grants-provided-by-the-gl.html

UNDP. (2015a). *Annual report on the implementation of UNDP project in support of the Government of the Kyrgyz Republic, funded by The Global Fund to Fight AIDS, Tuberculosis and Malaria—2014* (pp. 1–108). UNDP. Retrieved February 3, 2023, from https://www.kg.undp.org/content/kyrgyzstan/en/home/library/hiv_aids/gfatmannualreport_eng.html

UNDP. (2015b). *Newsletter: Grants on HIV, TB and malaria.* April 2014 (pp. 1–10). Retrieved March 3, 2023, from https://www.kg.undp.org/content/kyrgyzstan/en/home/library/hiv_aids/april-2014-newsletter%2D%2Dgrants-on-hiv%2D%2Dtb-and-malaria.html

UNDP. (2015c). *Newsletter: Grants for HIV, tuberculosis and Malaria.* January 2014 (pp. 1–10). Retrieved March 26, 2020, from https://www.undp.org/content/dam/kyrgyzstan/Publications/hiv-tb-malaria/2014/kgz_Newsletter%20UNDP%20GF_January%202014_ENG.pdf

UNDP. (2015d). *Newsletter Grants for HIV, tuberculosis and Malaria.* July 2014 (pp. 1–11). Retrieved March 26, 2020, from https://www.undp.org/content/dam/kyrgyzstan/Publications/hiv-tb-malaria/2014/kgz_Newsletter%20UNDP%20GF_July%202014_ENG.pdf

UNDP. (2015e). *Newsletter: Grants on HIV, TB and malaria.* June 2014 (pp. 1–10). Retrieved February 7, 2023, from https://www.kg.undp.org/content/kyrgyzstan/en/home/library/hiv_aids/june-2014-newsletter%2D%2Dgrants-on-hiv%2D%2Dtb-and-malaria.html

UNDP. (2020a). *Proekt PROON\Global'nogo fonda—Novosti za maj 2020: Adaptiruâ naš otvet k novomu kontekstu [UNDP\Global Fund project—News for May 2020: Adapting our response to a new context].* Retrieved February 20, 2023, from https://www.undp.org/sites/g/files/zskgke326/files/migration/kg/2020.05.RUS.pdf

UNDP. (2020b). *Proekt PROON\Global'nogo fonda—Novosti za noâbr' 2020: Prava pacientov—na pervoe mesto! [UNDP/global fund project—November 2020 Newsletter: Rights patients come first!]* (p. n.p.). Retrieved February 17, 2023, from https://www.undp.org/sites/g/files/zskgke326/files/migration/kg/2020.11.RUS.pdf

UNDP. (2021a). *Proekt PROON / Global'nogo fonda: novosti za mart: Novyj grant Global'nogo fonda dlâ kompleksnoj bor'by s VIČ i tuberkulezom [UNDP/Global fund project: March update: New Global fund grant for comprehensive response to HIV and TB]* (p. n.p.). Retrieved February 17, 2023, from https://www.undp.org/sites/g/files/zskgke326/files/migration/kg/2021.03.RUS.pdf

UNDP. (2021b). *Proekt PROON / Global'nogo fonda: novosti za sentâbr': Zâšita uâzvimyh grupp naseleniâ [UNDP/global fund project: September update: Protecting vulnerable populations].* Retrieved February 17, 2023, from https://www.undp.org/sites/g/files/zskgke326/files/migration/kg/2021.09.RUS.pdf

UNDP. (2021c). *Proekt PROON / Global'nogo fonda: novosti za avgust: Graždanskogosudarstvennoe partnerstvo dlâ lučšego kontrolâ nad VIČ i tuberkulezom [Project UNDP/Global background: News for August: Civil-state partnership for better control of HIV and tuberculosis].* Retrieved February 16, 2023, from https://www.undp.org/sites/g/files/zskgke326/files/migration/kg/2021.08.RUS.pdf

UNDP. (2023). *Effective TB and HIV control project in the Kyrgyzstan.* Retrieved February 15, 2023, from https://www.kg.undp.org/content/kyrgyzstan/en/home/projects/effective-tb-and-hiv-control-project-in-the-kyrgyzstan/

van den Boom, M., Mkrtchyan, Z., & Nasidze, N. (2015). *Review of tuberculosis prevention and care services in Kyrgyzstan 30 June–5 July 2014 Mission report* (pp. 1–95). Retrieved February 17, 2023, from http://www.euro.who.int/__data/assets/pdf_file/0010/287803/Review-of-tuberculosis-prevention-and-care-services-in-Kyrgyzstan.pdf?ua=1

WHO/Europe. (2011). *Tuberculosis country work summary.* Retrieved February 17, 2023, from http://www.euro.who.int/__data/assets/pdf_file/0004/185890/Kyrgyzstan-Tuberculosis-country-work-summary.pdf

WHO/Europe. (2014). *Regional joint WHO/GFATM TB priority investment setting and technical assistance mechanism* (TBTEAM) meeting (pp. 1–20). https://apps.who.int/iris/bitstream/handle/10665/129635/Regional%20joint%20WHO%20GFATM%20TB%20Priority%20Investment%20Setting%20and%20Technical%20Assistance%20Mechanism%20%28TBTEAM%29%20meeting.pdf?sequence=1&isAllowed=y

Wolfe, D. (2005). *Pointing the way: Harm reduction in Kyrgyz republic* (pp. 1–60). Harm Reduction Association of Kyrgyzstan. Retrieved February 17, 2023, from https://core.ac.uk/download/pdf/11872287.pdf

Zardiashvili, T., & Garmaise, D. (2017). *Kyrgyzstan's program continuation funding request to the Global Fund provides little information on the proposed program.* Retrieved April 19, 2019, from http://www.aidspan.org/gfo_article/kyrgyzstan%E2%80%99s-program-continuation-funding-request-global-fund-provides-little

# 10

# Aid Relationships and Power Dynamics in the Global Fund Grants

This chapter discusses the types of relationships between stakeholders involved in the Global Fund to Fight AIDS, Tuberculosis and Malaria (the Global Fund) grants to the Kyrgyz Republic. This discussion builds on the findings of the previous chapters. Chapter 2 outlined the general analytical framework used to analyze the relationships between providers and recipients of aid. It also introduced the analytical categories used to delineate the stakeholders: "donors" or aid providers, "recipient state" or state organizations receiving the assistance, and "civil society organizations" (CSOs) or nongovernmental organizations (NGOs) involved in health aid. Chapter 4 further elaborated on how the structural factors relevant to relationships and sustainability, namely, aid predictability and flexibility on the providers' side, as well as capacity and dependency on the recipients' side, evolved in the Global Fund grants. Chapter 8, in its turn, disentangled the roles of stakeholders in reference to the abovementioned analytical categories in initiating, designing, implementing, and monitoring the grants. Building on the findings of these chapters, this chapter discusses the power dynamics among stakeholders and defines the following types of relationships between actors involved in the grants (Table 10.1).

© The Author(s) 2024
G. Isabekova, *Stakeholder Relationships And Sustainability*, Global Dynamics of Social Policy, https://doi.org/10.1007/978-3-031-31990-7_10

**Table 10.1** Relationships between stakeholders involved in the Global Fund grants to the Kyrgyz Republic

| Actors | Reference | Type of relationships |
|---|---|---|
| The Global Fund—grant-recipient NGOs | Donor–civil society organizations | "Utilitarian" approach |
| The Ministry of Health, represented by the National Center of Phthisiology, and the Republican AIDS Center—NGOs | Recipient state–civil society organizations | "Utilitarian" approach |
| The Global Fund—other development organizations, working in tuberculosis and HIV/AIDS | Donor–donor | Coordination |
| The Global Fund—Ministry of Health, represented by the National Center of Phthisiology and the Republican AIDS Center | Donor–recipient state | "Unequal" cooperation |

# 10.1   Donor–CSOs: "Utilitarian" Approach

I define the relationships between the Global Fund and NGOs as evincing a "utilitarian" approach because of the equal participation of both actors throughout the project, structural factors favorable to hierarchical relations, and power dynamics between these stakeholders.

First, the participation of grant-recipient NGOs through the grant realization process was uneven. Civil society participation and empowerment are at the cornerstone of the Global Fund's mission, and this emphasis also found its reflection in the grant design and implementation phases. The organization was critical to NGOs' engagement in designing the grant applications and implementing them on equal terms with state organizations. However, the NGOs' involvement in the grant monitoring process was limited to data provision, filling no decision-making functions, reaffirming the conventional provider-recipient relations in which the local NGOs felt that they were relegated to being themselves mere grant implementers.

Second, the structural factors accommodated hierarchical relations between stakeholders. The Global Fund strives to provide predictable assistance, but the organizational dependence on replenishment cycles, the guaranteed financing is confined to three years. The organization also aims to ensure the compliance of its grants with recipient countries' priorities, among other things by ensuring the broader engagement of stakeholders in designing the grants. Still, the organizational structures seem to hinder the adaptability of grants to changes and suggestions, as alluded to in the case of the limited adjustment to NGO suggestions. Though the capacities of grant-recipient organizations greatly vary depending on their size, experience, and the areas they are working in, the services provided by NGOs appear highly dependent on external assistance.

Though surely case-dependent, a combination of the four structural factors, together with the NGOs' limited engagement in the monitoring process, lay down the basis for power dynamics to unfold the way they did:

The relationship between the financier and local NGOs vividly demonstrates the dilemma presented by, but also the interrelation between the "power to" and "power over." The former emerged through structural bias favoring the roles of NGOs in health aid and the strong organizational support for it. By contrast, the latter came into being through social order and discipline favoring the predictability of outcomes over flexibility in the grant realization process.

The structural bias and constraints promoted by the Global Fund in its grants created the "power to" for the local NGOs. Following Haugaard (2003, p. 107), structural biases occur through specific social order, which creates possibilities that empower or disempower actors through structural constraints. The Global Fund's emphasis on civil society participation (social order) provided a window of opportunity (empowerment) for the local NGOs to participate in the decision-making and implementation of grants. This social order was further supported by structural constraints. The Global Fund's requirements concerning the establishment of the Country Coordinating Mechanism (CCM) and the incorporation of a human rights perspective into the grants contributed to the involvement of local NGOs and persons affected by diseases in designing the country's applications.

Similar constraints were applied to the implementation of grants. The Global Fund facilitated NGOs' involvement in the project implementation process through its "dual-track" financing or channeling of funds via state and non-state actors. Though arduous, this collaboration between actors nevertheless set a precedent for joint lobbying for the continuity of tuberculosis (TB) and human immunodeficiency virus infection and acquired immune deficiency syndrome (HIV/AIDS) services beyond the duration of the grants.

Notably, the deference to the abovementioned structural bias was backed up by sanctions. The Global Fund rejected the country's HIV proposal due to the noncompliance of the Country Coordinating Mechanism Country with the "minimum requirements" of civil society representation. This gatekeeping action signaled national stakeholders to take this requirement seriously in order to access the grants. The Global Fund financing available for the capacity building offered means for "correcting" this situation and ensuring sufficient civil society participation in drafting the country's grant applications. This access to additional financing represented what Baldwin (1971, p. 23) would call a "positive sanction" aimed to reward a stakeholder for its acquiescence to the social order. Still, this positive sanctioning operated together with gatekeeping the Global Fund exercised to support the structural bias in favor of the roles of civil society organizations in health aid and, more specifically, its grants.

At the same time, the grant realization process suggests that the Global Fund has also exercised the "power over" local NGOs through social order and discipline to ensure the predictability of grant outcomes. In the implementation stage, the grants demonstrated limited openness to NGO suggestions. The indicators and objectives stated in the country's applications constituted structural constraints that disregarded recommendations made beyond the design phase. This closedness aimed to ensure the predictability of outcomes in relation to grant objectives.

Though logical in an organizational setting, this predictability considerably hinders responsiveness to the changing environment in which health aid is implemented. The "power over" created through predictability here was the outcome of a social order in which actors built structures concerning specific meanings (Haugaard, 2003, p. 107), in this case in the form of grants targeted at fighting certain diseases. The limited openness to NGO suggestions brought specific associations with the division of labor corresponding to the hierarchical relations between the "provider" and the "recipient" of aid. In these relationships, the recipient is fully aware of and limited by what a civil society representative recalled as functionary duties that were bestowed upon them. This hierarchy is further strengthened by NGOs' limited English proficiency and limited awareness of Global Fund regulations. In these circumstances, the local NGOs became "passive" recipients of assistance (Rasschaert et al., 2014, p. 7) and are reduced to the status of being the "means" for implementing it (Morgan, 2001, p. 221).

Another source of predictability was discipline in defining the roles and responsibilities of actors in the monitoring process. Following Haugaard (2003, p. 108), "practical consciousness knowledge" and "socialization through discipline" can be used to ensure the reproduction of existing power structures. As Sub-Recipients, the local NGOs do not directly participate in the monitoring process but rather report to the primary recipient, Country Coordinating Mechanism, and Local Fund Agent (discipline). This socialization of NGOs is based on the idea that by assigning specific roles to actors involved in grants, the Global Fund can avoid conflicts of interest. This idea is further supported by practical knowledge the NGOs apply in their reports. Sub-Recipients report using grant indicators stated in the country's application and following administrative and financial regulations and changes (if any) that the Global Fund representative communicates during her visits to the country. Though affected by the financier's regulations, Sub-Recipients have little say in the monitoring process. This power created through discipline reiterates hierarchical relations between the donor and the recipient. NGOs working with multiple donors spend extensive time reporting to each using different templates and complying with at times contradictory requirements.

Overall, the relationship between the Global Fund and the Sub-Recipient NGOs vividly demonstrates the affinity between the "power to" and "power over." It also shows that both conflict (limited flexibility) and consensus (promoting civil society) were integral to this relationship. Reiterating the hierarchy between the provider and the recipient of aid, this relationship is still based on freedom, an essential element of power relations (Foucault, 2002). Both providers and recipients of aid have the freedom to choose to (dis)engage in relationships with each other.

What incentives does the Global Fund have to utilize the "utilitarian" approach toward grant-recipient NGOs? First, civil society involvement corresponds to its organizational objectives and provides access to the groups targeted by the grants (e.g., commercial sex workers, men who have sex with men, injecting drug users, and others). These groups are often close to the state health care system due to stigma and discrimination in society. The expertise and context-awareness (Pape, 2014) make local NGOs essential to the provision of health services and health promotion among vulnerable groups (IO Partner 4). Furthermore, CSOs may advocate for a broader range of issues. For instance, NGOs can raise public awareness and demand action on issues that ministries and state agencies would not prioritize due to budget deficits (CSO 8). Popularity concerns and voters' support also restrict state actions on issues, such as the rights of sexual minorities, in conservative contexts.

Why do local NGOs engage in the "utilitarian" approach with the Global Fund? Participation in the Global Fund grants, even in terms of the "utilitarian" approach, offers capacity building and involvement in decision-making. Thus, through their relationships with the Global Fund, local NGOs have the possibility to advocate for the interests of their organizations and the groups they claim to represent. As Sub-Recipients, they also have access to financing. In the context of increasing competition among NGOs due to decreasing assistance for TB and HIV/AIDS, access to financing allows Sub-Recipient NGOs to continue their activities and ensure their own survival. In this way, the interaction with the Global Fund provides NGOs access to resources. Closely associated with power (Giddens, 1984), resources are crucial to understanding it. However, as the analysis of donor–CSO relations in the "Community Action for Health" project shows, resources alone do not define the power, nor does having similar access to resources mean that actors necessarily exercise similar power.

## 10.2 Recipient State–CSOs: "Utilitarian" Approach

The Ministry of Health, the National Center of Phthisiology, and the Republican AIDS Center pursue a "utilitarian" approach toward their collaboration with local NGOs, primarily driven by an interest in securing donor funding rather than a genuine perception of NGOs as equal partners.

Interestingly, the structural factors could have equally laid down the basis for the state organizations' "empowerment" approach toward local NGOs. The CSOs are financially independent of state agencies, which also explain their ability to raise "uncomfortable" issues, such as the rights of commercial sex workers and men who have sex with men. Furthermore, local NGOs seem to have greater capacity in terms of human resources than the Ministry of Health and its agencies. Thus, though the services of both actors are dependent on external aid, the actors themselves are inter-related but financially independent from each other. This situation may change as the country progresses with social contracting for NGOs, which will make them accountable to state agencies (Chap. 9). However, within the framework of the Global Fund grants, the structural factors did not favor hierarchical relations between the state and nongovernmental organizations.

Similarly, stakeholder participation in grants did not favor hierarchy among stakeholders. Both actors equally participated in the grant realization process, and both had limited roles in the monitoring process. Still, the project life cycle showed that NGO engagement in health aid was imposed on state organizations by the conditions set by the Global Fund. This involuntary engagement found its reflection in the power dynamics between the state and civil society organizations, as discussed below.

The accountability of public services, promoted by the Government of the Kyrgyz Republic and development partners, allowed for civil society scrutiny over state institutions. As demonstrated in the project cycle, local NGOs scrutinize the government in terms of use of funds. More specifically, they can send requests to a relevant state institution to obtain information on financing and other matters. The state organizations are expected to respond to public requests (including NGOs) within two weeks. In this way, the government aims to ensure the openness and responsiveness of state institutions to public concerns.

This social order, created by the government, opens up local NGOs access to necessary data. However, the form of reply and information is not necessarily straightforward. One civil society interviewee noted that her organization had to hire an external consultant to comprehend the information provided by the Ministry of Finance. Nevertheless, it did gain access to data necessary for analyzing the use of finances, with the goal of pointing at possible areas for rearrangement to ensure additional funding for the areas the organization advocated for. The interviewee highlighted that, if previously the state officials could refer to the budget deficit, now NGOs could show that the required funding was available by referring to the data the Ministry provided as evidence for it (CSO 8). In addition to the regulation mentioned above, this NGO scrutiny over state agencies is now possible thanks to the Sector Wide Approach. During the meetings with donors and civil society organizations, the Ministry of Health reports on the achievement of indicators stated in the national program and the use of funding. Thus, the NGOs can obtain data at the national level and on matters of particular interest to them.

At the same time, the project life cycle showed that state agencies contended with civil society participation in decision-making. Promoted by the Global Fund, this social order aimed to empower persons affected by diseases and local NGOs representing them to ensure their participation in drafting and implementing grants (see the previous subsection). The Global Fund initially rejected the country's application as the CCM did not comply with "minimum requirements." State organizations outnumbered the civil society representatives who had limited capacity to fully participate in designing the grants (see Chap. 8). This situation, along with the issues between state and civil society actors during the grant implementation process, hints at the state organizations' unwillingness to accept the social order promoted by the Global Fund. This unwillingness also relates to the government's perception of its role as the leading actor in health care.

In addition to opposing the social order on civil society participation, the state partners have aimed to exercise their "power over" local NGOs by creating a favorable system of thought. The state institutions, particularly the Ministry of Health, advocate for the central role of the government in health care. As demonstrated in the project life cycle, the former Minister of Health has repeatedly questioned the expertise and ability of NGOs to provide health care services. He also advocated for scrutinizing their use of finances by highlighting the leading role of the Ministry in the health sector. These remarks were not limited to a single politician. In both the grant implementation and monitoring stages, state officials viewed the NGOs as "grant eaters" (Spicer et al., 2011, p. 1750) rather than equal implementation partners (Murzalieva et al., 2009). These systemic biases about stakeholders and their roles are based on two premises. First, health in the post-Soviet region is viewed as purely medical and not a social phenomenon. Second, health care remained the state domain, which is also reflected in the leading role of state institutions in the regulation and provision of health care services. Both interpretations correspond to the *Semashko* health care system present in the former Soviet Union, in which the government was the main financier, regulator, and service provider. Due to budget deficits, state organizations gave up on the financial part of this obligation, but seemed to be keen on keeping their authority in the two other areas.

Notably, the remarks about the use of funding by NGOs and the role of the state were limited to individual figures during the data collection process for this book in 2018. However, on June 26, 2021, these statements materialized into a new law necessitating NGOs to report on the sources of their financing and the use of these funds (Government of KR, 2021). Accordingly, the state organizations gained access to the financial data they had longed for.

What interests did stakeholders have in the selected form of aid relationships? The interaction of the Ministry of Health and its agencies with local NGOs is largely driven by access to donor financing. Although openly disagreeing with the work of the CSOs, the state institutions continued to follow the Global Fund's requirements because incompliance would have resulted in a rejection of the country's grant application. A similar logic lay behind the Ministry's collaboration with NGOs during the negotiations with the Ministry of Finance and Parliament. NGOs' advocacy was the key to increasing the TB and HIV/AIDS financing necessary for the gradual transition of the country from the Global Fund's assistance. In both cases, the Ministry and other state agencies seem to perceive the local NGOs as a means to an end, not as equal partners. Furthermore, the NGOs provided access to groups, such as commercial sex workers, men who have sex with men, and injecting drug users, that are typically beyond the outreach of state health care organizations. In so doing, they offer expertise and skills necessary to combatting HIV/AIDS (Pape, 2014).

What are the NGOs' interests in engaging in a "utilitarian" approach with state organizations? NGOs interact with state institutions largely due to their dominant role in health care. As one civil society representative noted, donors cover some activities, but the government is still responsible for regulating health care facilities, providing social benefits, and the rule of law—all relevant to the NGOs' work (CSO 6). Through collaboration with state health institutions, NGOs gain access to public resources and infrastructure critical to achieving sustainable results (Pape, 2014). The role of the government can grow only further in the context of decreasing donor financing for TB and HIV/AIDS and the introduction of social contracting to ensure continuous funding for NGO services (see Chap. 9). In these conditions, collaboration with state organizations, particularly on the terms of a "utilitarian" approach, becomes even more sensible.

## 10.3   Donor–Donor: Coordination

Based on the actors' roles throughout the project life cycle and the lack of a hierarchy and power dynamics, the relationships between the Global Fund and other donors can be qualified as coordination.

The project life cycle showed the continuous involvement of development organizations working in tuberculosis and HIV/AIDS in the realization of the Global Fund grants. The engagement seems to work well, particularly in the design and implementation phases. However, the relationships among actors are somewhat limited in the monitoring process, which causes duplication of efforts and an additional burden on national stakeholders having to report to different aid providers.

In contrast to donor–recipient relations, there is no explicit hierarchy in the relationship among donors. Therefore, the structural factors are not prone to a ranking among donors. For instance, although leading in financial terms, the Global Fund adheres to standards set by other organizations that have established themselves in particular niches (e.g., the United Nations organizations).

In terms of power dynamics, relationships of the Global Fund with other development actors in tuberculosis and HIV/AIDS combine the attributes of both "power over" and "power to." The former is related to the preeminent position of some organizations as norm-setters in health, whereas the latter concerns the ability of organizations to work with each other.

First, the World Health Organization (WHO) and the Joint United Nations Programme on HIV/AIDS (UNAIDS) exercise "power over" other organizations working in health through their expertise. Explicitly devoted to health, the WHO has established itself as a norm-setter in health (Kaasch, 2015). Its recommendations are equally followed by the state, civil society, and donor organizations. For instance, in Kyrgyzstan, the WHO recommendations provided the basis for the clinical protocols on methadone substitution therapy (Subata et al., 2016), HIV treatment (Murzalieva et al., 2009), and treatment of TB/HIV coinfection (Government of KR, 2012). In addition to the recipient state, the WHO recommendations are equally followed by donor organizations in the health sphere. The Global Fund, for instance, may specify the procurement of medical products accredited by the WHO and compliance of treatment activities with WHO standards, as it did in the grant to Armenia (see Global Fund, 2009, pp. 9–12). Similarly, the Country Coordinating Mechanism introduced through the Global Fund grants connates with the "Three Ones" principles (one national AIDS framework, one national AIDS authority, and one system for monitoring and evaluation—all categories are listed verbatim) promoted by UNAIDS (2005, p. 8). Though UNAIDS is less salient in comparison to its counterpart, its regulations are equally followed in HIV/AIDS.

Why do other stakeholders adhere to the WHO and UNAIDS regulations and suggestions? Again, following Haugaard (2003, pp. 104–105), this compliance could be on the grounds that actors perceive a proposed system of thought more than a "simply arbitrary convention." The WHO positions itself as an "evidence-based multilateral agency" (Kaasch, 2015, p. 27) and promotes a typical "evidence-based" approach to health. Though less assertive, a similar system of thought, based on evidence, could be attributed to UNAIDS. This reference to the evidence suggests that the non-arbitrariness of norms suggested by these organizations has a scientific underpinning, which serves as a basis for reification (see Haugaard, 2003, pp. 104–105). In other words, by following the WHO and UNAIDS guidelines and recommendations, organizations, in a way, comply with the scientific evidence.

The "power to" among donors manifests itself through their coordination with each other. This coordination follows the social order outlined by the Paris Declaration on Aid Effectiveness (2005) and the following Accra Agenda for Action (2008) (hereinafter "Paris Agenda") outlined the five principles of development assistance that became the synonym for effectiveness and guiding norms for aid in the twenty-first century (Brown, 2020). The five principles are ownership, alignment, harmonization, managing for results, and mutual accountability (OECD, n.d.).

The Paris Agenda set the social order recognized and reproduced by development actors. The analysis of the Global Fund grants to Kyrgyzstan demonstrated that the multilateral organization closely coordinated its activities with other donors in the design and implementation phases. At the core of this coordination lies the principle of harmonization of donor activities, aimed at avoiding the duplication of efforts to ensure the greater effectiveness of aid. Similarly, donors jointly supported the Ministry of Health, its agencies, and NGOs in designing the grant applications to the Global Fund. This support complied with the principles of ownership. By building the capacity of national stakeholders, donors aimed to support their ownership over development aid. The division of labor among donors was intended to avoid duplications and ensure the complementarity of their support (harmonization). By following these principles, donors confirmed the meaning of aid effectiveness in the social order promoted by the Paris Agenda. This recognition and reproduction of meaning are at the core of the social order (Haugaard, 2003, pp. 90–93).

However, during the grant monitoring process, the donor coordination cracks due to each donor's visibility concerns. The duplication problems in monitoring indicate the limits of donors' adherence to the harmonization principle. In reference to the social order, Haugaard (2003, p. 96) notes that the structures accepted and taken for granted today were fought for in the past. In this way, the limits of harmonization in monitoring may suggest that the social order has not fully established itself. Furthermore, donors' accountability to their financiers additionally hinders the realization of harmonization principles. The Global Fund is expected to demonstrate the result of its activities by specifying the number of patients treated, health products distributed, and training sessions organized. This health impact of the Global Fund is essential to its continued funding by donor countries (see Chap. 4). Other donor organizations have similar concerns.

Despite the consensus over the harmonization principles, there is conflict regarding its implementation. Both consensus and conflict are integral to the social order (Haugaard, 2003, p. 90). Notwithstanding the issues observed during the monitoring phase, the relationships among donors still qualifies as coordination due to the visible adherence to non-duplication in other stages of the grant realization process.

What interests did stakeholders have in the form of aid relationships selected? The abovementioned social order on aid effectiveness is essential to understanding the actors' interests in coordinating their activities with each other. There are no explicit sanctions for noncompliance spelled out in the Paris Agenda, but rather peer pressure standing behind this Agenda, supported by the global call for the sustainable use of resources. The project cycle shows that as the share of its grants to the country decreases, the Global Fund has intensified its coordination with other donors to ensure the sustainability of its TB and HIV/AIDS activities. Other donors have similar concerns. Yet, power dynamics among donors have remained relatively equal throughout the project life cycle.

## 10.4   Donor–Recipient State: Unequal Cooperation

The relationships between the Global Fund, the Ministry of Health, the National Center of Phthisiology, and the Republican AIDS Center qualifies as unequal cooperation.

The domination of the Global Fund is visible throughout the project cycle, except for during the initiation phase. Thus, the Global Fund project unequivocally increased the type and breadth of services offered to TB and HIV patients affected by TB and HIV. Still, the initiative behind the TB and HIV/AIDS services was already in place before this project. For this reason, the grant activities and objectives corresponded to the issues present and pressing to the country. However, during other phases, the recipient state complied with the Global Fund recommendations and regulations with few reservations.

Additionally, the structural factors remained in favor of hierarchical relations. The Global Fund attempted to increase the predictability of its assistance by introducing continued financing for well-performing projects and announcing the list of countries eligible for grants. However, grant disbursements are guaranteed for only three years, due largely to the organizational dependence on replenishments by its financiers every three years. Although relatively independent from the Global Fund's technical assistance, government institutions largely rely on financing for prevention and treatment programs. The Global Fund project also provides limited space for change during the implementation process. Time- and effort-consuming bureaucratic processes discourage state agencies from suggesting any revisions to the initial grant agreed to with the Global Fund. All these factors, namely, the Global Fund's limited flexibility, aid dependency, and capacity issues of government institutions, contributed to the situation in which the aid recipient fully complied with the terms established by the aid provider as long as the donor controlled the finances.

The combination of stakeholders' roles through the project life cycle and structural factors in the grants laid down the basis for power dynamics contributing to unequal cooperation. Overall, the power relations between the Global Fund, the Ministry of Health, the National Center of Phthisiology, and the Republican AIDS Center were probably the most comparatively complex. Combining the "power to" and "power over," the Global Fund has opted for a more diverse array of sources of power, including social order, structural bias/constraints, discipline, coercion, and systems of thought.

The Global Fund empowered the recipient state ("power to") through social order. Ownership, or compliance of development aid with the needs and structures of aid recipient countries, is one of the five norms promoted by the Paris Agenda discussed in the previous section. The support for the existing structures is inherent to the effective development assistance promoted by the Paris Agenda. This social order, reproduced and confirmed by donor organizations, empowered the recipient state by providing financial and technical assistance to the national monitoring and evaluation systems. As discussed in the project life cycle, the Global Fund integrated its monitoring indicators into national systems and assigned a part of grant finances to strengthen them (Chap. 8). This assistance did not solve structural issues, but still advanced parts of the health care monitoring system relevant to grants. Through its support, the Global Fund reproduced and confirmed the meaning of "ownership" stated in the Paris Agenda and, in so doing, confirmed the social order on aid effectiveness empowering the grant-recipient state.

At the same time, the Global Fund exercised the "power over" the recipient state through structural biases/constraints, empowering NGOs and, in so doing, challenging the dominant role of state organizations in health care. It also turned to discipline, limiting the roles of state agencies involved in grants in the monitoring stage. The organization also resorted to coercion in response to grant misappropriation, combined with a justification to keep an international organization as the primary recipient of its grants.

First, the Global Fund exercised the "power over" the recipient state through structural biases. According to Haugaard (2003, p. 107), structural biases occur when social order produces power through structural constraints that eventually (dis)empower others. As the project cycle shows, the Global Fund regulations on co-financing, human rights, and CCM considerably shaped the content of grant applications, along with its recommendations for a joint application for two diseases. As noted earlier, although not obligatory, the recommendations nevertheless were followed by grant applicants, most likely in order to secure positive feedback from a financier (see Chap. 8). Both recommendations and regulations represented structural constraints that intended to ensure the predictability and stability of a system by enabling desired outcomes (Haugaard, 2003, p. 94). Thus, they intended to demonstrate growing state funding for target diseases, support for human rights, and inclusion of civil society organizations in the decision-making process.

Not necessarily "repressive," these structural constraints may be enabling to some stakeholders but disabling to others (ibid.). For the recipient state, the regulations and recommendations were rather disabling as they supported the role of CSOs in health, both in decision-making and service provision, which are traditional state domains. The authorities were also compelled to increase their financial commitments and introduce changes regarding the rights of vulnerable groups of society. In the context of the continuous budget deficit and rather a conservative attitude toward reproductive health and sexual rights, these changes did not necessarily correspond to voters' or politicians' agendas.

The second source of the Global Fund's "power over" the recipient state was discipline. In both the implementation and monitoring phases, state agencies comply with indicators and activities indicated in the country's proposal. This practical knowledge provides for the "socialization through discipline" that secures existing power relations (Haugaard, 2003, p. 108). Furthermore, the discipline establishes a routine which ensures the predictability of an outcome, as opposed to irregularities unwanted by the existing social order (Haugaard, 2003, p. 106). Grant agreements spell out the responsibilities and rights of all parties. As long as stakeholders comply with these agreements, there is a sense of predictability and foreseen achievement of stated goals.

Nonetheless, the grant implementation and monitoring phases demonstrated the limits of power created by discipline. According to Haugaard (2003, p. 107), compliance with discipline depends on the extent routine is internalized by stakeholders. Implementation and monitoring routines outlined in the agreement and supported by the Global Fund regulations and recommendations intended to preempt irregularities. Yet, the misappropriation of grant finances by the National Center of Phthisiology is an irregularity it did not prevent. The Global Fund audit and investigation outcomes indicated limited internalization of the "routine" by some stakeholders. It also pointed to the mismatch between personal and organizational interests, highlighting the relevance of individual and organizational perspectives on actors.

Further non-reimbursement of missing finances demonstrated a similar limitation of discipline. As noted in the project cycle, the Global Fund repeatedly requested state authorities to repay unaccounted-for finances. The Ministry of Health ignored these requests on the grounds of not having access to the finances, or (allegedly) not having received these requests in the first place (see Chap. 8). The money that had been misappropriated by state official(s) was deducted from the following grant to the country, but this has probably affected TB and HIV patients much more than the Ministry itself. The reaction from the Global Fund was remarkable in that it did not halt the project funding. Instead, it continued its project in the country while combining multiple means of creating "power over" the recipient state.

The third source of power, coercion, followed the delinquency of discipline. Following the misappropriation and mismanagement of grants, the United Nations Development Programme (UNDP) became the Principal Recipient (PR). The organization was proposed by the Country Coordinating Mechanism and approved by the Global Fund. One may disagree with my attribution of coercion here due to the fact that the Country Coordinating Mechanism is composed of national stakeholders, and the Global Fund merely confirmed the choice made by these stakeholders. However, in relation to the recipient state, this decision was indeed coercion. The project cycle shows that the reassignment of the PR functions was a remerging issue repeatedly brought up by state organizations and discussed in the Country Coordinating Mechanism already in 2014. Continuous discussions resulted in the establishment of the Project Implementation Unit under the Ministry of Health. This continuity of discussions and content raised by state officials interviewed for this research points to the presence of a conflict and the fact that the decision to keep the UNDP as the Principal Recipient of grants was made against their will. The notion of willingness is critical to defining the activities of aid providers as coercion.

At the same time, consonant with Arendt's (1970) notion of power, coercion did not represent its strongest form but was used as a measure of last resort (Barnes, 1988, p. 15). Maybe for that reason, to create power, coercion was not applied alone, but rather in combination with the systems of thought, discussed below.

The fourth source of power is the system of thought related to the recipient state's capacity. As discussed in the project cycle, both state and non-state actors may become Primary Recipients of grants as long as they have the necessary capacity to fulfill the related functions. The lack of this capacity was the main justification for the donor's decision to keep the UNDP as the Principal Recipient. Indeed, the Project Implementation

Unit failed to meet the minimum criteria to demonstrate its ability to implement the grants. This failure showed that the Ministry of Health was "not ready" to take over the PR responsibilities (see Chap. 8).

Following this line of argumentation, compliance with specific criteria demonstrates the "capacity" and "readiness" necessary to become the PR. In power created by systems of thought, specific meanings are not just "out there," but instead are the results of knowledge based on "particular interpretative horizons" (Haugaard, 2003, pp. 107–108). Thus, the interpretation of capacity and fulfillment of criteria as necessary preconditions for resuming PR functions supports the decision of the Global Fund by making it non-arbitrary and based on reasoning. It also creates a relevant perception among the state, civil society, and international actors working in the country that the lack of Ministry of Health capacity is the reason for reassigning the UNDP as the Principal Recipient ("social consciousness-sustaining structural practices" in Haugaard's words) (2003, p. 108).

Overall, the relationships between the Global Fund and state organizations combined both "power to" and "power over," generated through the multiple sources discussed above. This multiplicity also points to the fact that the aid relationships did not solely rely on the premise that one actor had resources another wanted to access, but also on how stakeholders used these resources to create power (e.g., coercion and sanctions). The power to and the power over also occurred in a combination of conflict and consensus among actors, vividly demonstrated in the project cycle.

What interests did stakeholders have behind the selected form of aid relationships? It should be noted that despite the "power over," actors still have freedom. For instance, structural bias (here in the form of donor regulations and recommendations) can be changed but may require changing the "rules of the game" (donor–recipient hierarchy), which may be costly for some actors, and therefore they resist doing so (Haugaard, 2003, p. 95). Free to choose, the recipient state opts for compliance with recommendations and regulations as compliance offers access to grant finances and technical assistance the actor would forego otherwise.

The access to resources is essential to understanding the recipient states' interests in unequal cooperation. During an interview in 2018, a state representative noted that 95% of financing for preventive activities, including syringes, condoms, methadone, and lab tests, came from international organizations. For this reason, "willingly or not," the government worked with them as "one team" (State Partner 2). Though decreasing with time, the Global Fund remained critical to TB and HIV/AIDS activities (Chap. 9), which explains the Ministry and its agencies' readiness to engage in unequal cooperation with this organization.

In turn, the Global Fund was interested in working with government institutions due to its key role in regulating and providing health care services. Government authorities are essential to accessing the country's health care system and ensuring the sustainability of health care provision beyond the duration of the grants. Moreover, cooperation with state institutions allows donors to influence national policy (Ancker & Rechel, 2015). For the Global Fund, it meant the ability to advance its agenda on the rights of groups vulnerable to TB and HIV/AIDS.

# References

Ancker, S., & Rechel, B. (2015). HIV/AIDS policy-making in Kyrgyzstan: A stakeholder analysis. *Health Policy and Planning, 30,* 8–18. https://doi.org/10.1093/heapol/czt092

Arendt, H. (1970). *On violence.* Houghton Mifflin Harcourt.

Baldwin, D. A. (1971). The power of positive sanctions. *World Politics, 24*(1), 19–38. https://doi.org/10.2307/2009705

Barnes, B. (1988). *The nature of power.* University of Illinois Press.

Brown, S. (2020). The rise and fall of the aid effectiveness norm. *The European Journal of Development Research, 32*(4), 1230–1248. https://doi.org/10.1057/s41287-020-00272-1

Foucault, M. (2002). The subject and power. In J. D. Faubion (Ed.), *Power essential works of Foucault 1954–1984* (Vol. 3, pp. 326–348). Penguin Books.

Giddens, A. (1984). *The constitution of society. Outline of the theory of structuration.* University of California Press.

Global Fund. (2009). *Amended and restated grant agreement for the rolling continuation channel ('RCC') program.* Retrieved May 10, 2020, from https://data.theglobalfund.org/investments/grant/ARM-202-G06-H-00/2

Government of KR. (2012). *Gosudarstvennaya programma po stabilizatsii epidemii VICH-infektsii v Kyrgyzskoy Respublike na 2012–2016 gody: Postanovlenie*

*Pravitel'stva Kyrgyzskoy Respubliki ot 29 dekabrya 2012 goda N 867.* Retrieved February 23, 2023, from http://cbd.minjust.gov.kg/act/view/ru-ru/93959/20?cl=ru-ru

Government of KR. (2021). *Zakon Kyrgyzskoj Respubliki ot 15 oktâbrâ 1999 goda № 111 "O nekommerčeskih organizaciâh" [Law of the Kyrgyz Republic dated October 15, 1999 No. 111 "On non-profit organizations"].* Retrieved February 4, 2023, from http://cbd.minjust.gov.kg/act/view/ru-ru/274

Haugaard, M. (2003). Reflections on seven ways of creating power. *European Journal of Social Theory, 6*(1), 87–113. https://doi.org/10.1177/1368431003006001562

Kaasch, A. (2015). *Shaping global health policy: Global social policy actors and ideas about health care systems.* Palgrave Macmillan.

Morgan, L. M. (2001). Community participation in health: Perpetual allure, persistent challenge. *Health Policy and Planning, 16*(3), 221–230.

Murzalieva, G., Aleshkina, J., Temirov, A., Samiev, A., Kartanbaeva, N., Jakab, M., Spicer, N. & Network, G. H. (2009). *Tracking global HIV/AIDS initiatives and their impact on the health system: The experience of the Kyrgyz Republic: Final report* (pp. 1–89). Royal College of Surgeons in Ireland. Retrieved March 4, 2023, from https://repository.rcsi.com/articles/report/Tracking_Global_HIV_AIDS_Initiatives_and_their_Impact_on_the_Health_System_the_experience_of_the_Kyrgyz_Republic/10776524/1

OECD. (n.d.). *The Paris declaration on aid effectiveness (2005) and Accra agenda for action (2008).* Retrieved February 15, 2023, from http://www.oecd.org/dac/effectiveness/34428351.pdf

Pape, U. (2014). *The politics of HIV* (Vol. 92). Routledge.

Rasschaert, F., Decroo, T., Remartinez, D., Telfer, B., Lessitala, F., Biot, M., Candrinho, B., & Van Damme, W. (2014). Sustainability of a community-based anti-retroviral care delivery model—A qualitative research study in Tete, Mozambique. *Journal of the International AIDS Society, 17*(18910), 1–10. https://doi.org/10.7448/IAS.17.1.18910

Spicer, N., Harmer, A., Aleshkina, J., Bogdan, D., Chkhatarashvili, K., Murzalieva, G., Rukhadze, N., Samiev, A., & Walt, G. (2011). Circus monkeys or change agents? Civil society advocacy for HIV/AIDS in adverse policy environments. *Social Science & Medicine, 73*(12), 1748–1755. https://doi.org/10.1016/j.socscimed.2011.08.024

Subata, E., Moller, L., & Karymbaeva, S. (2016). *Evaluation of opioid substitution therapy in Kyrgyzstan* (pp. 1–10). WHO/Europe. Retrieved February

17, 2023, from https://www.researchgate.net/publication/307966135_
Evaluation_of_Opioid_Substitution_Therapy_in_Kyrgyzstan

UNAIDS. (2005). *The global task team on improving AIDS coordination among
multilateral institutions and international donors* (pp. 1–34). UNAIDS. Retrieved
February 3, 2023, from https://www.theglobalfund.org/media/1393/repleni
shment_2005romegtt_report_en.pdf?u=636727910200000000

# 11

## "Missing Link"

This book aims to make a theoretical contribution to understanding the interaction between the relevant actors and the impact of that interaction on the sustainability of development assistance for health care. The notion of *impact* in the research question presumes a causal relationship between interaction and sustainability. For this reason, this book refers to causal inferences in qualitative research for guidance. This chapter expands on the formulation of causal inferences within cases covered in this book by referring to causal mechanisms formed in the "Community Action for Health" and the Global Fund grants to Kyrgyzstan. This section is followed by hypotheses that help us understand how the relationships between stakeholders influence the sustainability of health aid. Although developed on the basis of causal mechanisms, these inferences are more general and applicable across cases. This chapter concludes with limitations of causal inferences made in this book.

The intention of the detailed analysis of selected project phases is to provide a basis for causal inferences within these cases. A within-case analysis is essential for the identification of mechanisms linking the cause (aid relationships) and the outcome (aid sustainability) and the factors relevant to these mechanisms (Rohlfing, 2012, p. 12). The project-level analysis offers sufficient context specificity, which is essential for

© The Author(s) 2024
G. Isabekova, *Stakeholder Relationships And Sustainability*, Global Dynamics of Social Policy, https://doi.org/10.1007/978-3-031-31990-7_11

developing causal mechanisms. It provides the necessary environment for the mechanisms to function (see Falleti & Lynch, 2009; Hedström & Ylikoski, 2010), because depending on the context, the same mechanism may produce different outcomes (Beach & Pedersen, 2019). I use the concept of a social mechanism as "a constellation of entities or activities that are linked to one another in such a way that they regularly bring about a particular type of outcome" (Hedström, 2005, p. 11). This approach to mechanisms as units composed of "entities" and "actions" is also in accord with scholars working in this field (see Beach & Pedersen, 2019, p. 70; Hedström & Swedberg, 1998; Rohlfing, 2012, p. 35).

## 11.1  The "Community Action for Health" Project

Based on the relationship between the actors throughout the project and the power dynamics between them, I have defined the following types of aid relationships: "empowerment" approach (donor–civil society organizations [CSOs]), "utilitarian" approach (recipient state–CSOs), and (un)equal cooperation (donor–recipient state; donor–donor).

### 11.1.1  Impact of an "Empowerment" Approach on Sustainability

I argue that the "empowerment" approach by the donor influenced the sustainability of the health care program in two ways, namely, through the mechanisms of "ownership" and "learning."

First, community engagement throughout the duration of the project, following an "empowerment" approach by the donor, influenced the sustainability of the project by developing a sense of ownership in the VHCs. This contributed to community capacity building, and continuity of project activities (Diagram 11.1):

The "empowerment" approach had a considerable impact on community capacity building, by contributing to the continued survival of community-based organizations beyond the end of the "Community

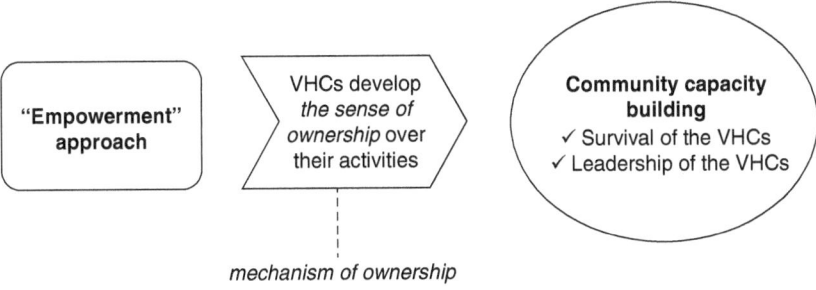

**Diagram 11.1** The impact of the donor's "empowerment" approach on sustainability

Action for Health" project (CAH). It did so by contributing to their leadership skills, and their ability to mobilize resources. I visited the Village Health Committees (VHCs) almost a year after the end of the project, and the community-based organizations were still functioning. The VHCs continued their work and were "thinking of extending it" (CSO 5). A literature review of studies of sustainability of development programs emphasizes the involvement of volunteers as being vital to program sustainability (Scheirer, 2005). Yet, volunteerism alone does not guarantee sustainability (Glenton et al., 2010), as high attrition among health volunteers is common (e.g., Khetan et al., 2017; Sivaram & Celentano, 2003). For this reason, it was not volunteering that ensured the organizations' continuity beyond the end of the project, but, rather, the sense of ownership the VHCs developed throughout the CAH.

The VHCs raised their own initiatives and worked on solving broader social issues, which contributed to their leadership skills. This included the construction of a bridge so that children would no longer have to go through water on their way to school (CSO 4) or solving residents' waste management problems (CSO 5). The VHCs I visited showed me a notice board of their initiatives, such as support for poor households (CSO 1), for the elderly, for persons with disabilities (CSO 5) and others. One interviewee showed me the photos of the sports competition the VHC had organized to raise funds for a villager in need of surgery, and they had collected about 27,000 KGS (about €290)[1] as a result of this event (CSO

---

[1] The exchange rate, as of March 17, 2023, was applied throughout this book.

2). In this way, the VHCs not only identified problems, but also sought solutions, which contributed to their leadership capacity.

The Swiss Red Cross (SRC) encouraged the "self-initiatives" and included such initiative-taking in the evaluation criteria.[2] The VHCs were expected to suggest and implement activities based on their own initiative (CSO 5). This encouraged the freedom of the community-based organizations to initiate their own activities. According to one of my interviewees, at some point there was "a fear" that VHCs were "leaving the health care" (IO Partner 5), since the scope of their activities was very broad. Following the SRC's suggestion, the VHCs adopted a mission statement, which described the organizational goals and their focus on health care.[3] This mission statement was intended to emphasize the VHCs' activities in health care, but not limit it to this area, as the community-based organizations still continued their work on solving broader social problems.

The community-based organizations look for various resources with which to conduct their self-initiatives. The VHCs also use their organizational funds to finance their self-initiatives. Thus, one of my interviewees conducted the self-initiatives for 25,000 KGS (around €268) grant the VHC received from the SRC during the CAH to finance various initiatives in the village (CSO 2). It should be noted, however, that the size of the VHCs' budget varies greatly between 2000 and 3000 KGS (€21–32) in the case of the smallest budget, and 100,000 and 150,000 KGS (€1072–1608) in the case of the largest budget (CSO 4). According to the interviewee who was working closely with the VHCs, the size of the budget depends largely on the VHC leaders and their ability to work with local actors and donors to increase the size of their organizational budget (CSO 4). The VHCs write appeals to local self-governments to solve residents' problems (CSO 5). Although not offering financing, the local authorities provide in-kind support to the VHCs (State Partner 12). The community-based organizations also write project applications to donors, which they learned how to do in the training courses provided by the CAH (CSO 7). However, the VHCs do not seem to simply sit back

---

[2] These criteria are used by the VHCs and HPUs to assess the VHCs and their activities.

[3] The author has the sample of the statement in Kyrgyz.

and wait for donor support (CSO 2); rather, they try to use other means to mobilize resources instead.

The "empowerment" approach between the Swiss organizations (e.g., the Swiss Agency for Development and Cooperation and the Swiss Red Cross) and the VHCs, also reflected in the encouragement of "self-initiatives," contributed to community capacity building by developing a sense of "ownership" among the community-based organizations. The VHCs did not just define the issues on their own, but also looked for the solutions. As one development partner noted, "from a passive [role of] providing information," they transformed themselves into organizations seeking solutions to the issues pressing their communities "at this point at the local level" (CSO 1).

Secondly, the VHCs' sense of ownership, developed through the "empowerment" approach, contributed to the continuity of project activities on tuberculosis (TB) and HIV/AIDS, after the project ended. The VHCs defined the issues targeted by the CAH, either by surveying the local population or by suggesting their own initiatives. Andrews (2013) suggests that aid recipients tend to take ownership of development programs which are driven by local problems and solutions, rather than the ones guided by a global agenda. At the same time, as described in Sect. 5.2, Design, reproductive tract infections were among the issues identified by the local population, while tuberculosis was not (see Schüth et al., 2014). Nevertheless, the VHCs continued their activities around TB and HIV/AIDS, beyond the end of the CAH. Based on this, I argue that the community-based organizations continued their activities targeting both diseases because of the sense of "ownership" they developed through the "empowerment" approach. It should be noted that this sense of ownership was not limited to a specific activity or area of health care, but extended to the health of the communities as a whole:

> Since we have collected all this information, well, our village needs it; the Swiss Red Cross or Tobias [Dr. Schüth] does not need all this, [but] we need [it] ourselves, to preserve our health, to maintain the health of our village, [these] were the reasons for us to learn all that. (CSO 2)

Thus, TB and HIV/AIDS prevention continued along with the other activities, due to the VHCs' sense of ownership, and responsibility for the health of the local population.

The VHCs use various means to continue their health-related activities. The VHC member I interviewed reported that she contacts the *feldsher-midwife [akusher]* point to get up-to-date information on diseases and their prevalence, and she disseminates this information in her village (CSO 2). The VHC also uses the brochures available in the organization to "refresh" the knowledge of the local population about certain diseases from time to time (ibid.). The VHCs do not limit themselves to the training courses provided by the Health Promotion Units (HPUs), and try to attend other events and training courses to learn more about health issues and their prevention (ibid.), even where travel costs are not covered by the organizers (CSO 5). All these attempts to continue health-related activities point to the sense of ownership the VHCs have developed toward all of their activities, which has also resulted in continuity of TB and HIV/AIDS-related activities, more specifically (Diagram 11.2).

Furthermore, the "empowerment" approach contributed to community capacity building through the mechanism of "learning." According to the VHC member, the SRC stressed learning throughout the project, and community-based organizations were aware of its importance for the continuation of their activities beyond the CAH (CSO 2). The SRC provided extensive training to support the VHCs' organizational capacity (book-keeping, budgeting, how to organize the seminars, write appeals, etc.) and health-specific activities (essential information about the

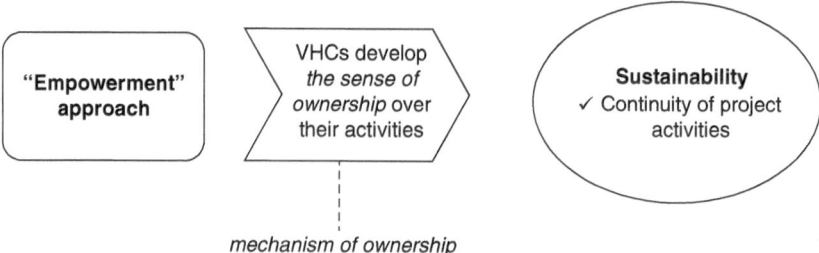

mechanism of ownership

**Diagram 11.2** The impact of the donor's "empowerment" approach on sustainability

diseases, their prevention, and health promotion). Using this knowledge, the VHCs overcame structural barriers, such as illiteracy, disinterest on the part of the local authorities and local population, to pursue their activities. Although important (Walsh et al., 2012), training alone is not sufficient for community capacity building. The VHCs learned extensively through their participation in implementing and evaluating the project-related activities, and by exchanging their experiences with each other. I argue that this involvement of community members through the "empowerment" approach contributed to their learning, and resulted in community capacity building. Through engagement during the period of the CAH, the VHCs developed their expertise and planning skills, which contributed to their organizations' survival.

Firstly, as they work closely with communities, the VHCs are well aware of community issues, which make them the first point of contact for local authorities and donor organizations. During their initial dissemination campaigns, members of community-based organizations visited local households:

*VHCs know more about the problems in their village ... because they make the rounds in the village [and] visit the households to disseminate their information.* (CSO 4)

Through this close contact with the villagers, the community-based organizations became aware of the living conditions and concerns of most of the households in their villages. It was this awareness which eventually contributed to the acknowledgement of the VHCs by local authorities and donor organizations. Being well aware of the problems of the local population, the VHCs serve as mediators between the villagers and the local authorities (CSO 4). The community-based organizations support the villagers in their claims to the local self-governments by helping them write petitions for example, but also assist the authorities with outreach in their community by mobilizing the villagers to meet the authorities. The VHCs are also the first point of contact for donors who are willing to work with communities. The local authorities refer any development partners looking for local initiative groups to the VHCs, emphasizing the fact that there is no need to establish any new groups, when the VHCs already exist (ibid.) (Diagram 11.3).

Secondly, by implementing and evaluating the project-related activities, the VHCs started planning their activities and their connection to other organizations. The community-based organizations had the freedom to organize their activities as they saw fit. They defined the timing and the frequency of their meetings (Schüth, 2011) and activities, with no intervention from the SRC. This freedom contributed to improving the VHCs' planning skills. One community member I interviewed explained that in order to manage their household responsibilities and project-related activities, the VHC members started to divide their labor and plan their activities (CSO 5). This planning also allowed them to distribute the villages and households among each other to ensure a broader coverage of their seminars (ibid.).

The VHCs' self-assessment strengthened this planning further. The community-based organizations compared their current performance to the previous years, identified the issues and the possible ways for improvement, which were then included in the organizational work plan (CSO 4). The VHCs also enlisted the key local organizations they sought to cooperate with, including local self-governments, schools, associations, the court of elders, and others (CSO 2). Some of these organizations approached the VHCs themselves, proposing to develop a joint plan of activities (ibid.).

In this way, by implementing and evaluating the project-related activities, the VHCs planned their activities and their links to other

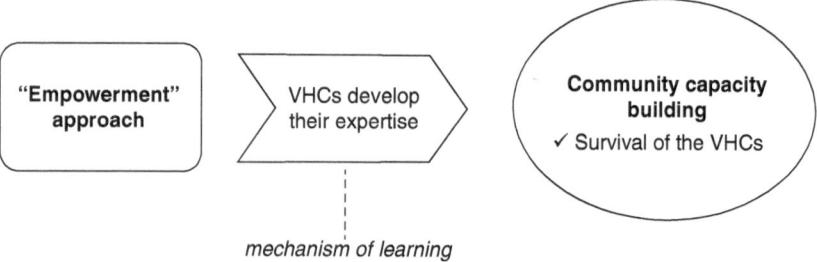

**Diagram 11.3** The impact of the donor's "empowerment" approach on sustainability

institutions, which is essential for the survival of the community-based organizations. The VHCs collaborate with multiple local actors, including the family medicine centers, *feldsher-midwife (akusher)* points, youth councils, local association of women, and so on (CSOs 2, 4, and 5). Collaboration with local actors provides access to technical (in the case of health care organizations) and administrative support (from local self-governments, schools, etc.), which is essential for the VHCs and the continuation of their activities beyond the end of the CAH. This link to local actors is important for the survival of community-based organizations (see Glenton et al., 2010) (Diagram 11.4).

Thirdly, the VHCs continued to meet and share their experiences, which also contributed to the continued survival of the organizations beyond the end of the CAH. During the CAH, there were monthly meetings at district level (Rayon Health Committees), where the VHCs shared their experiences and learned from each other (Schüth, 2011, p. 44). The associated costs were covered by the project (CSO 4). One interviewee, closely working with the community-based organizations, suggests that the exchange of experiences during these meetings stimulated competition between the VHCs and contributed to their performance (ibid.).

Since the end of the project, the frequency of the VHC meetings has decreased to a quarterly basis (CSO 2), with travel costs being covered by funds from the Rayon Health Committees (CSO 7) or the VHCs (CSO 4). One representative of the Rayon Health Committee reported that where, before, the community-based organizations had waited for the CAH to gather them together for a meeting, now they initiated the

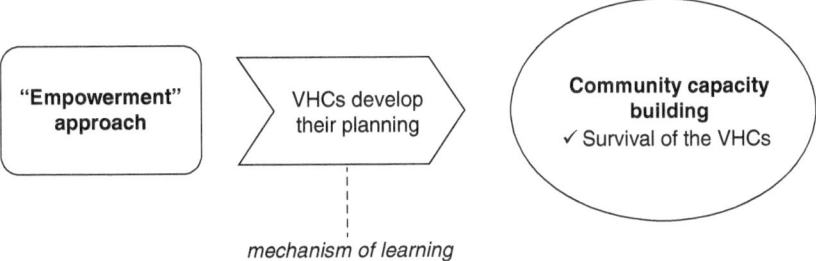

Diagram 11.4  The impact of the donor's "empowerment" approach on sustainability

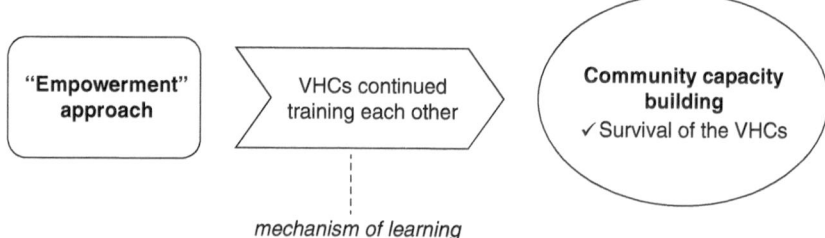

**Diagram 11.5** The impact of the donor's "empowerment" approach on sustainability

gathering themselves, even if it was at their own expense (CSO 7). Although the frequency of the VHC meetings has decreased, nevertheless, the meetings are still continuing, and therefore so, too, is this exchange of experiences between organizations, which has contributed to their continued survival beyond the end of the project (Diagram 11.5).

In addition to community capacity building, the mechanism of learning, developed through the "empowerment" approach, also contributed to the maintenance of benefits. In addition to attending training courses, the VHCs learned extensively by carrying out project-related activities, and through the evaluation of the VHCs. This learning was essential for the maintenance of benefits, namely, the survival and quality of the information provided by the VHCs. In addition to evaluating their own organizations, the VHCs I have met supported others by organizing "nomadic seminars" and training courses.

During these "nomadic seminars" the VHCs share their experiences following the principle of "nondominance," which encourages the organizations to continue their work. During the annual self-assessment, the VHCs take note of the organizations which are having problems with their documentation. For instance, if there was a problem with documentation in one of the VHCs, the organizations offering the "nomadic seminars" organized joint visits for the heads of organizations to a VHC, which performed well in this regard (CSO 5). The example of an organization encouraged others to improve their documentation accordingly (ibid.).

The exchange of experience follows the principle of "nondominance," and instead of pointing out problems, the organizers of the seminars appeal to the consciousness of the heads of the VHCs. Organizers of the "nomadic seminars" use their organizational funds to cover their travel expenses (CSO 5). My interviewee suggested a growing interest in their initiative, which encouraged the VHCs to continue their support to other organizations (ibid.). She stressed that the seminars encourage the VHCs to continue their work, because the organizations that discontinue their work miss out on opportunities to collaborate with development projects coming to their villages (ibid.).

Furthermore, some organizations share their knowledge with other organizations that did not have access to the same training. Since the end of the CAH, there is no longer any donor covering all of the VHCs; instead donor organizations provide specific training courses for community-based organizations in certain regions, depending on the project-specific objectives and tasks. However, community-based organizations which are not covered by donor organizations are also interested in learning (CSO 1). The VHCs have solved this inequality in access to training by sharing their knowledge with each other. One development partner notes that the community-based organizations covered by the project started training the organizations in the neighboring villages and regions (ibid.). Thus, one of my interviewees visited two other villages at her own expense to provide training in the areas covered by the development project she was working on (CSO 7). Related to this, the above-mentioned development partner stressed the motivation of the VHCs to learn and continue learning. The interviewee noted that as unpaid VHC members, they were not interested in a mere formal existence of their organizations; they were "interested in changing something" instead and, in so doing, gaining "some authority" (CSO 1).

This motivation to learn contributes to further exchanges of experience and learning; despite the inequality in their access to training, the VHCs share their knowledge with other organizations. Similar to the "nomadic seminars" the training courses organized by the community-based organizations for each other evolved during and beyond the CAH and contributed not just to the continued survival of the VHCs, but also to their organizational capacity and their ability to conduct their awareness-raising activities through continued learning.

## 11.1.2    Impact of a "Utilitarian" Approach on Sustainability

In this section, I argue that the interaction between the recipient state and the CSOs had differing influences on sustainability. To examine the influence of the interaction between the recipient state and the CSOs, I differentiate between the interaction between the Ministry of Health and the VHCs, and the interaction between local self-governments and the VHCs.

In the case of the interaction between the Ministry of Health and the VHCs, the Ministry's "utilitarian" approach on its own did not influence sustainability. Certainly, the HPUs continued to provide training for the community-based organizations after the end of the CAH, which contributed to the maintenance of benefits. However, the Ministry established the HPUs in response to the *SRC's* request to provide the health care workers, and not in response to the VHCs. Although the interaction between the VHCs and the Ministry of Health contributed to changing the perspective of state officials, this was not the reason for the Ministry to provide the HPUs. For this reason, I propose that the interaction between the Ministry and the community-based organizations on its own did not influence sustainability, but did so in a combination with the relationship between the Ministry of Health and the SRC.

At the same time, the "utilitarian" approach taken by local self-governments toward the VHCs influenced sustainability by contributing to the continued survival of the community-based organizations beyond the end of the CAH. The local authorities provide administrative support to the community-based organizations by offering office space, and referring any donor organizations that approach them on to the VHCs. Furthermore, the local authorities involve the community-based organizations in decision-making, and, in doing so, recognize their activities and their authority in the village. Collaboration with and recognition by the local authorities is essential to the activities of community-based organizations (Glenton et al., 2010), and for this reason I would argue that the interaction between the local authorities and the VHCs has contributed to the continued survival of community-based organizations beyond the end of the CAH.

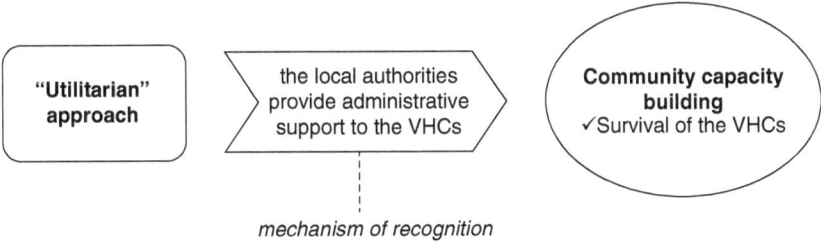

**Diagram 11.6**  The impact of the local self-governments' "utilitarian" approach on sustainability

Although both levels of government in the recipient state, national and local, have pursued a "utilitarian" approach to the community-based organizations, the interaction of the Ministry of Health with the VHCs and the interaction of the local authorities with the VHCs have had different impacts on sustainability. Both have contributed to sustainability by supporting the maintenance of benefits and the ongoing survival of the organizations. However, the impact in the case of the Ministry was the outcome of the interaction of the Ministry with the SRC, whereas in the case of the local self-governments, the impact on sustainability was the result of the direct interaction with the VHCs (Diagram 11.6).

## 11.1.3  Impact of (Un)equal Donor-Driven Cooperation on Sustainability

First of all, I argue that (contingent) equal cooperation between the Ministry of Health and the SRC contributed to the long-term sustainability of the Community Action for Health project, namely, by contributing to the maintenance of benefits through the mechanism of "institutionalization." In the literature on development, the term "institutionalization" is frequently used interchangeably with the term "sustainability" (see Chap. 3); however, in the framework of this book, by institutionalization I refer to the Ministry of Health's formalization of its commitments by including them in the Sector Wide Approach, and establishing the Health Promotion Units (Diagram 11.7).

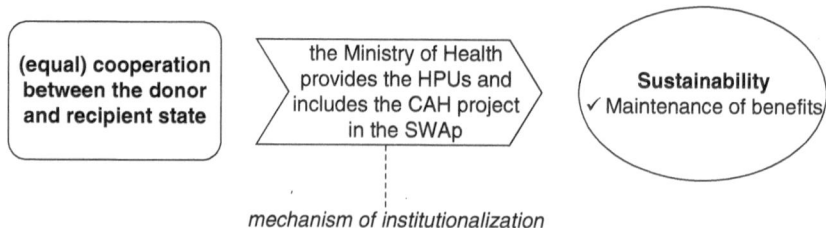

Diagram 11.7   The impact of (contingent) equal cooperation on sustainability

By including the CAH in the Sector Wide Approach, the Ministry authorized the extension of the project throughout the country, and committed its resources to ensuring continuity of the VHCs beyond the end of the CAH. By including the CAH in the SWAp in 2005, the Ministry encouraged its nationwide rollout (IO Partner 11). This acknowledgement and commitment from the Ministry was essential for the extension of the CAH, and for the commitment from the United States Agency for International Development (USAID) and the Swedish International Development Cooperation Agency (Sida) to support the extension process (ibid.). In addition to the rollout, inclusion in the SWAp also provided resources for the continuity of the VHCs beyond the CAH. More specifically, the commitment of the Ministry of Health to commit its resources to this initiative facilitated further support from other donors (ibid.).

In this way, the VHCs became part of the national health care program: *"Manas Taalimi"* (2006–2011), *"Den Sooluk"* (2012–2018), and the "Healthy Person—Prosperous Country" programs (2019–2030) mention the VHCs (see Government of KR, 2006, 2012, 2018). The Ministry of Health established the HPUs, and assigned salaries for them from the state budget, as part of primary health care. Even so, as one of my interviewees noted, because of underfinancing, the majority of the expenses in the area of health care promotion are still covered by donors, and not by the recipient state (IO Partner 5). Another interviewee however, emphasized the fact that the ministerial support to the VHCs will continue, as the activities of community-based organizations in health promotion comply with the interests of the Ministry (IO Partner 11). HPUs remained among the key sources of training for community-based organizations, particularly within the framework of the *"Den Sooluk"* program. Similarly, the ongoing (2019–2030) program stipulates the development of training modules for the VHCs (Chap. 6). For this reason, the mechanism of institutionalization triggered by the donor-driven cooperation was the key to continuous training of the VHCs and maintenance of benefits beyond the duration of the CAH.

Furthermore, the unequal cooperation between the SRC, Sida, and USAID influenced sustainability by contributing to the survival of the community-based organizations beyond the end of the CAH through a process of "uniformity." The national rollout of the CAH resulted in the establishment of the network of VHCs throughout the country, and the establishment of the Association of VHCs to support this network. The financing from Sida and USAID was essential for the establishment of the HPUs by the Ministry of Health. However, it was not just their financing alone which contributed to the continued survival of the VHCs beyond the project, but rather the compliance of these two donors with the SRC's approach to community capacity building. The presence of the SRC as "lead" donor ensured the uniformity of the donor relationship with the VHCs, and the process of establishing the community-based organizations. Thus, the VHCs in Issyk-Kul region had similar structures and received similar training to the VHCs in Batken or Talas regions. This uniformity was essential for the interaction between the community-based organizations during their joint meetings, and their ability to share their experiences and issues, based on the similarity of the activities they were all conducting. In 2010, the network of VHCs was strengthened further with the establishment of the Association of VHCs, which provides supervision (IO Partner 5) and support to the community-based organizations throughout the country.

Though it did contribute to the survival of the VHCs beyond the CAH, unequal cooperation between the donors, in itself, does not necessarily result in sustainability. It was the mechanism of "uniformity" which was the key to the expansion of the "empowerment" approach the SRC pursued with the communities. A similar expansion under a "utilitarian" approach, however, would not necessarily have contributed to sustainability to the same extent as did the "empowerment" approach'. However, the presence of the "lead" donor would nevertheless ensure the expansion

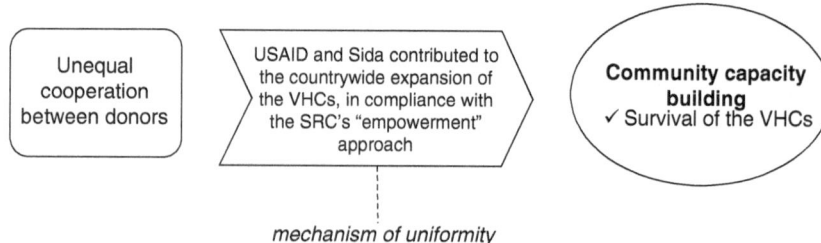

mechanism of uniformity

**Diagram 11.8** The impact of the unequal cooperation between donors on sustainability

of community-based organizations according to the approach pursued by the lead donor, as other donors would comply with its approach (Diagram 11.8).

To summarize, in the case of "Community Action for Health," the missing link unfolded in the following way:

1. The "empowerment" approach between the SRC and the VHCs influenced the sustainability of the CAH in two ways, namely, through the processes of ownership and learning. The mechanism of ownership contributed to the continued survival of the community-based organizations beyond the end of the project, as well as to the continuity of the VHCs' activities, including those targeting TB and HIV/AIDS. The mechanism of learning similarly contributed to the VHCs' survival beyond the CAH and maintenance of benefits, or the presence and quality of information on disease prevention and health promotion provided by the VHCs.

2. The (contingent) equal cooperation between the Ministry of Health and the SRC resulted in the maintenance of benefits through the mechanism of institutionalization. The HPUs, established, by the Ministry, have continued to provide training in the four areas prioritized in the national health care program, which contributes to the quality of the relevant information provided by the VHCs.

3. The "utilitarian" approach of the Ministry of Health and the local self-governments toward the VHCs had different impacts on sustainability. The interaction of the Ministry with the community-based organizations affected the maintenance of benefits only in combination with the interaction between the Ministry and the SRC. The "utilitarian" approach of the local self-governments, however, contributed to the continued survival of the VHCs beyond the end of the project, due to the dependence of the local authorities on the expertise and the authority of the VHC members in their villages.

4. The unequal cooperation between the SRC, USAID, and Sida contributed to the ongoing survival of the community-based organizations through the process of uniformity. In combination with the "empowerment" approach of the SRC toward the community members, the mechanism of 'uniformity' resulted in the establishment of the network of VHCs. This was essential for their unity and exchange of experience. It should be noted, however, that the "empowerment" approach of the "lead" donor (the SRC) was essential to this outcome.

## 11.2 The Global Fund Grant to Kyrgyzstan

Based on the relationship between the actors throughout the project and the power dynamics between them, I have defined the following types of aid relationships: "utilitarian" approach (donor–civil society organizations [CSOs]; recipient state-CSOs), unequal cooperation (donor–recipient state), and coordination (donor–donor).

### 11.2.1 Impact of a "Utilitarian" Approach on Sustainability

The "utilitarian" approach of the Global Fund toward grant-recipient NGOs contributed to community capacity building by ensuring the CSOs' survival beyond the grants. The Global Fund grants to the country increased the number of NGOs and facilitated their competition over

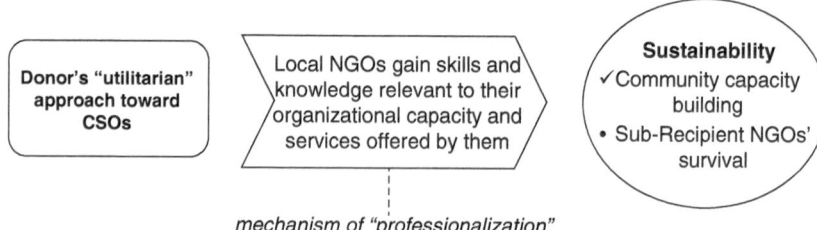

*mechanism of "professionalization"*

**Diagram 11.9** The impact of the donor's "utilitarian" approach toward CSOs on sustainability

resources. According to the NGO representatives interviewed for this research, civil society needs to continuously increase its capacity and expertise in all areas, including treatment, medication supply, procurement, and budgeting (CSO 8). The organizations also improved their advocacy skills by arguing from legal and health care perspectives and not just going on strike and demanding a revolution (CSO 6). The competition between the NGOs contributed to their development and selection of well-performing organizations in the Global Fund project. These organizations had to comply with the Global Fund's standards and requirements for project management, accounting, monitoring and evaluation (M&E), and reporting, facilitating the introduction of new positions and the recruitment of additional personnel in the NGOs and subsequently contributing to their "professionalization" (Harmer et al., 2013, p. 304). The skills obtained during the design and implementation of the Global Fund grants contributed to the NGOs' survival by advancing their negotiation skills essential to fundraising (Diagram 11.9). However, similar to other sustainability components, NGOs' survival beyond the duration of the Global Fund project depends on broader political and economic factors.

## 11.2.2   Impact of Unequal Donor-Driven Cooperation on Sustainability

Unequal cooperation, formed between the Global Fund, the Ministry of Health, and its agencies, contributed to sustainability through the mechanism of institutionalization. The Ministry and its agencies comply with the Global Fund's requirements throughout the project life cycle by establishing the CCM and increasing the share of government co-financing of the grants. Government institutions also took over first-line TB medications and increased their financing for antiretroviral therapy (see Chap. 9). All these commitments contributed to the continuity of treatment activities beyond the duration of the Global Fund project (Diagram 11.10):

In addition, "unequal" cooperation between the Global Fund, the Ministry of Health, and its agencies contributed to community capacity building by supporting the local NGOs' mobilization of resources. The Global Fund facilitated civil society involvement in the design and implementation of its grants. As an NGO representative interviewed for this research noted, "willingly or not," the recipient state was open to civil society participation (CSO 8). Furthermore, following the Global Fund conditions, social contracting featured the country's joint application for TB/HIV to Fund 2017–2019 (Zardiashvili & Garmaise, 2017). By incorporating social contracting, the government committed itself to financing NGOs. As a state official interviewed for this research noted, the donors "come and leave," and social contracting is the only possibility for the continuity of NGO activities (State Partner 2). The Ministry of Health agreed to sign a contract with two NGOs in 2018 and six NGOs in 2020 (see Government of KR, 2017). The Ministry also committed to developing the normative-legal basis for social contracting by 2021 (ibid.). In this way, the unequal cooperation between the Fund, the Ministry, and its agencies ensured continuous financing of the NGOs

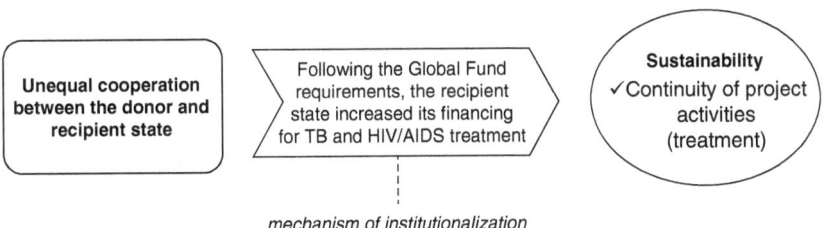

**Diagram 11.10** Donor–recipient state: the impact of unequal cooperation on sustainability

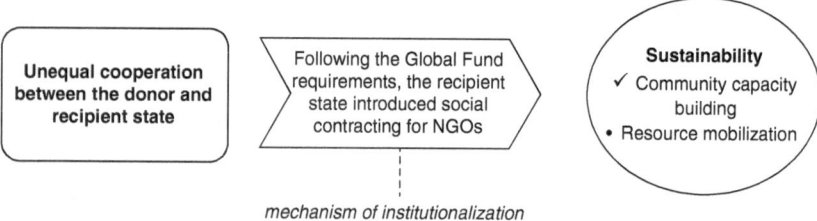

Diagram 11.11   Donor–recipient state: the impact of unequal cooperation on sustainability

working on TB and HIV/AIDS beyond the duration of the Global Fund project (Diagram 11.11).

Overall, the interaction that developed between the actors involved in the Global Fund project in Kyrgyzstan had various impacts on sustainability and its components. As the interaction between the recipient state and the local NGOs was the result of the Global Fund's condition, I suggest that it did not impact the sustainability of the Global Fund project in Kyrgyzstan.

## 11.2.3   Impact of Coordination on Sustainability

The Global Fund is the largest financier of TB and HIV/AIDS programs, and the country depends on its contributions in these areas. The financial gap in the national HIV/AIDS program demonstrates the government's inability to meet the ongoing and increasing demand for antiretroviral therapy and other services. Furthermore, Kyrgyzstan is among the countries with a large number of multidrug-resistant TB, which requires long and expensive treatment. In this way, the government's ability to provide these services after the country's transition from Global Fund grants is questionable (see Chap. 9). However, in addition to the state budget deficit, the country's dependence on the Global Fund is also the outcome of its coordination with other donors. Throughout the project life cycle, except for the monitoring phase, the Global Fund demonstrated extensive coordination with other donors working on TB and HIV/AIDS.

However, this coordination was not limited to the duration of the Global Fund project in Kyrgyzstan. Before the award of the Global Fund grants to Kyrgyzstan in 2004, the government had already collaborated with multiple donors. These are the Soros Foundation Kyrgyzstan, German Development Bank (*die Kreditanstalt für Wiederaufbau*—KfW), World Health Organization (WHO), Joint United Nations Programme on HIV/AIDS (UNAIDS), United Nations Development Programme (UNDP), the United Kingdom's Department for International Development (DFID),[4] International Committee of the Red Cross (ICRC), Doctors Without Borders/*Médecins Sans Frontières* (MSF), and the World Bank.

Upon the commencement of the grants, these donors gradually discontinued their TB and HIV/AIDS-related activities. Some respondents in the study by Ancker and Rechel (2015, pp. 822–823) connected the World Bank and DFID's retrenchment from HIV/AIDS to nonduplication of efforts, rather than donor "fatigue." A state representative interviewed for this research similarly pointed to continuous cuts of HIV funding in the Sector-Wide Approach due to Global Fund grants (State Partner 9). The Global Fund is the largest international initiative against TB, HIV/AIDS, and Malaria, financed by multiple countries, including Germany, the United States, France, and the United Kingdom, among others (see Global Fund, 2023). For this reason, the countries financing the Global Fund decreased their bilateral assistance in the area of TB and HIV/AIDS to avoid the duplication of activities with the grants. In doing so, however, they have also contributed to the dependency of the country on a single donor—the Global Fund.

The coordination between donors activates the mechanism of "replacement," which contributes to the continuity of TB and HIV/AIDS activities after the country's transition from Global Fund grants. Currently, no international organization, except for the MSF, who made an oral commitment, can guarantee the stock of TB drugs in the case of interruptions in the Global Fund grants to the country. Similarly, the continuity of HIV/AIDS-related prevention and treatment is uncertain. However, the lack of commitments does not necessarily hint at donors' unwillingness

---

[4] DFID was replaced by Foreign, Commonwealth and Development Office in 2020.

to support the government in its fight against the two diseases; instead, it points to their inability to make long-term commitments. Aid predictability is a general problem in development assistance, and the Global Fund is among the few donors, along with Swiss aid agencies, offering longer commitments (see Isabekova, 2019). According to my interviewees, the national and international actors in Kyrgyzstan attempt to avoid the situation of all donors leaving the country at once (IO Partner 3). As the Global Fund grants to the country decrease, other donors, such as USAID and the President's Emergency Plan for AIDS Relief (PEPFAR), increase their contributions (State Partner 9). This tendency is not limited to Kyrgyzstan. In Sub-Saharan Africa, USAID and PEPFAR took over most of the activities previously provided by the Fund. Coordination among the donors triggers the mechanism of "replacement," according to which an area left by one donor is taken over by another actor or other actors working in the same area. Continued provision of TB medications in Kyrgyzstan is another example of the mechanism of "replacement" in practice. The German government provided first-line medications against drug-resistant TB between 2002 and 2004 based on the agreement that the Government of Kyrgyzstan would take over financing these medications in 2005 (Government of KR, 2001). However, with the commencement of Global Fund grants to Kyrgyzstan in 2004, all costs of TB medications were transferred to these grants, not to the state budget.

Nevertheless, the mechanism of "replacement" does not necessarily guarantee the same level of assistance, which affects the sustainability of the Global Fund project in Kyrgyzstan. The Global Fund, unlike other donors, was explicitly established to combat TB, HIV/AIDS, and Malaria. Other donors approach TB and HIV/AIDS but not as the central parts of their aid portfolio. As a result, their financial contribution to these areas will be significantly lower than that offered by the Global Fund. Although contributing to the continuity of some activities after the country's transition from the Global Fund grants, other donors are unlikely to provide the same level of services. Despite the continuity of preventive and treatment activities, financial incentives to patients are likely to be discontinued (see Chap. 9). Lower donor financing also implies less funding to the local NGOs working on TB and HIV/AIDS that are dependent on the Global Fund grants. Since the Global Fund

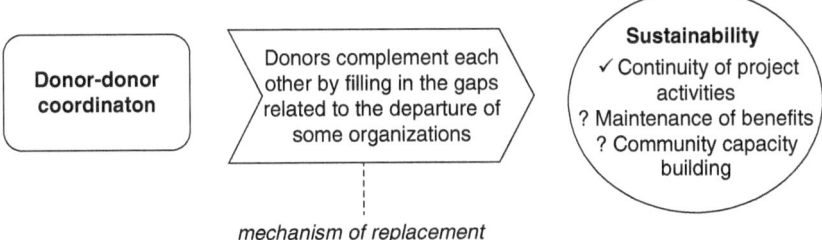

Diagram 11.12   The impact of coordination among donors on sustainability

project is still ongoing and due to the high level of uncertainty in Kyrgyzstan's economic and political situation, the impact of coordination among donors on sustainability is uncertain. Though one is clear, the donor that steps in following the Global Fund will not offer the same level of services unless several organizations take over the TB and HIV/AIDS programs in Kyrgyzstan (Diagram 11.12).

To summarize, in the case of the Global Fund grants to Kyrgyzstan, the missing link unfolded in the following way:

1. The "utilitarian" approach of the Global Fund toward the local NGOs contributed to community capacity building through the mechanism of "professionalization." Through participation in the grants, the NGOs developed skills and knowledge in service provision and other areas, which are essential to their survival of the organizations beyond the grants.

2. The "utilitarian" approach of the Ministry of Health and its agencies on TB and HIV/AIDS toward local NGOs did not affect the sustainability of grants. Notably, the interaction between the state and civil society was triggered by the conditions established by the Global Fund. Although contributing to collaboration between actors, the joint realization of grants has demonstrated continuous disagreement and conflict between the state and non-state sectors.

3. The unequal cooperation of the Global Fund with the Ministry of Health and its agencies contributed to the continuity of grant-related activities and community capacity building. Following the Global

Fund's requirements, the recipient state gradually took over the treatment of TB and HIV/AIDS, which contributed to their continuity beyond the grants. Furthermore, the recipient state committed to concluding social contracting to continue the work of local NGOs with the key groups. This work provided additional funding to the NGOs (e.g., mobilization of funds).

4. The coordination between the Global Fund and other donors contributed to sustainability through the mechanism of replacement. The amount of the Global Fund grants to Kyrgyzstan has decreased, but the project is still ongoing. Due to donor organizations' inability to provide long-term commitments, epidemiological situations, and the political and economic instability in the country, estimating the exact impact of this coordination on specific components of sustainability is not feasible.

## 11.3  Cross-Case Causal Inferences

At the same time, this exploratory study aims to build (but not test) hypotheses about the potential impact of the interaction between actors on the sustainability of development aid. Therefore, a cross-case analysis of the findings from within-case analysis to explore the possibilities for theorizing these mechanisms beyond the selected cases is foreseen. The generalization of causal mechanisms to a "population of causally similar cases" is possible (Mahoney, 2008, pp. 413–420), also via simplification of specific boundaries of cases or "layered generalization" (Rohlfing, 2012, pp. 209–212). In other words, the less specific the case, the more generalizable the causal mechanism, which, however, also means that the claims are less specific as well. Following this logic of the "layered generalization," I conduct a cross-case analysis, which offers the possibility of examining and theorizing causal effects, or what Rohlfing calls "theoretically intelligible and systematic" relationships (Rohlfing, 2012, p. 12). These effects provide the foundation for the formulation of hypotheses about the causal impact of interaction on sustainability.

By relying on cross-case analysis of how the actors' relationships in the selected countries might have causally shaped the sustainability of the relevant health care programs, it is possible to formulate the following findings:

First, an "empowerment" approach of a donor toward CSOs contributes to community capacity building by improving and streamlining the leadership of the CSOs, and their capacity to mobilize resources, as well as by facilitating their survival beyond the duration of development assistance projects. As demonstrated in the case of the CAH project, an "empowerment" approach may also develop in the environment of unequal power dynamics between a provider and a recipient of development aid. Illiteracy (Jana et al., 2004), gender-related biases (WHO, 2008), the political situation in the country, and poverty (Fawcett et al., 1995) may prevent the CSOs from initiating a development project without external assistance. For this reason, the Swiss Agency for Development and Cooperation (SDC) and the SRC were essential to the initiation of the CAH. However, the design, implementation, and evaluation phases of the project largely depended on community members' leadership and their consent to engage in voluntary work. The Village Health Committees' extensive role throughout the project, along with the SRC and SDC's emphasis on community empowerment, flexibility, and predictability of development assistance, was critical to altering the initially unequal power dynamics between the actors. This changing nature of power identified in the CAH corresponds to the findings in the literature on aid relationships (e.g., Andrews, 2013; Swedlund, 2017). In addition, it points to the possibility of changing the power dynamics despite the aid recipient's capacity issues and dependence on development assistance. Furthermore, the CAH project demonstrated a causal link between the donor's "empowerment" approach and community capacity building. The impact of this approach was clearly seen in the continued survival of the majority of the Village Health Committees beyond the end of the CAH and their mobilization of resources through fund-raising, member contributions, and donor support.

Second, a "utilitarian" approach of a donor toward the CSOs contributes to their survival beyond the duration of development assistance programs, but it does not affect the quality of the leadership of these CSOs. A "utilitarian" approach may also facilitate resource mobilization for CSOs, but only in a situation in which the relevant donors cooperate on an unequal basis with the authorities in the recipient country. The cases of the Global Fund grants vividly illustrate a "utilitarian" approach of a donor toward the CSOs engaged in the design and implementation of grants, though on the terms defined by the donor, and not by the CSOs themselves. The unequal power dynamics established between the Global Fund and grant-recipient NGOs, due to the latter's aid-dependence and capacity issues, intensified in the course of the grant realization process. The Global Fund used disbursements as leverage to ensure the grant recipients' performance and their compliance with its regulations. The limited flexibility of the assistance precluded responsiveness to changing needs and approaches of NGOs that followed the activities and indicators stated in the projects instead. This compliance contributed to NGOs' awareness of the Global Fund's procedures and other technical skills, but not their leadership, which was not stimulated through their implementation of grants. At the same time, the Global Fund's "utilitarian" approach toward the NGOs contributed to their existence beyond the projects by providing access to resources (i.e., social contracting) through authorities of grant-recipient countries, which cooperated with the Global Fund on an unequal basis.

Third, unequal cooperation between a donor and the relevant authorities of aid-recipient countries does in fact contribute to the continuity of project activities. However, the extent of the services that might continue beyond the period of the development assistance is highly dependent upon decision-makers' priorities, the presence of stigma and discrimination against groups targeted by assistance, as well as the epidemiological, political, and economic situation in the aid-recipient countries. The case of the Global Fund grants illustrated unequal cooperation between the donor and the recipient state. Government authorities' participation in the design and implementation of grants did not change the unequal power dynamics. Limited flexibility of the Global Fund in regard to the grant activities and indicators, along with the state authorities' aid-dependence and capacity issues, strengthened the unequal power dynamics. The inequality between the actors intensified further as the Global Fund used the grant disbursements to impose its conditions and regulations on the recipient governments. To access the financing, the state authorities designed the grant applications and established the institutions (i.e., Country Coordinating Mechanism) according to the focus and procedures of the Global Fund. Moreover, a prolongation of a standard three-year-long project cycle also depended on grant recipients' compliance with the Global Fund's regulations and achievement of the objectives stated in the grant agreements. As part of the Global Fund's conditionalities, the recipient governments also increased their contribution to the treatment of TB and HIV/AIDS, which contributed to the continuity of these activities beyond the duration of the projects. At the same time, the long-term commitment of the government remains unclear.

Fourth, coordination between donors decreases aid fragmentation and contributes to the sustainability of benefits and activities resulting from sponsored health care programs, as long as these activities and benefits comply with the donors' objectives and priorities in the aid-recipient countries. The analysis of the Global Fund grants unveiled a curious feature of the national actors' dependence on the Global Fund by linking it to the donors' coordination with each other. The arrival of the Global Fund grants to a certain degree resulted in the retrenchment of other donors from TB and HIV/AIDS. In addition to reinforcing the financial dependence of state authorities and CSOs on the Global Fund, this retrenchment evidently pointed at coordination among the donors, which was also visible throughout the realization (i.e., design, implementation, and evaluation) of grants. The Global Fund took over the activities previously provided by other donors, such as the Soros Foundation, ICRC, and others. This book suggests that the transition of Kyrgyzstan from the Global Fund grants does not necessarily presume discontinuity of the project activities currently financed by the Fund; it rather suggests the transfer of these activities to the account of other donors, who will replace the Global Fund. Clearly, this replacement strongly depends on the presence of a donor whose objectives and priorities include TB and HIV/AIDS. The range of benefits and activities may also change depending on the priorities and financial means of this donor.

## 11.4   Methodological Limitations

I acknowledge the limits of causal effects and causal mechanisms defined through exploratory research. The mechanisms identified in this book may be only a part of many alternatives leading to the same outcome (for the problem of indeterminacy, see Rohlfing, 2012, p. 7). While emphasizing the significance of the necessary and sufficient conditions behind the causal effects, this book nevertheless acknowledges the uncertainty regarding the inclusion of all possible relevant conditions. Moreover, multiple conditions or their combinations may lead to the same outcome (the problem of equifinality, see King et al., 1994; Mahoney, 2008), and some conditions may produce the outcome "only if they are simultaneously present" (conjunctural causation) (Rohlfing, 2012, p. 56). Furthermore, in defining the causal mechanisms and causal effects behind interactions and sustainability in development assistance, this book may have unwittingly overlooked other conditions of equal relevance to the outcome and included only those known to the literature (Gerring, 2010, pp. 1508–1512).

Other than acknowledging the exploratory limits of this study and incorporating the causes relevant to interaction and sustainability into the "scope conditions," there are no known solutions to the problems of equifinality and conjunctural causation. The "scope conditions" are the "boundary conditions, delineating the domain within which a specific causal relationship is expected to exist" (Rohlfing, 2012, p. 9). An increase in the number of boundary conditions contributes to the validity of identified mechanisms by specifying the population of cases and decreasing the number of nonexamined cases (Rohlfing, 2012, p. 8). However, increased case specification also decreases the generalizability of the identified mechanisms (Rohlfing, 2012, pp. 147–148). The specificity of boundary conditions relates back to the "layered generalization" indicated above.

I do not pretend to solve the issues associated with causal effects and causal mechanisms, most of which go beyond the scope and interest of this book. However, acknowledging the limits of the selected methodology, it nevertheless aims to make a meaningful contribution to understanding the link between the interaction among stakeholders and the sustainability of health aid.

# References

Ancker, S., & Rechel, B. (2015). Policy responses to HIV/AIDS in Central Asia. *Global Public Health, 10*(7), 817–833. https://doi.org/10.1080/1744169 2.2015.1043313

Andrews, M. (2013). *The limits of institutional reform in development: Changing rules for realistic solutions* (Illustrated Ed.). Cambridge University Press.

Beach, D., & Pedersen, R. B. (2019). *Process-tracing methods: Foundations and guidelines* (2nd ed.). University of Michigan Press.

Falleti, T. G., & Lynch, J. F. (2009). Context and causal mechanisms in political analysis. *Comparative Political Studies, 42*(9), 1143–1166. https://doi.org/10.1177/0010414009331724

Fawcett, S. B., Paine-Andrews, A., Francisco, V. T., Schultz, J. A., Richter, K. P., Lewis, R. K., Williams, E. L., Harris, K. J., Berkley, J. Y., Fisher, J. L., & Lopez, C. M. (1995). Using empowerment theory in collaborative partnerships for community health and development. *American Journal of Community Psychology, 23*(5), 677–697. https://doi.org/10.1007/BF02506987

Gerring, J. (2010). Causal Mechanisms: Yes, But…. *Comparative Political Studies, 43*(11), 1499–1526. https://doi.org/10.1177/0010414010376911

Glenton, C., Scheel, I. B., Pradhan, S., Lewin, S., Hodgins, S., & Shrestha, V. (2010). The female community health volunteer programme in Nepal: Decision makers' perceptions of volunteerism, payment and other incentives. *Social Science & Medicine (1982), 70*(12), 1920–1927. https://doi.org/10.1016/j.socscimed.2010.02.034

Global Fund. (2023). *Government and public donors.* Retrieved February 3, 2023, from https://www.theglobalfund.org/en/government/

Government of KR. (2001, January 1). *Natsional'naya programma Kyrgyzskoy Respubliki "Tuberkulez-II" na 2001–2005 gody: Postanovlenie Pravitel'stva Kyrgyzskoy Respubliki ot 6 iyunya 2001 goda N 263.* http://cbd.minjust.gov.kg/act/view/ru-ru/6838?cl=kg-kg

Government of KR. (2006). *Nacional'naâ programma reformy zdravoohraneniâ Kyrgyzskoj Respubliki "Manas taalimi" na 2006–2010 gody [National Health Care Reform Program "Manas Taalimi" for 2006–2010]: Utverždena postanovleniem Pravitel'stva Kyrgyzskoj Respubliki ot 16 fevralâ 2006 goda № 100 [Approved by the Decree of the Government of the Kyrgyz Republic dated February 16, 2006 No. 100].* Retrieved March 3, 2023, from http://cbd.minjust.gov.kg/act/view/ru-ru/57155

Government of KR. (2012). *Postanovlenie ot 24 maya 2012 goda № 309 O Natsional'noy programme reformirovaniya zdravookhraneniya Kyrgyzskoy Respubliki "Den sooluk" na 2012–2016 gody [Decree dated May 24, 2012 No. 309 On the National Healthcare Reform Program of the Kyrgyz Republic "Den Sooluk" for 2012–2016].* http://cbd.minjust.gov.kg/act/view/ru-ru/93628? cl=ru-ru. Accessed 2 February 2023.

Government of KR. (2017). *O Programme Pravitel'stva Kyrgyzskoj Respubliki po preodoleniû VIČ-infekcii v Kyrgyzskoj Respublike na 2017–2021 gody [The program of the Government of the Kyrgyz Republic sight to overcome HIV infection in the Kyrgyz Republic for 2017–2021]: Postanovlenie Pravitel'stva Kyrgyzskoj Respubliki ot 30 dekabrâ 2017 goda № 852 [Decree of the Government of the Kyrgyz Republic dated December 30, 2017 No. 852].* Retrieved February 3, 2023, from http://cbd.minjust.gov.kg/act/view/ru-ru/11589

Government of KR. (2018). *Programma Pravitel'stva Kyrgyzskoj Respubliki po ohrane zdorov'â naseleniâ i razvitiû sistemy zdravoohraneniâ na 2019–2030 gody "Zdorovyj čelovek—procvetaûsaâ strana" [Program of the Government of the Kyrgyz Republic for the protection of public health and the development of the healthcare system for 2019–2030 "Healthy Person—Prosperous Country"]: Priloženie 1 (k postanovleniû Pravitel'stva Kyrgyzskoj Respubliki ot 20 dekabrâ 2018 goda № 600) [Annex 1 (to the Decree of the Government of the Kyrgyz Republic dated December 20, 2018 No. 600)].* Retrieved February 2, 2023, from http://cbd.minjust.gov.kg/act/preview/ru-ru/12976/10?mode=tekst

Harmer, A., Spicer, N., Aleshkina, J., Bogdan, D., Chkhatarashvili, K., Murzalieva, G., et al. (2013). Has global fund support for civil society advocacy in the former Soviet Union established meaningful engagement or "a lot of jabber about nothing"? *Health Policy and Planning, 28*, 299–308. https://doi.org/10.1093/heapol/czs060

Hedström, P. (2005). Dissecting the social: On the principles of analytical sociology (Illustrated Ed.). Cambridge University Press.

Hedström, P., & Swedberg, R. (1998). Social mechanisms: An introductory essay. In P. Hedström & R. Swedberg (Eds.), *Social mechanisms. An analytical approach to social theory* (pp. 1–31). Cambridge University Press.

Hedström, P., & Ylikoski, P. (2010). Causal mechanisms in the social sciences. *Annual Review of Sociology, 36*(1), 49–67. https://doi.org/10.1146/annurev.soc.012809.102632

Isabekova, G. (2019). *The relationships between stakeholders engaged in development assistance: Towards an analytical framework.* SOCIUM SFB 1342 Working Papers, 3, 1–28.

Jana, S., Basu, I., Rotheram-Borus, M. J., & Newman, P. A. (2004). The Sonagachi project: A sustainable community intervention program. *AIDS Education and Prevention, 16*(5), 405–414. https://doi.org/10.1521/aeap.16.5.405.48734

Khetan, A. K., Purushothaman, R., Chami, T., Hejjaji, V., Madan Mohan, S. K., Josephson, R. A., & Webel, A. R. (2017). The effectiveness of community health workers for CVD prevention in LMIC. *Global Heart, 12*(3), 233–243. https://doi.org/10.1016/j.gheart.2016.07.001

King, G., Keohane, R. O., & Verba, S. (1994). *Designing social inquiry: Scientific inference in qualitative research.* Princeton University Press.

Mahoney, J. (2008). Toward a unified theory of causality. *Comparative Political Studies, 41*(4–5), 412–436. https://doi.org/10.1177/0010414007313115

Rohlfing, I. (2012). *Case studies and causal inference: An integrative framework.* Palgrave Macmillan.

Scheirer, M. A. (2005). Is sustainability possible?: A review and commentary on empirical studies of program sustainability. *American Journal of Evaluation, 26*(3), 320–347. https://doi.org/10.1177/1098214005278752

Schüth, T. (2011). *Appreciative principles and appreciative inquiry in the Community Action for Health Programme in Kyrgyzstan.* Tilburg University, n.p. https://pure.uvt.nl/ws/portalfiles/portal/1359087/Schueth_appreciative_07-11-2011.pdf

Schüth, T., Jamangulova, T., Aidaraliev, R., Aitmurzaeva, G., Iliyazova, A., & Toktogonova, V. (2014). Community Action for Health in the Kyrgyz Republic: Overview and Results. Sharing Experiences in International Cooperation. *Issue Paper on Health Series*, (3a), 1–31.

Sivaram, S., & Celentano, D. D. (2003). Training outreach workers for AIDS prevention in rural India: Is it sustainable? *Health Policy and Planning, 18*(4), 411–420. https://doi.org/10.1093/heapol/czg049

Swedlund, H. J. (2017). *The development dance: How donors and recipients negotiate the delivery of foreign aid* (1st ed.). Cornell University Press.

Walsh, A., Mulambia, C., Brugha, R., & Hanefeld, J. (2012). "The problem is ours, it is not CRAIDS' ". Evaluating sustainability of community based organisations for HIV/AIDS in a rural district in Zambia. *Globalization and Health, 8*(1), 40. https://doi.org/10.1186/1744-8603-8-40

WHO. (2008). *Community involvement in tuberculosis care and prevention Towards partnerships for health: Guiding principles and recommendations based on a WHO review.* Retrieved February 28, 2023, from http://apps.who.int/iris/bitstream/10665/43842/1/9789241596404_eng.pdf

Zardiashvili, T., & Garmaise, D. (2017). *Kyrgyzstan's program continuation funding request to the Global Fund provides little information on the proposed program.* Retrieved April 19, 2019, from http://www.aidspan.org/gfo_article/kyrgyzstan%E2%80%99s-program-continuation-funding-request-global-fund-provides-little

# 12

# Conclusion and General Implications of This Study

Uncovering the "missing link" between aid relationships and sustainability was one of the major reasons for conducting this research. During the extensive analysis of the literature on aid and public health programs, I came across the association of project sustainability with stakeholder relationships. This relationship seemed intuitive, particularly given the Paris Agenda and its emphasis on ownership and harmonization. Nevertheless, in light of the lack of a systematic analysis of the two phenomena and the link between them, I aimed to develop an analytical framework by synergizing the findings of other researchers. This academic curiosity and endeavor found further support in practice during the first fieldwork conducted in 2016, during which multiple actors expressed their concerns over the sustainability of initiatives beyond donor funding. Similarly, inequality among stakeholders and broader structural issues were repeatedly recalled by stakeholders, along with collaboration. These considerations reiterated broader concerns in the literature. The sections below outline the major findings concerning aid relationships, the sustainability of selected initiatives, and the possible link between these two phenomena. This chapter also outlines the academic and practical benefits of this research, along with limits and directions for further studies.

© The Author(s) 2024
G. Isabekova, *Stakeholder Relationships And Sustainability*, Global Dynamics of Social Policy, https://doi.org/10.1007/978-3-031-31990-7_12

## 12.1   Aid Relationships

This book aimed to synergize the discussion of aid relationships in the development aid literature with a discussion of power and its sources in political theory to provide a more refined analytical framework for analyzing aid relationships (Chap. 2). It differentiated between the conventional and alternative perspectives on stakeholder relationships and power and expands on the latter. More specifically, the analytical framework aimed to enable further examination of recipients' roles, actors' interdependence, and the changing nature of power throughout the assistance. This book pursued four steps in analyzing these aspects.

The first step commenced with a reflection on the meaning of power and the common terms associated with it, such as resources, consensus/conflict, and interests (Chap. 2). Then, to more fully grasp the inequality present in development assistance, it followed the distinction between "power over," which is associated with a hierarchy between stakeholders, and "power to," which corresponds to changes in this hierarchy. Whether these two types of power are separate or merely represent distinct aspects of the same power is debated among scholars. Nevertheless, the distinction between the two types of power helped differentiate the power dynamics between stakeholders, establishing the basis for identifying the sources of power.

As a second step, this book emphasized the relevance of both stakeholders and the context in which they interact. It approached stakeholders as agents who act depending on incentives provided in the relevant structures and the roles assigned to these agents (Dowding, 2017, p. 22). Both individual and collective agencies were emphasized as being equally important. An empirical analysis confirmed this assumption. Individual agency featured in multiple instances, such as in the cases of the Global Fund to Fight AIDS, Tuberculosis and Malaria (the Global Fund) grants and the "Community Action for Health" (CAH) project. However, the empirics also demonstrated the equal importance of collective agency on behalf of organizations, as well as abstract categories, such as donors, recipient states, and civil society organizations.

In addition, this book emphasized the relevance of the context in which these agents operate by defining them as structures or "recursively organized sets of rules and resources" that enable and constrain stakeholders (Dowding, 2011, p. 10). An extensive literature review highlighted the significance of aid dependency, capacity, volatility, and flexibility. The empirical analysis of each of these aspects provided the following insights:

1. In both the Global Fund grants and the CAH, the aid recipients (state and civil society organizations) were dependent on the donors' financial and technical support. Following Lensink and White (Lensink & White, 1999, p. 13), this book attributed aid dependency if a country (in this framework, a recipient state or a civil society organization) could not "achieve objective X in the absence of aid for the foreseeable future." This book demonstrated the limits of the conventional quantitative indicators, which are normally used to illustrate the share of external assistance for health care; instead, it proposed using the sector-specific definition of aid dependence. Despite the relatively low share of donor contributions compared to the total share of health financing, this research provided compelling evidence for the extensive dependence of state and civil society organizations on the financing and technical assistance offered by donors. This was found to be particularly visible in the cases of human immunodeficiency virus infection and acquired immune deficiency syndrome (HIV/AIDS)-related preventive activities, antiretroviral therapy, and treatment of multidrug and extensively drug-resistant tuberculosis (TB).

2. Defining capacities as individual-, organizational-, and system-level abilities to implement functions (European Centre for Development Policy Management, 2008, p. 2), this book specifically focused on the structural issue, namely, the availability of human resources. Overall, staff turnover in ministries and state agencies, which is also due to the political situation in the country and low salary rates, was found to significantly jeopardize their abilities to perform their functions (see Isabekova & Pleines, 2021). Relevant to both cases, the problem of the limited capacity of state organizations, also in terms of monitoring, was particularly evident in the Global Fund grants, which resulted

in a corruption scandal. In addition, there were cases of former state officials working for international organizations and NGOs. Although the scale of this phenomenon requires further research, it resonates with the conclusions of other relevant studies (Swedlund, 2017; Toornstra & Martin, 2013). Notably, staff rotation and attribution in nongovernmental organizations (NGOs) varied, depending on organizations and specific positions. While it was found to be substantial in the case of outreach workers, it did not seem equally pressing in other positions of NGOs.

Furthermore, in contrast to the high level of attrition in community-based organizations, which is described at length in the literature (e.g., Glenton et al., 2010; Khetan et al., 2017), the CAH was actually characterized by a *low* level of attrition among the Village Health Committee (VHC) members, demonstrating the persistence of the VHC members. Labor migration and conventional gender roles contributed to the "female" profile of the volunteers in the VHCs, for whom VHC membership offered the possibility of participating in social life and decision-making processes. Most of those women did not migrate to other parts of the country or abroad, which contributed to the stability of VHC membership.

3. Acknowledging the uncertainty in which stakeholders operate because of aid appropriation procedures and the relatively short duration of development programs, this book focused on aid volatility as another structural factor. As demonstrated in Chap. 4, both the Global Fund and the Swiss Agency for Development and Cooperation (SDC) worked on increasing the predictability of their assistance. However, the SDC offered higher predictability than the Fund, which depended on its financiers' replenishment of its three-year cycle. The Global Fund's inability to provide long-term commitments was explicitly visible in the National Program for HIV/AIDS Prevention, where the financing from the Global Fund was only confirmed for the first three years, and the funding for the two remaining years was unknown. However, in contrast to the findings made by Swedlund (2017), this research did not find any evidence that the actors involved were unaware of upcoming assistance or that the volatility of the aid had any impact on the recipients' commitment to the aid. However,

despite this outcome, the long-term duration of the CAH (approximately 17 years) did contribute to the commitment of the recipient state and community members to the project.

4. Emphasizing the significance of aid flexibility to the relationships present among stakeholders, this book examined this phenomenon by associating it with the ability to adjust to local priorities and context (see Hirschhorn et al., 2013). An empirical analysis showed that the CAH was more flexible to the recipients' changing needs and priorities than the Global Fund grants. The SDC, similar to other Swiss development agencies, provides a high level of decision-making autonomy to its field offices, meaning that these offices can decide on important issues without prior approval from the head office. The Global Fund, in contrast, does not have branches in the recipient countries; rather, it has a designated officer who deals with specific regions. While this officer is closely involved in all the processes relating to the implementation of the grant, the key decisions are made by the Global Fund's head office in Geneva. Although it was open to making minor adjustments to the program, the Global Fund proved to be reluctant to consider more substantial changes. Meanwhile, the availability of senior staff from the SDC in Kyrgyzstan and the strong personality of the project manager from the Swiss Red Cross (SRC) contributed to the remarkable flexibility of the CAH toward the changing needs and priorities of the local communities. Based on these findings, this book would argue that the decentralization of donor organizations and a high level of decision-making autonomy of local field offices, along with a strong personality in the project manager, could contribute to the responsiveness of development projects to the recipients' changing needs and priorities.

Overall, the first two steps composed the initial level of the analysis, as they laid down the conceptual basis for understanding power, stakeholders, and the context in which they interact. The following steps linked this conceptual basis to the alternative perspective of stakeholder relationships that this book aimed to expand on. More specifically, the following steps offered the analytical depth and tools necessary to grasp recipients' roles, actors' interdependence, and the potentially changing nature of power in development assistance.

The third step called for a project-level analysis that differentiated among initiation, design, implementation, and evaluation phases. An empirical analysis at this level offers a detailed and yet standardized (in terms of a project cycle) analysis of development projects. Thus, Chaps. 5 and 8 in this book provide an elaborative overview of stakeholders' roles throughout the realization of the CAH and the Global Fund grants. In addition to understanding the agency (both individual and collective) of stakeholders, this depth of the analysis was critical to understanding the actors' interdependence. It also allowed a better grasp of the recipients' agency and a glimpse of the changes in stakeholders' roles throughout the project cycle.

The fourth step culminated the analytical framework by linking the empirical insights from step three and the conceptual basis defining stakeholders, power, and the context in the first two steps to a theorization of power dynamics and aid relationships. This step was necessary to understand the empirical cases by placing them in a broader theoretical framework. This step combined the seven ways of creating power suggested by Haugaard (2003) with the "ideal" types of aid relationships defined by the author of this book in Chap. 2. For simplicity of comprehension, this analysis was conducted in a dyadic manner by referring to broader analytical categories: donor–recipient state, donor–donor, recipient state–civil society organizations (CSOs), and donor–CSOs. This step synergized the findings of all three other steps.

The synergy of the four steps unfolded as follows in the case of the CAH. The donor pursued an "empowerment" approach toward community-based organizations (CBOs). In addition to continuous participation, the CBOs maintained a decision-making role throughout the project cycle (based on the project life cycle analysis in Chap. 5). Moreover, structural factors, including aid flexibility and predictability (Chap. 4), were favorable to altering the conventional stakeholder positions by ensuring continuous project responsiveness to community needs. Volunteering or unpaid roles of community members and their leadership (individual agency) were additional assets. These findings from two chapters strongly hinted at the feasibility of an "empowerment" approach of a donor toward CBOs. However, the theorization linking these empirical findings to types of power, stakeholders' interests (step one), and the ways of creating power (step four) was critical to the validity of my claims. The theorization showed that the donor primarily pursued the "power to" the CBOs for several reasons because of its emphasis on and belief in the decisive role of communities in aid ("system of thought" in Haugaard's terms). A relevant practice of nondominance followed by SRC staff members further supported this belief, contributing to community members' realization of their decisive roles (transformation of tacit knowledge into discursive, according to Haugaard). These ways of creating power vividly demonstrated that the SRC used its resources to produce "power to" CBOs rather than "power over" them. Additional reflections on donor and CBOs' interests in following the specific forms of power, namely, bringing change to a community, self-development, organizational perspective, and individual agency, further supported an "empowerment approach" (Chap. 7).

In contrast to the CAH, donor–CSO relationships in the Global Fund grants followed a "utilitarian approach." In addition to uneven levels of participation throughout the project life cycle (Chap. 8), the structural factors remained in favor of hierarchical relations (Chap. 4). Indeed, the Global Fund committed itself to ensuring the predictability of its assistance, which, nevertheless, in practice, remains dependent on its replenishment cycle. Furthermore, although driven by the grantee's needs in the design phase, the provider offered the limited adaptability of grants during the implementation stage (e.g., limited flexibility). In addition, local NGOs and their activities depended considerably on donor financing, which varied across organizations but strengthened the inequality between the provider and recipient of aid. Theoretically, the relationship between the Fund and NGOs was characterized by a combination of "power over" and "power to." Notably, the Global Fund's emphasis on civil society participation in decision-making (a structural bias in Haugaard's term) offered a window of opportunity for NGOs to strengthen their positions in the project life cycle. This opportunity was further reinforced by solid organizational support in the form of finances and technical assistance from the Global Fund. However, this "power to" came along with the aid provider's "power over" the recipient. One source of this form of power was the propensity to predict grants and their outcomes ("social order"), while another source was assigning specific roles and tasks to stakeholders in the project lifecycle (discipline in Haugaard's theory). Similar to the CAH, a "utilitarian approach" was found to have its underpinning in stakeholders' interests, including access to resources, reaching out to vulnerable groups, and organizational perspective.

The analysis of aid relationships between other stakeholders in the CAH and the Global Fund grants followed a similar logic (for more information, see Chaps. 7 and 10).

## 12.2   Sustainability of the Selected Health Projects

In addition to analyzing aid relationships, this book also offered a systematic analytical framework to assess the sustainability of health care interventions, in which various approaches to the operationalization of sustainability and relevant factors were given special attention. Following the most frequently used approach in the literature, namely, the approach developed by Shediac-Rizkallah and Bone (1998), this book defined sustainability as the continuity of project activities, the maintenance of benefits (e.g., services and infrastructure), and community capacity building. This book complemented this approach with three extensions.

First, it acknowledged the relevance of the analysis of both ongoing and complemented projects. In the former, it approached state commitment in terms of necessary legislative amendments and financing as the sign for sustainability of ongoing initiatives. The analysis of the Global Fund grants vividly demonstrated both the validity of and issues with this approach (Chap. 9). In contrast, the CAH presented the case for a completed health care project in which the actual fulfillment of obligations and activities upon the end of the donor funding could be assessed. However, both ongoing and completed initiatives are subject to continuous socioeconomic, political, and epidemiological changes (e.g., the coronavirus disease 2019 [COVID-19]), representing similar uncertainty and jeopardy to the sustainability of projects. Moreover, by acknowledging that sustainability does not automatically come at the end of aid but rather is built throughout its realization process, this book highlighted the equal validity of the analysis of ongoing and completed initiatives.

Second, this book complemented the operationalization of community capacity building with an adaptation of Laverack's framework by focusing on participation, leadership, and mobilization of resources (see Labonte & Laverack, 2001a, 2001b).[1] Furthermore, in contrast to the original framework by Shediac-Rizkallah and Bone (1998) and the operationalization of community capacity building suggested by Laverack (see Labonte & Laverack, 2001a, 2001b), the current research introduced a new category for assessment that is commonly highlighted by the interviewees but absent in the two older frameworks, namely, the survival of CSOs beyond the period of development assistance provision. This aspect is important, as, for instance, unlike the NGOs involved in and highly dependent on the Global Fund grants, the CBOs in the SDC's CAH demonstrated remarkable continuity beyond the development assistance. In this way, although financing was important for the functioning of the CSOs, it did not seem to be a necessary factor.

Third, the initial framework was complemented with a list of factors relevant to the sustainability of health care interventions developed by the author of this book through an extensive review of related literature. These factors included financing; accounting for the influence of general economic, social, and political situations in the aid-recipient country; integrating within context; and disentangling organizational factors into further categories (see Table 3.3 in Chap. 3).

Overall, the analytical framework and its three extensions found their reflection in the analysis of the selected health care programs.

The CAH demonstrated the continuity of tuberculosis and HIV/AIDS-related services previously pursued by the project beyond the duration of donor funding. Unpaid, the CBOs continued their awareness-raising activities in these areas (among others) by informing the local population, organizing community events and walking campaigns, and so on. Indeed, the means available to CBOs varied depending on their coverage with analogous donor programs, particularly given that the

---

[1] The original source is an unpublished Ph.D. thesis by Laverack (1999), which was expanded further by Labonte and Laverack (2001a, 2001b).

government training activities became uneven. Among others, the main issues thereof were found to be related to reimbursing the travel expenses of state trainers, which were also due to the ongoing optimization reforms and change in the national health care toward a systemic (and not disease-specific) target-setting. Similarly, the survival of CBOs beyond the CAH was uneven, but the majority continued due to members' leadership and organizational support from the Association of Village Health Committees. Indeed, resource mobilization remained a challenging task due to not only the socioeconomic situation in the country but also the lack of extensive training on this matter. However, CBOs continued and even expanded their work by overcoming challenges, such as COVID-19 implications, and exploring opportunities for their organizational growth.

In the case of the Global Fund grants, the government demonstrated an unprecedented commitment to continuing TB and HIV/AIDS-related activities, although with mixed success. In addition to legislative changes aimed at eliminating the discrimination of groups affected by TB/HIV, it also adopted a roadmap to optimize state health care services and facilities to provide additional financing for the areas funded by donors. The mobilization of resources at both the national and local levels was found to be challenging due to the socioeconomic situation in the country defining its gross domestic product (GDP). However, even with savings, state funding was found to be insufficient to cover medications for antiretroviral therapy, the treatment of drug-resistant forms of TB, and other areas. There was skepticism that state-level commitments highly depended on decision-makers' choices. However, other factors equally mattered to the fulfillment of these commitments, such as procurement costs and opportunities, the availability of medical professionals, the epidemiological situation in the country, and the COVID-19 implications on the already strained health care system. The maintenance of benefits received by patients with TB/HIV similarly demonstrated the commitment of the government, as reflected in the social support offered by the state that was found to be nevertheless insignificant. The survival of NGOs beyond the duration of the grants provided ambiguous answers depending on the capacities of the organization in question. Indeed, NGOs explored possible alternatives that were nevertheless scarce due to a decrease in external funding. The state's reform toward social contracting, following the Global Fund grant conditions, was a reasonable alternative, although with further implications with regard to accountability and the dependence of CSOs on the government.

Overall, both the theoretical and empirical discussions vividly showed that the sustainability of health care initiatives is a complex question requiring nuanced answers. Thus, a project may hypothetically perform well in terms of continuity of activities or the maintenance of benefits but not in terms of community capacity-building (or a specific aspect of it). Are such projects still sustainable, then? As noted in Chap. 3, I refrain from suggesting degrees of sustainability; however, I also argue that this is not a yes/no question. Furthermore, the analysis only reflects the state of affairs at a certain period of time. Consequently, sustainability, similar to power and aid relationships, evolves.

## 12.3   The "Missing Link" between Aid Relationships and Sustainability

The link between aid relationships and sustainability was explored at the project level and beyond. At the project level, this link embodied mechanisms or processes connecting the two phenomena (see Chap. 11 for more details). However, these mechanisms have limited implications for our understanding of the link beyond the selected cases. For this reason, I used broader causal links due to the level of abstraction in both wording and approach.

First, after providing a comprehensive picture of aid relationships and sustainability, this book assessed the impact of the different types of relationships formed between the actors over each component of sustainability (e.g., continuity of activities, maintenance of benefits, and community capacity building). Based on Rohlfing's (2012) integrative framework for case studies and causal inference, this research identified the following positive links between interaction practices among involved actors and the sustainability of the selected health care programs: ownership, learning, institutionalization, recognition, uniformity, replacement, and "professionalization"[2]:

- The mechanism of ownership—aid-recipient community-based organizations develop a sense of ownership and responsibility for their communities' health. Triggered by the donor's "empowerment" approach toward the CSOs, this mechanism affects the CSOs' survival and continuity of health activities beyond the duration of health aid.
- The mechanism of learning—through their extensive participation in the realization of development assistance, aid-recipient community-based organizations increase their awareness of local issues and links to local organizations. In doing so, the CSOs become the first point of contact for local authorities and donors, which ensures their survival beyond the development assistance. This mechanism is similarly generated by the donor's "empowerment" approach toward the CSOs.
- The mechanism of professionalization—"professionalization" of NGOs in specific areas takes place through their training, fulfillment of donor requirements, and implementation of project activities. This contributes to their survival beyond the duration of health aid. This mechanism evolves through the "utilitarian" approach of a donor toward the CSOs.
- The mechanism of institutionalization is characterized by a recipient state's formalization of its commitments, leading to the continuity of project activities and CSOs' survival beyond the duration of donor assistance. This mechanism develops through unequal cooperation between the donor and aid-recipient authorities.
- The mechanism of recognition occurs when state authorities approach community-based organizations in order to achieve their own objectives. This cooperation provides the CSOs with additional means for their survival beyond the donor funding. This mechanism evolves through the aid-recipient government's "utilitarian" approach toward the CSOs.

---

[2] The author of this book identified and names these mechanisms according to the process they trigger.

- The mechanism of uniformity, under certain conditions, contributes to the expansion of development assistance and continuity of its activities. This mechanism develops through unequal cooperation among donors.
- The mechanism of replacement—donors take over each other's activities to ensure their continuity. This mechanism is triggered by the coordination among donors.

In addition to identifying these mechanisms, this book also outlines the conditions under which these mechanisms may take place beyond the context of the selected health care programs and country. These include aid dependency and limited capacity of government authorities and CSOs, precarious economic and political situation in the aid-recipient country, and the structure of development assistance, defining its flexibility and predictability. Furthermore, labor migration, conventional gender roles, stigma, discrimination against sexual minorities (i.e., lesbian, gay, bisexual, trans, intersex, and queer [LGBTQ]), and the personalities of decision-makers and project implementers are essential to the realization of these mechanisms in the case of health projects.

Furthermore, theory-centered, the aim of this book was to make a general theoretical contribution toward evaluating the impact of relationships between relevant actors on the sustainability of development assistance for health care. For this reason, in addition to the causal mechanisms mentioned above, identified through the intensive analysis of the selected TB and HIV/AIDS programs, the other aim of this research was to formulate tentative results in the form of causal links between the cause (aid relationships) and the outcome (sustainability). These causal links require a definition of causal effects or "theoretically intelligible and systematic" relationships (Rohlfing, 2012, p. 12) defined through cross-case analysis of the selected health care programs. Though context-specific, these effects provide the theoretically grounded claims about the link between the cause and the outcome:

1. An "empowerment" approach of a donor toward CSOs contributes to community capacity building by improving and streamlining the leadership of the CSOs, and their capacity to mobilize resources, as

well as by facilitating their survival beyond the duration of development assistance projects.

2. A "utilitarian" approach of a donor toward the CSOs contributes to their survival beyond the duration of development assistance programs, but it does not affect the quality of the leadership of these CSOs.

3. Unequal cooperation between a donor and the relevant authorities of aid-recipient countries does, in fact, contribute to the continuity of project activities. However, the extent of the services that might continue beyond the period of the development assistance is highly dependent upon decision-makers' priorities, the presence of stigma and discrimination against groups targeted by assistance, as well as the epidemiological, political, and economic situation in the aid-recipient countries.

4. Coordination between donors decreases aid fragmentation and contributes to the sustainability of benefits and activities resulting from sponsored health care programs, as long as these activities and benefits comply with the donors' objectives and priorities in the aid-recipient countries.

As it has provided herein comprehensive analytical frameworks together with detailed case studies, this book is of interest to academics and practitioners working in areas related to development and public health, as well as area studies and regional specialists. Despite its very specific focus on the health care programs financed by the Global Fund and SDC in Kyrgyzstan, the intention of this book was to provide a general perspective on types of aid relationships, components of sustainability, and the link between these two phenomena. Certainly, the abovementioned causal mechanisms and effects have been identified in the specific context of Kyrgyzstan.

However, by easing region-specific characteristics, it is possible to generalize the causal links between aid relationships and sustainability stated above beyond the selected health care programs. There are different "layers of generalization" (Rohlfing, 2012) for the causal mechanisms and causal effects identified in this research. The country (Kyrgyzstan), policy areas (TB and HIV/AIDS), and donors (the Global Fund and SDC) represent three major "scope conditions" defining the context for specific causal relationships (ibid., p. 9).

Kyrgyzstan offers interesting observations of a lower-middle-income country that inherited a state-dominated health care system from the Soviet Union. Indeed, the country's epidemiological, cultural, historical, and other aspects are country-specific, which may require some caution in interpreting the results provided in this book. Nevertheless, some issues, including the discrimination of groups affected by TB/HIV, aid dependency, conventional gender roles in society, and other aspects, are not unique and are equally present in other settings.

Furthermore, TB and HIV/AIDS are specific policy areas requiring continuous access to quality medications and health care personnel to ensure timely detection and uninterrupted treatment of persons affected by the diseases. As shown in the analysis, health care programs' sustainability depends on political engagement, financing, training, and awareness not restricted to specific country borders (e.g., the rise of HIV infection among labor migrants). Furthermore, unlike TB, for instance, HIV requires lifelong treatment, and in contrast to other health care areas (mother and child health, cardiovascular diseases, etc.), HIV is burdened with a high level of stigma and discrimination, not just on the grounds of the disease itself, but also the "moral issue" attached to it (see Chap. 3). Therefore, the causal links and results identified using the examples of TB and HIV/AIDS may vary in the case of other diseases.

Moreover, this book focused on "bottom-up" health projects designed by aid-recipients using the examples of the programs financed by the Global Fund and SDC. Equally stressing the recipient's ownership over development aid and civil society involvement in it, these donors varied in terms of their structures and approaches (Chaps. 1 and 4). While the Global Fund grants may have comparable outcomes in other countries with similar epidemiological profiles, the SDC's CAH represents a country-specific project, which can nevertheless be applicable to other settings willing to apply a similar approach.

Causal mechanisms and effects presented in this chapter are specific to the cases examined in this book. However, by easing the "scope conditions" stated above, it is possible to generalize these causal links, connecting different types of relationships relevant to the sustainability of health care assistance into more general conclusions about the way in which key actors and the relationships between them might affect the sustainability of development assistance.

## 12.4 Further Findings and Limitations of This Research

In line with the existing literature, this book finds unequal donor-driven cooperation as the most common form of interaction between the donor and aid-recipient government due to the limited capacity and aid dependency of state authorities. The analysis of the Global Fund project in Kyrgyzstan also shows that donor conditionalities did not end with the World Bank and International Monetary Fund's Structural Adjustment Loans in the 1980s–1990s[3] and continued up until nowadays.

Nevertheless, multiple findings of this book (in addition to those mentioned in the previous subsections) are new to our understanding of the relationship between the interaction among stakeholders and the sustainability of health projects.

First, initially unequal power dynamics between providers and recipients of aid may change during the development assistance. The analysis of the two projects in Kyrgyzstan shows that aid flexibility and donor's inclination to aid recipient's empowerment contribute to changing. In contrast, the financial requirements and the threat of withdrawal of funds strengthen the unequal power dynamics. These findings refine the suggestions in the existing literature, which suggests increased power of donors at the beginning and recipients—at the end of the assistance (e.g., Andrews, 2013; Swedlund, 2017).

Second, this book offers a new perspective on aid dependency. It argues for a sector-specific definition beyond the quantitative indicators. In 2018, the share of external health expenditure (% of current health expenditure) in Kyrgyzstan was about 5% (World Bank Group, 2023). However, the analysis shows that 60–90% of tuberculosis and HIV/AIDS prevention and treatment services depend on the Global Fund. The country also relies on the technical assistance offered by the World Health Organization, the German Corporation for International Cooperation, SDC, and others. Furthermore, the empirical analysis

---

[3] For more information on adjustment policies and outcomes, see Cornia et al. (1987, 1988).

shows that aid dependency may be an outcome of donor coordination. Several donors, including the World Bank and German Development Bank, discontinued their tuberculosis and HIV/AIDS programs shortly after the commencement of the Global Fund projects in Kyrgyzstan, which resulted in the country's dependence on a single donor.

Third, in addition to the inclusion of all relevant stakeholders, this book emphasizes further differentiation of actors within the categories of "donors," "recipient state," and CSOs to understand their interaction with each other better. The empirical analysis vividly demonstrates the contrast between a Global Health Initiative pursuing a "standardized" approach across the aid-recipient countries irrespectively of the SWAp and a traditional bilateral donor providing country-specific allocations driven by geopolitical interests. National and local state authorities also differ in their interests and capacities, similar to community-based organizations, local and international NGOs. Differentiation of actors grouped into one category allows for a better understanding of power dynamics, interaction, and its implications for the sustainability of aid.

Fourth, in contrast to the existing literature on development aid, including health aid, this book does not find high staff attrition in the case of CSOs, though high staff rotation in government organizations complies with the findings in development aid literature. The analysis of community-based organizations in Kyrgyzstan shows that conventional gender roles in the family and society ensured male migration and retention of women in households. As a result, there was no high staff attrition among the local female volunteers working in health organizations. It should be noted that health is primarily viewed as a "female" responsibility.

Fifth, this book also makes some empirical findings new to the literature on development aid and its sustainability, along with the abovementioned theoretical contributions. It contains unique primary material and thus offers new knowledge of complex processes inherent to development assistance implemented in the region, mostly neglected in development studies. Shortly after the collapse of the Soviet Union in 1991, newly independent countries received significant financial and technical assistance from international organizations. However, except for the number of articles discussing the conditionality (e.g., Pleines, 2021; Stubbs et al., 2020), assumptions (Wilkinson, 2014), and implications of international support (Kim et al., 2018), the post-Soviet region is overlooked in the literature on development aid (Leitch, 2016), which largely focuses on Sub-Saharan Africa, Latin America, and Southeast Asia.

In addition to multiple academic and empirical contributions mentioned above, this book has far-reaching policy implications. The Global Fund project is still ongoing, although the share of donor funding has considerably decreased during the last ten years. Furthermore, the national health care system of Kyrgyzstan, as elsewhere in the world, is burdened by the COVID-19 pandemic. In these conditions, the sustainability of the Global Fund project is essential to ending the epidemics of AIDS and tuberculosis in the country. Similarly, the sustainability of the SDC's completed primary health care project is critical to achieving universal health coverage in rural regions of the country. Although context-specific, the analysis of two projects in Kyrgyzstan nevertheless demonstrates the issues common to other developing countries, distressed by the shortage of finances and human resources. Therefore, the issues and opportunities for the sustainability of health aid presented in this book will benefit decision-makers working in the relevant areas in Kyrgyzstan and beyond.

At the same time, this research has several limitations. Firstly, this book focused primarily on the organizational level, and did not elaborate on the interconnection between individual and organizational levels and on how this could be relevant to understanding the actors that were involved in the development assistance. Furthermore, in discussing only dyadic relationships (meaning between two actors at a time), it did not focus on the interdependence of actors' choices, for instance. Thirdly, this research defined the causal mechanisms and causal effects linking the different types of relationships in the development assistance programs under study, with the sustainability of those programs. As an exploratory study, it did not further test the causal inferences and the tentative results in the form of causal links between aid relationships and sustainability identified in this research. For this reason, this book did not elaborate on sufficiency, necessity, equifinality, conjunctural causation, and other issues pertinent to further understanding the relationship between the cause and the outcome. Moreover, because of the fragmentation of the relevant literature, this study acknowledges the limits of the theoretical basis of the book, and uncertainty about the inclusion of all relevant conditions into the analysis. These and other areas could be possible directions for future research on this topic.

# References

Andrews, M. (2013). *The limits of institutional reform in development: Changing rules for realistic solutions* (Illustrated Ed.). Cambridge University Press.

Cornia, G. A., Jolly, R., Stewart, F., Cornia, G. A., Jolly, R., & Stewart, F. (Eds.). (1987). *Adjustment with a human face: Volume 1, protecting the vulnerable and promoting growth.* Oxford University Press.

Cornia, G. A., Jolly, R., Stewart, F., Cornia, G. A., Jolly, R., & Stewart, F. (Eds.). (1988). *Adjustment with a human face: Volume 2, ten country case studies.* Oxford University Press.

Dowding, K. (2011). Agency-structure problem. In *Encyclopedia of power* (pp. 10–11). SAGE Publications.

Dowding, K. (2017). *Social and political power.* Oxford Research Encyclopedia of Politics. https://doi.org/10.1093/acrefore/9780190228637.013.198.

European Centre for Development Policy Management. (2008). Capacity change and performance: Insights and implications for development cooperation. *Policy Management Brief, 21,* 1–12.

Glenton, C., Scheel, I. B., Pradhan, S., Lewin, S., Hodgins, S., & Shrestha, V. (2010). The female community health volunteer programme in Nepal: Decision makers' perceptions of volunteerism, payment and other incentives. *Social Science & Medicine (1982), 70*(12), 1920–1927. https://doi.org/10.1016/j.socscimed.2010.02.034

Haugaard, M. (2003). Reflections on seven ways of creating power. *European Journal of Social Theory, 6*(1), 87–113. https://doi.org/10.1177/1368431003006001562

Hirschhorn, L. R., Talbot, J. R., Irwin, A. C., May, M. A., Dhavan, N., Shady, R., Ellner, A. L., & Weintraub, R. L. (2013). From scaling up to sustainability in HIV: Potential lessons for moving forward. *Globalization and Health, 9*(57), 1–9. https://doi.org/10.1186/1744-8603-9-57

Isabekova, G., & Pleines, H. (2021). Integrating development aid into social policy: Lessons on cooperation and its challenges learned from the example of health care in Kyrgyzstan. *Social Policy & Administration, 55*(6), 1082–1097. https://doi.org/10.1111/spol.12669

Khetan, A. K., Purushothaman, R., Chami, T., Hejjaji, V., Madan Mohan, S. K., Josephson, R. A., & Webel, A. R. (2017). The effectiveness of community health workers for CVD prevention in LMIC. *Global Heart, 12*(3), 233–243. https://doi.org/10.1016/j.gheart.2016.07.001

Kim, E., Myrzabekova, A., Molchanova, E., & Yarova, O. (2018). Making the 'empowered woman': Exploring contradictions in gender and development programming in Kyrgyzstan. *Central Asian Survey, 37*(2), 228–246. https://doi.org/10.1080/02634937.2018.1450222

Labonte, R., & Laverack, G. (2001a). Capacity building in health promotion, part 1: For whom? And for what purpose? *Critical Public Health, 11*(2), 111–127. https://doi.org/10.1080/09581590110039838

Labonte, R., & Laverack, G. (2001b). Capacity building in health promotion, part 2: Whose use? And with what measurement? *Critical Public Health, 11*(2), 129–138. https://doi.org/10.1080/09581590110039847

Leitch, D. (2016). *Assisting reform in post-communist Ukraine 2000–2012: The illusions of donors and the disillusion of beneficiaries.* Ibidem Press.

Lensink, R., & White, H. (1999). Aid dependence. Issues and indicators. *Expert Group on Development Issues, 2*, 1–86.

Pleines, H. (2021). The framing of IMF and World Bank in political reform debates: The role of political orientation and policy fields in the cases of Russia and Ukraine. *Global Social Policy, 21*(1), 34–50. https://doi.org/10.1177/1468018120929773

Rohlfing, I. (2012). *Case studies and causal inference: An integrative framework.* Palgrave Macmillan.

Shediac-Rizkallah, M. C., & Bone, L. R. (1998). Planning for the sustainability of community-based health programs: Conceptual frameworks and future directions for research, practice and policy. *Health Education Research, 13*(1), 87–108. https://doi.org/10.1093/her/13.1.87

Stubbs, T., Reinsberg, B., Kentikelenis, A., & King, L. (2020). How to evaluate the effects of IMF conditionality. *The Review of International Organizations, 15*(1), 29–73. https://doi.org/10.1007/s11558-018-9332-5

Swedlund, H. J. (2017). *The development dance: How donors and recipients negotiate the delivery of foreign aid* (1st ed.). Cornell University Press.

Toornstra, F., & Martin, F. (2013). Building country capacity for development results: How does the international aid effectiveness agenda address the capacity gaps? In H. Besada & S. Kindornay (Eds.), *Multilateral development cooperation in a changing global order. // multilateral development cooperation in a changing global order* (pp. 89–114). Palgrave Macmillan.

Wilkinson, C. (2014). Development in Kyrgyzstan: Failed state or failed state-building? In A. Ware (Ed.), *Development in difficult sociopolitical contexts: Fragile, failed, pariah* (pp. 137–162). Palgrave Macmillan.

World Bank Group. (2023). *External health expenditure (% of current health expenditure)—Kyrgyz Republic*. Retrieved February 28, 2023, from https://data.worldbank.org/indicator/SH.XPD.EHEX.CH.ZS?locations=KG

# Appendix

The Global Fund Grants to Kyrgyzstan (2004–Present)[1]

| # | Period | Grant | Area | PR | Disbursed |
|---|--------|-------|------|----|-----------| 
| 1. | 2004–2009 | Development of preventive programs against TB, HIV, and malaria (KGZ-202-G02-T-00) | TB | National Center of Phthisiology | US $2,771,070 |
| 2. | 2004–2009 | Development of preventive programs against TB, HIV, and malaria (KGZ-202-G01-H-00) | HIV | National AIDS Center | US $17,073,306 |
| 3. | 2007–2012 | Enhancing DOTS implementation by strengthening strategic planning and management of the National TB Program (NTP) under the Manas Taalimi National Health Care Reform Program and by its further integration into health care services, scaling-up DOTS-plus implementation beyond the pilot phase, and reducing the burden of TB, TB/HIV, and MDR-TB in the penitentiary system (KGZ-607-G04-T) | TB | National Center of Phthisiology | US $6,212,840 |

*(continued)*

© The Author(s) 2024

G. Isabekova, *Stakeholder Relationships And Sustainability*, Global Dynamics of Social Policy, https://doi.org/10.1007/978-3-031-31990-7

(continued)

| | | | | | |
|---|---|---|---|---|---|
| 4. | 2009–2011 | Increasing universal access to prevention, detection, treatment, care, and support for key population groups in the Kyrgyz Republic (KGZ-708-G05-H) | HIV | National AIDS Center | US $11,020,755 |
| 5. | 2011–2015 | Consolidation and expansion of the "Directly Observed Treatment, Short Term" (DOTS) program in Kyrgyzstan by providing access to diagnostics and treatment of drug-resistant tuberculosis (KGZ-910-G07-T) | TB | Project HOPE | US $5,849,523 |
| 6. | 2011–2016 | Promoting accessibility and quality of prevention, treatment, detection, and care services for HIV among the most vulnerable populations in the Kyrgyz Republic (KGZ-H-UNDP) | HIV | UNDP Kyrgyzstan | US $31,893,603 |
| 7. | 2011–2016 | Consolidation and expansion of the "Directly Observed Treatment, Short Term" (DOTS) program in Kyrgyzstan by providing access to diagnostics and treatment of drug-resistant tuberculosis (KGZ-S10-G08-T) | TB | UNDP Kyrgyzstan | US $23,349,032 |
| 8. | 2016–2023 | Effective HIV and TB control project in the Kyrgyz Republic (KGZ-C-UNDP)[2] | TB, HIV | UNDP Kyrgyzstan | US $67,184,606 |

[1]Titles of the grants are listed verbatim according to the relevant Grant Agreements. Disbursements, areas, and organizations are listed according to the information provided at https://data.theglobalfund.org/location/KGZ/grants/list.

[2]This grant has been excluded from the analysis, as there are no documents available on this grant; see https://data.theglobalfund.org/grant/KGZ-C-UNDP/3/overview.

All other grants were analyzed by accessing the documentation available on the Global Fund website before it moved to a new data explorer platform. Titles of the grants are listed verbatim according to the relevant Grant Agreements.

# Index[1]

---

[1] Note: Page numbers followed by 'n' refer to notes.

© The Author(s) 2024
G. Isabekova, *Stakeholder Relationships And Sustainability*, Global Dynamics of Social
Policy, https://doi.org/10.1007/978-3-031-31990-7